to René-Louis Vallée, with my gratitude and my admiration.

As a Frenchman and as a physicist, I cannot write in English like Shakespeare. I guess this book must be full of blunders and syntax errors, but this work is the result of forty years of research and its content is what is most important to me. It also stands as my scientific testament.

Readers, be generous and forgive me, I did my best. Corrections will come later, progressively.

Cover : painting by Gérard Derrez

By the same author :

-*Multicoupleurs et filtres VHF/UHF*, Hermes, 2007

-*VHF/UHF Filters and Multicouplers*, ISTE-Wiley, 2010

-*L'univers de Maxwell,* Lulu, 2013

Bernard PIETTE

THE UNIVERSE OF MAXWELL

2014
Edition 1

Table of Contents

Foreword

Intense pleasures are rare in a lifetime: a tramp who finds a wallet full of banknotes, an athlete who breaks a world record, winning the jackpot at the lottery, and whatnot.

Now think of Marie Curie when she extracts her first gram of radium after having toiled her soup of pitchblende for years? What about Hertz, who justifies the well-founded theory of Maxwell through magical experiments? What about Tombaugh, when he discovers Pluto at the place computed by Lowell? What about Marconi, who creates the first transcontinental radio link?

Physics is the foundation of our technological western civilisation. It becomes concrete in industrial applications, which change our lives progressively, helps us to get out of darkness and from time to time brings us a breath of passion and happiness, maybe too scarcely. With its daughter, chemistry, and its cousins, natural sciences, it is the only human activity which really makes our knowledge of the world progress and, as a consequence, our intelligence.

But though, what sadness, what rigidity, what pretentious academism in the tuition and the approach of physics! What badly-deserved purism, what stiffness in schoolbooks as well as in specialized periodical publications, what pedant style used to teach a subject which should be offered to every one as a pleasure but which has become an instrument of suffering and elitism, like mathematics, and partly because of it and its abuse.

According to Einstein, physics should belong to everybody; in this case, it should be taught or explained so that everybody can understand it. If we want it to be well assimilated, it must necessarily be attractive. So why not speak about physics like we do at home, at the pub, on the bus or in the underground, in all common places, with an appropriate common language, and in order to improve its transmission, why not use a few swear words, or a touch of vulgarity from time to time ?

This book is not for the narrow-minded ones who shoot up with transcendent equations, with the old perfumes of the academic encaustic and naphthalene, the ones who venerate Einstein, who believe that physics is a sanctuary reserved for an elite, that you have to deserve it, that it belongs to mathematicians and the brightest students, to workaholics. May they keep on suffering from their own sophisticated science and congratulate themselves among geniuses in their pretentious nebula.

This book is for ordinary people, who deserve physics as well, for goodness sake!

Introduction

"It is important that the great majority should be able to know the driving forces and the conquests of science with clarity and precision. It would be deplorable if only a restricted circle of specialists appropriated these conquests to improve and apply them in a concrete domain. If the circle of people who has access to knowledge should be reduced to a small group of initiates, this would mean the decay of the philosophical spirit in the people and the advent of intellectual poverty".

What a beautiful message, so generous, so altruistic, except that the use of the conditional is not appropriate any more: the worst has happened. Science well and truly belongs to a restricted circle of individuals, whether we like it or not. The esoteric language of theoretical physics has become incomprehensible to the general public for a long time, and intellectual poverty has invaded public space, in the same way and with the same density as the mobile phone. However, behind this warning message, we can guess the troubled consciousness of a disillusioned and anxious scientist, who may have been overtaken by unforeseen events for which he realized that his own responsibility was involved after all, against his will and in spite of his warning.

The twilight thoughts of Albert Einstein -because this is him- show us once more that if the supposed big brains of the human species are gifted for the prevision of disasters, for which they are sometimes responsible, they also are often incapable of preventing them. Nevertheless, voices have always been raised to warn us against an irrelevant use of our discoveries, and Einstein was neither the first nor the last. Three centuries earlier, doctor Rabelais had also had the premonition of a coming catastrophe: “Science without conscience is nothing but the ruin of the soul”, Pantagruel said. The quotation is well-known, but does it still have meaning today? Though the current environmental issues, which evidently threaten our future or even our survival on earth, perhaps begin to give rise to a general awareness. This issue of collective irresponsibility may require to be somewhat developed, and it is probably not pointless to make a short way back to a recent past, to try to understand why and how we have reached that stage.

At the end of the 19th century and at the beginning of the 20th, it was the triumph of mechanics. Not theoretical mechanics, which nowadays belongs to mathematicians, but rather practical, industrial mechanics, the one used for constructions, engines and means of transport. In less than fifty years, everything was invented: motorcars, planes, buildings, metallic bridges, pneumatics, trains, internal combustion engines... At the same time, at the instigation of Gramme, electro-mechanics allowed to generate electricity from steam and mainly to transport it from a distance. Between 1875 and 1905 not only did we go from the 19th to the 20th century, but in thirty years we, Westerners, changed from the age of stone to the age of bronze, that is from animal to electric traction, from the ox-cart to the plane. Thirty years is nothing. It is hardly the time elapsed today since the first time we heard about controlled fusion, this gigantic joke which evidently is one of the biggest technological fiascos of modern times and the biggest scientific swindle after the Big Bang, of the same nature as the mystic “sniffer planes” during the tenure of French President Giscard d'Estaing. For those who don't know or who were too young in the late 70s, these planes were supposed to detect oil fields by the use of a high-frequency transmitter-receiver installed in their noses. We are still laughing at it.

In those days, this triumphant advent of mechanics was widely shared, followed and supported by a large public, enthusiastic, fully spoilt, filled up with discoveries, expecting the announcement of a new technological revolution at any moment, rushing towards World Fairs in which France was one of the leaders. People believed in the future, things were evolving quickly, and science was correctly doing its educative job: little mathematics, many pictures, reviews published weekly, monthly and yearly. It was quite easy to keep in touch with the discoverers, all you had to do was regularly buy some issues among plenty of scientific journals like *La Science pour Tous*, *Sciences et Voyages*, *La Revue Rose*, *La Technique Moderne*, *Le Monde et la Science*, *Omnia*, *La Vie Scientifique*, *La Science Illustrée* by Figuier, and more particularly *La Nature* by Gaston Tissandier, a weekly in which it was then absolutely natural to include the digests of the reports of the Academy of Science in its columns, for the attention of a wide readership. Can you imagine a weekly scientific journal giving us information on what happens at the Academy of Science today? In this connection, I would like to welcome the remarkable initiative of the Conservatoire National des Arts et Métiers for having put the archives of *La Nature*, from the first issue of 1873, at the disposal of everyone on the Internet: it's a mine of information that anybody can get by simply exploring the website of the CNUM CNAM and which I have widely used to find and extract some references for this book. This journal is also a nostalgic journey deep into the world of our grand-parents, the pages of which we often browse through with a smile on your face, but which also teaches a lot about the evolution of technologies and of currents of thought, by reminding us of forgotten landmarks, with many surprises in store about the supposed huge progression made in the domain of knowledge since then.

What remains today of that feverish urge to know of the early 19th century? Actually, it may still exist, and we hope so, but scientific journals are less attractive than before, and they are now published every month instead of every week. The disappearance of scientific weeklies is the undeniable sign that there is less to write today, for the three reasons that have already been mentioned: the rhythm of inventions in fundamental physics has lowered, their nature has deeply changed and the language used to present them is no more the common language. Whom is a monthly like *La Recherche* – the top reference among scientific journals in France - intended for? For searchers, for graduate students, let's say for informed readers, but not for any reader, not for ordinary people. Moreover, even more popular publications like

Science et Vie, born in 1913, or *Science et Avenir*, the last avatar of *La Nature*, despite valuable efforts at being clear, have much difficulty to appeal to the common reader in search of novelties. Of course, there are still technological evolutions to feed one's curiosity, but it is not the same: it is not real science any longer, but already business.

Mechanics, in comparison with the other branches of science, possesses something particular: it is part of us in a way. Since man exists as the dominant species on earth, he has been, despite this self-attributed status, so much bound to Nature that it is evident that the latter has transmitted him, for centuries, and without his being really aware of it, the basis of his abstract, say mathematical thought, the thought with which he has taken the habit of representing reality to himself by the process of modelling: for example, a quiet lake has become the horizontal plane in our brains, the trajectory water dripping from the roof of a cave or of a suburban cottage - depending on the era - has become the vertical one. Joining and comparing the two of them has induced perpendicularity, and a stone thrown into the air has printed the parabolic trajectory into our eyes, long before we put it into equation. This is the reason why mechanics can be assimilated without much difficulty: it has somehow gone into our genes, it is part of our nature. Actually any inhabitant of the suburbs could have been a geometrician, but the fact that he doesn't know that he has inherited a fabulous skill from the past generations and that he cannot use it because he has not been taught to, is tragic after all. It is tragic because of the consequences on the average standard of the average individual, in other words all of us. This is probably due to our so imperfect education system, and therefore constitutes a recurrent problem which completely gets away from us, as well as a key to our gregarious behaviour, irrational and sometimes even definitely imbecile.

As for modern sciences, which roughly date back to the second quarter of the 20th century, this is another story. Understanding such specialities as electronics, atomistic, the Quantum Theory, or Relativity requires an appropriate mathematical background, that is of high standard, which only a small minority of people possesses. And this automatically leads us to the heart of one of the major problems of our education system: the preponderant role of mathematics which in fact proves to be excessive and too systematic. In the 80s, Jack Lang, the French Minister of Education at that time, declared: "*we must get rid of the dictatorship of mathematics...*" A beautiful sentence, a beautiful project that we would like to support, a beautiful resolution in principle, but as inefficient as Einstein's warning against scientific sectarianism. However, Jack Lang was right: mathematics has actually taken the power in teaching, it is set up everywhere, including in non-exact sciences, like sociology or economy to mention just a few, which it tries to make exact. Moreover, and above all, it has become the intelligence test par excellence, the first selection criterion which will determine the good students, the brain-boxes so to speak, who, as a reward for their "gifts" and hard work, will open the door of the Grandes Écoles[1] – those prestige university-level colleges we drag like millstones around our necks - then reach positions of great responsibility in which they will eventually prove that, despite their brilliant studies, they are neither better nor brighter than others.

Here is another observation of modern life: we are in a society which is completely governed by international trade, where the business world totally controls scientific research, which only gets its vital credits if research managers accept to be framed in action lines leading to something profitable. Today they must spend a good part of their time on administrative tasks, and permanently check the balance in their budget. But above all, they must prove that their studies will both lead to marketable products and generate new activities which will bring enough money not to regret the investments, generally granted painfully and mistrustfully.

The notion of profitability is something which undermines Western society from within. Montesquieu used to say that international trade gathered peoples but divided individuals; he was certainly right, and the fact that our relationships with the Third World and emerging nations are getting more and more difficult is only one outward sign of it. In this context, the average citizen feels as badly as science, and we eventually wonder what has be-

1 High schools in France

come of it in the collective consciousness, inasmuch as there is still one. Contrary to what most of us think, science is not a crystal sanctuary impervious to human ups and downs. It is an activity which has become, like everything else, completely controlled by a finance company in which the only decision-maker is money. Pure research doesn't exist any longer, and sponsors prefer to invest in professional sport rather than in knowledge.

Just like the boiler of a steam locomotive, the human brain is a blazing fire that must be constantly fed if we don't want it to stop working. The essential difference between man and the animal lies in man's irrepressible desire to know and learn, to penetrate the secrets of the Universe, to perpetually pursue a quest the final goal of which he doesn't even know, and there is only one means to advance into the unknown: science. And when science hesitates, when it sinks into the quagmire of mathematics, when it ceases finding anything any longer, when it stops fulfilling its educative role, all the dangers that threaten the human spirit, crouching in the darkness, crop up again from nobody knows where. Religion, all the religions, are there, ready to offer their soft, protective and hypocritical cocoon, with answers for everything which don't stain the brain, which anaesthetize the freedom to think and the critical mind, and reduces the intellectually fragile populations to disciplined and brainless herds. It is quite sure that tackling the mysteries of Nature and the Universe head-on demands a certain amount of courage, because becoming gradually aware of our smallness next to the cosmos, of our precariousness in an environment the fragility of which we assess progressively, of our helplessness in the face of natural forces, don't usually contribute to the blind optimism and the feeling of well-being that faith can bring. But the emancipation of Man as a species has to go through this necessary awareness, and greatness of man is to keep standing, to refuse ignorance, to reject the mollycoddling of ready-made thoughts, to always try to increase his meagre knowledge and to live with dignity, even if he has known, since he was born, that he must eventually die. Albert Camus[2] wrote that there lied man's greatness.

2 French Nobel Prize of literature.

But let us forget about those philosophic reflections to come back to the problem of science in our everyday lives, and ask ourselves the following question: how could scientific studies appeal to young people nowadays? The media have already given to a 15-year-old teenager a complete view of what society can offer him as a life plan, and at the same time have made him shape the consumer society. Why then strain himself doing mathematics until the age of 25, when he perfectly knows what will happen to him? Why study over five years after high school if it doesn't prevent him from being unemployed? To have the same life as his parents', who do their best in a future-less world, who work frantically before going home late in the evening, who curse employers and politicians, and when the vital necessity has been done, sit down in front of the TV to watch the procession of the clowns or the news, the buffoons of the weather forecast, the liars of the government, the pretentious of the sitcoms, while slimy commercials trickle down the screen?

One day, the Earth will disappear, dragged into the heart of the solar whirlwind. Mercury, then Venus, will have known the same fatal outcome earlier and to make up for it, one or several newly captured planets will perhaps appear beyond Pluto. The engulfing of Mercury by the solar blaze will trigger off such a cataclysm that every sign of life on Earth will vanish, at least for some time. It will happen in such a long time that the human species, as we know it today, will have had hundreds of opportunities to commit suicide by destroying its environment until then. Perhaps it will be the price to pay so that, in the same way as Cro-Magnons replaced Neanderthals, Homo rationalis will succeed the ill named Homo sapiens to correctly assume his responsibility towards Mother Earth, at long last. But the survival of mankind will have to lie in the capacity Man will have acquired, thanks to scientific and technological knowledge, to leave a planet where life will have become impossible and migrate to another place, obviously Mars to start with. This prediction, founded on all the arguments put forward in this book, also implies that we, or more exactly our remote ancestors, would have already made the same journey from Venus to Earth, thus giving the myth of Noah's Ark new resonance, both unexpected and unconventional: we don't perfectly know yet the cosmic avatars of the planet we may come from, but we can easily imagine that the state in which Venus is, as we can observe it today, with its temperature of 400° and its atmosphere constituted of carbonic gas, could be the result of a previous episode of our saga in the solar system and of our extraordinary gift for systematically destroying our living conditions, wherever we are. It is well-known that Man's worse enemy is himself.

But the consumer society, closed in on itself, cares little and even ignores that. The financiers, the accountants, the economists on one side and the religions on the other side have taken the power and impose their peremptory and hypocritical speeches on us, and it is not the few so-called ecologists, with their hazy ideas, who will awaken the world's conscience or who give the means to improve the situation.

However, we have all we need to progress and get out of the rut we are in. In 1970 René-Louis Vallée, an engineer at the CEA - the French Atomic Energy Commission – elaborated an extraordinarily well-thought-out theory, based on a new and modern conception of the ether. By simply eliminating the basic hypothesis of the restricted Theory of Relativity, which consists in denying the existence of a propagation medium for electromagnetic waves and in considering the speed of light as a universal constant, he demonstrated that, on the contrary, it was eminently variable and directly linked to the gravitation field, of which it is the image and the indicator in a way. The ensuing model of photon is astoundingly beautiful and fecund, and it is the only one which makes the corpuscular and vibratory aspects of light compatible. But above all, it shows that the ether is structurally a gigantic and inexhaustible source of energy which can be converted into heat or electricity. The experiments René-Louis Vallée conducted on the Tokamaks - i.e. fusion tori – of Fontenay-aux-Roses seem to prove he was right, and yet he was crushed by the nuclear Establishment, who couldn't bear seeing someone, however gifted he was, challenge their lines of study, and above all, who were probably terrorized by the disruptive economic implications of his discovery. Nevertheless, a few hundred engineers and worthy scientists studied the Synergetic Theory -that's how its inventor called it-, and testified to its validity. But this restricted group of supporters was meticulously rendered harmless and all the documents relating to the discovery were destroyed or banned from publication. As for Vallée, he was forced to resign and finally exiled himself to the USA, where more intelligent people gave him the possibility to carry on his work and where he became a member of the National Academy of Sciences.

We have all we need to be successful and happy, to take full advantage of this temporary space oasis that Earth is, but through what has been jotted down so far, we can better understand why there is little hope of avoiding a world-wide catastrophe - which has finally little to do with global warming - if we don't quickly become aware of the situation and react. We are constantly caught between the two mortal enemies of our culture that we have created ourselves: money, which corrupts everything when it departs from its original function, and religion, always in the waking state when it doesn't hold power, and always ready to seize it back when it has lost it. Because it is all about power after all. Power in all its forms, social or societal when it is a question of money, intellectual or spiritual when it is a question of religion. In between, science is quite unhappy, when it should be our driving force. Oddly dressed, rotten from the inside, overtaken by the technology it has created, science is no longer attractive to young people looking for exciting careers. It has lost its soul.

Does that mean that we are doomed to helplessly witness the uncontrolled proliferation and the moral and intellectual degradation of an animal species which, though holding all the aces, is stupidly destroying its vital environment? To see the youths turn away from the fundamentals and only dream of suicidal comfort, blinded by the illusion of the consumer society? To listen to narrow-minded economists boring us with growth, when programming a population decrease is the only reasonable option to at least delay, if not avoid, the tragic destiny that we are collectively building?

Instead of this picture of no future, we could rather imagine Man as a sickly little being, assaulted by bad weathers, diseases and calamities, a wretched reed attacked from all sides by a Nature which ignores it, but still clinging to the rock, embittered, vindictive, cantankerous, dishevelled, livid in the storm, but also intelligent, wilful and tenacious, pointing an angry finger at the sky while trembling with anger and obstinacy and muttering: "One day, I will understand all that mess!". There is a long way to go, but we must keep hope alive, or it is not worth living. We must join our wits and our wills, or what is left of them, to try to get out of the rut of the consumer society and get rid of this bloody civilization as soon as possible.

Long live physics!

Chapter 1 : the Ether (1)

1-1 : Observation and analogies

"Curiosity killed the cat", we often say to children who behave badly. But scientific curiosity is an acknowledged quality, though not so frequent. It is rarely something that one acquires with the passing years, like wisdom or seasoning, it is more like an innate ability which some people possess, but which most people don't, as it seems. You can't be a good physicist if you don't belong to the first category, if you don't have the habit, the reflex which consists in suddenly stopping and having a look at an object or a phenomenon which puzzles you, without knowing why at first glance, while others will move on without noticing or paying attention, because it seems ordinary and consequently uninteresting to them. The former reaction is the first sign of a scientific vocation among children. For example, when we look at the sky at night, we can either simply see a permanent and unique show and just enjoy it, or let our unsatisfied brain begin to wonder, insidiously first and then more and more acutely, about very simple questions for which there are no ready-made answers: how many stars are there? What is infinity? Was there a beginning? Does the Universe have limits? and so on.

These are the unanswered questions which kids ask and which embarrass their parents so much. Other questions come later, more precise, more targeted, the ones we ask ourselves during physics courses when we acquire a minimum of knowledge, a minimum of tools, the ones we put to the test of reality. As far as the astronomical sky is concerned, the true scientifically-based questions usually arise as soon as we learn that it takes light thousands, millions or billions of years to come to us from the most remote stars. This is a first shock, some information that takes time to be digested before it starts meaning something: the cosmic time scale is not the same as our time scale at all, and it is necessary to make quite a consequent intellectual effort to be able to go

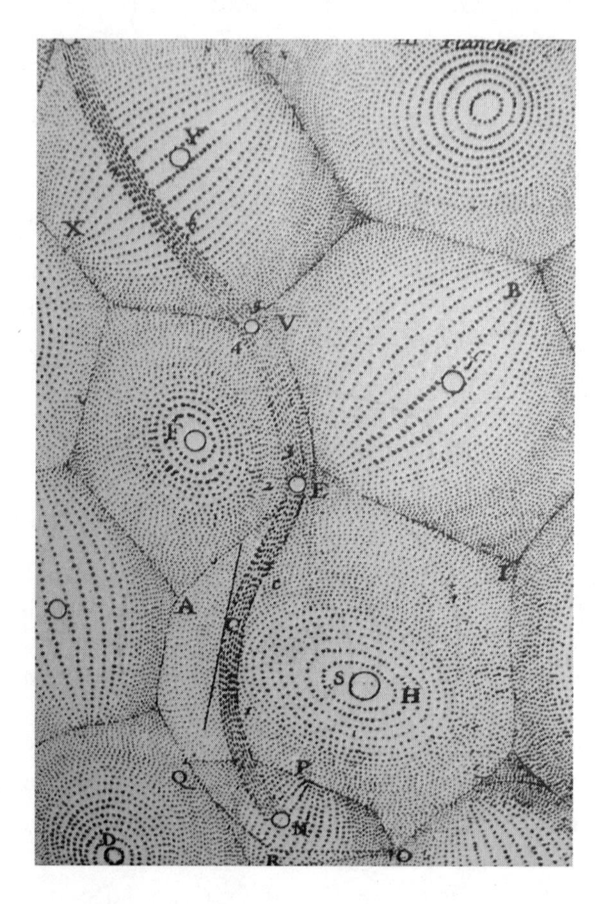

figure 1-1 : Solar system seen by Descartes

from one to the other, and then to practice the exercise in order to master it. And indeed, being aware of seeing a picture of the past while looking at the stars is something not easy to assimilate when we are young. After a while, it all ties up and things start to clear up, but as we browse through cosmography books for new information in vain, the fundamental questions remain unanswered: we have the strange feeling that the farther the astronomers see and the more new objects they discover, the more question marks it raises.

Since astronomy exists, popularization works have always sold well. It's the kind of present that we easily offer to children old enough to understand, it's also an instinctive appeal to a domain about which we confusedly feel that we have to get interested in, and which is a window to the great mystery of the World. The covers of these books, of which one issue at least should be on display in every good library, very often contain photographs on four traditional themes: the astronomical telescope, the observatory dome, the night sky and the Andromeda Nebula. The latter, together with all other spiral galaxies, sooner or later provokes the basic wonder to any physicist, either professional or amateur, through comparison with other natural phenomena presenting the same aspect: this particular shape irresistibly reminds the shape of other dynamic phenomena on earth. All of them are vortices appearing in fluids, reminding us either of the autumn wind which blows up dead leaves and dust to reveal to our blind eyes a vaguely circular and upward movement, otherwise invisible, or of water covered with a layer of floating particles which shows us an eddy when situated next to a lock gate, or of a whirlwind over the plains of Texas, or of a cyclone seen from a satellite....

While making the morning tea or coffee, when we add one or two sugar lumps into the mug and stir it with a spoon, and then carefully pour a few drops of milk right in the centre of the gyratory motion that we have initiated, we can see the birth of a miniature spiral nebula, which surprisingly looks like the illustrations on astronomy books. The resemblance is so strong that it always provokes the same reaction in all of us, which consists in first having a hesitant smile, halfway between perplexity and inspiration, because we suspect something without really understanding, and then, after a few minutes of contemplation, we inevitably wonder if it is only a mere similarity, a trick of nature, or if it is the same phenomenon but on a different scale. We are all the more tempted to believe in the latter hypothesis as the pictorial techniques of our audiovisual century give us other points of comparison which didn't exist in the previous century, like weather forecasts, scientific documentaries and reports on events related to hurricanes. And then we are in the heart of the matter: a spiral nebula may have the shape we know it has because it is situated in a whirling fluid medium which, though it remains invisible to us, reveals its existence. Unfortunately, like Saint Thomas, we only believe in what we can see. If we suppose that it is a real fluid, the most difficult for us is to define the physical characteristics compatible with our comprehension. That is the reason why Einstein decided to forget about this hypothesis, to go back to theoretical physics and publish his theory known as the Special relativity in 1905, which turned the ether into a cursed corpse, after three centuries of various and vain attempts to give it a face, with René Descartes as a forerunner.

1-2 : Descartes and the ether

We don't know whether Descartes used to have tea or coffee - coffee was new in England at that time- but he may also have observed this phenomenon like anybody else, and it could have been for him the beginning of deep meditation and the prelude to his cosmological theory. In the same way, the legend says that Newton would have elaborated his Theory of Universal Gravitation by seeing an apple falling from a tree and interpreted it thanks to his exceptional gift for analysis. As a matter of fact, Descartes could not know the existence of spiral nebulas, the first one of which was discovered in 1780 by Pierre Méchain, but a particular intuition made him aware of the hypothesis that the sky was "liquid", as he said, and that each celestial body was moving inside an etheric vortex. Allegedly, he would have had this inspiration while contemplating a torrent. This is both logic and possible, but impossible to check. Anyway it will live on as a symbolic picture of the history of science, like Newton's falling apple or Einstein's elevator.

If the name of Descartes comes first, it is mainly because we cannot speak of the ether without mentioning him. Not because he was the first to put forward this concept, as we can always find past records on the philosophical problem of vacuum, more or less precise, by searching hard in the history of science, but especially because he devoted an important part of his works to it, with a profusion of details, thus becoming an indisputable reference point. To the neophytes who want to take an interest in the subject, I recommend *Principia Philosophiae* (*Principles of Philosophy*), which is not a book on philosophy, contrary to what its title may suggest, but truly on physics. This work is of paramount importance as far as the orientation of the thoughts on the existence of the ether is concerned, because it is the first one which describes it in so many aspects and with so much precision, though some points have become indefensible today. But it matters little: *Principia Philosophiae* is a solid and rather undisputed starting point. At the beginning of the 20th century, H. Parenty, the author of *Les tourbillons de Descartes*, deplored the fact that no publisher had popularized Descartes's *Principles of Philosophy*, except for a few deluxe editions ordinary people could not afford. So it would be proper to thank Vrin bookshops in Paris for enabling amateurs and searchers to get an affordable new edition of both the original Latin version and the French version translated by Abbé Picot and brilliantly deciphered by Adam and Tannery. For those who are not really into this kind of studies, let's say that Abbé Picot was to Descartes what Solovine was to Einstein or Abbé Moigno to Tyndall, that is the guarantee of a high-quality and faithful translation. Besides, Descartes had approved Abbé Picot's translation at the time. Do many physicists, professors or students take advantage of this opportunity to go back to the source of the ether? That's another question, which it is up to them to answer.

What does Descartes tell us about the "subtle matter", as he names it? The essential idea for him is that vacuum cannot exist, because his mind cannot imagine it -that's the way Descartes was!-.

So we must replace vacuum by something else, by some substance present everywhere, that is inside the matter, impregnating and filling it, but also where there is no matter.

		1550	1600	1650	1700
Cassini	1625-1712				
Copernic	1473-1543				
Descartes	1596-1650				
Fermat	1601-1665				
Galilee	1564-1642				
Gassendi	1592-1655				
Grimaldi	1618-1663				
Hooke	1635-1703				
Huygens	1629-1695				
Kepler	1571-1630				
Leibniz	1646-1716				
Mariotte	1620-1684				
Mersenne	1588-1648				
Newton	1642-1727				
Pascal	1623-1662				
Roberval	1602-1675				
Römer	1644-1710				
Toricelli	1608-1647				
Louis XIII	1601-1643				
Louis XIV	1638-1715				

Main 17th-century personalities

And where there is no matter, there is the ether, which could only be imagined as a material fluid, for its nature not to be incompatible with the appearance of things and with the working of the cosmic machine. The argument may seem very lightweight today, but in those times it was not shocking, insofar as science was quite widely supported by the existence of God, the Saviour of the scientist who had run out of arguments, in a research circle still controlled by the Church. In the 17th century, the Church had a major role in the circulation of scientific ideas, in which it always saw a potential danger, because they may have led to a dangerous emancipation of the intelligences towards rationality and intellectual free will, the worst disaster for any religion. Similarly, we must keep in mind that the last book by Descartes, *Le Monde* (*The World*), also called *Traité de la Lumière* (*Treatise on Light*), was published posthumously, the author not being keen to attack the religious authorities head-on at a time when the trial of Galileo was taking place, in which his judges are well-known for their equity and impartiality today. In order to best control an activity which was finally and fortunately inevitable, bound to human curiosity and the desire to know, and to keep it at the service of its vital interests, the Church constantly dispatched its own members into the scientific circles, who by the way quite often contributed towards the advance and the diffusion of knowledge in physics, it's fair to admit it. Among them was father Marin Mersenne, of the order of the Minims, placed at the centre of the international group of intellectuals specialized in philosophical research that we gladly consider the seed of the future Academy of Sciences, and where we could find Descartes, Roberval, Huygens, but also Malebranche - the “Great Orator”, both a disciple and a critic of his mentor Descartes - Abbot Picot, Father Noël (Reverend Estienne Noël, of the Company of Jesus, a student of Descartes a bit too zealous whom the King affectionately called his “father of the whirlwinds”), and many others whom I will refer to later if necessary. But let's get back to the ether.

For the philosophers of that time, the world was made of four elements, and Descartes would reduce the number to three in his testimonial book, while keeping their agreed names. The first element is Fire, the second Air and the third Earth. It's quite difficult to establish a correspondence between the vision of the World, seen by Descartes and his predecessors, and our present knowledge, but there cannot be any doubt on the name that we would give today to the first element: it's the Ether. The following extract is from Chapter 5 of *Treatise on Light* (translated by Michael S. Mahoney, Professor of the History of Science at Princeton University):

The philosophers assure us that there is above the clouds a certain air much subtler than ours. That air is not composed of vapors of the earth as it is, but constitutes an element in itself. They say also that above this air there is still another, much subtler body, which they call the element of fire. They add, moreover, that these two elements are mixed with water and earth in the composition of all the inferior bodies. Thus~ I am only following their opinion if I say that this subtler air and this element of fire fill the intervals among the parts of the grosser air we breathe, so that these bodies, interlaced with one another, compose a mass as solid as any body can be.
But, in order better to make you understand my thought on this subject, and so that you will not think I want to force you to believe all the philosophers tell us about the elements, I should describe them to you in my fashion.

I conceive of the first, which one may call the element of fire, as the most subtle and penetrating fluid there is in the world. And in consequence of what has been said above concerning the nature of liquid bodies, I imagine its parts to be much smaller and to move much faster than any of those of other bodies. Or rather, in order not to be forced to admit any void in nature, I do not attribute to this first element parts having any determinate size or shape; but I am persuaded that the impetuosity of their motion is sufficient to cause it to be divided, in every way and in every sense, by collision with other bodies and that its parts change shape at every moment to accommodate themselves to the shape of the places they enter. Thus, there is never a passage so narrow, nor an angle so small, among the parts of other bodies, where the parts of this element do not penetrate without any difficulty and which they do not fill exactly.

As for the second, which one may take to be the element of air, I conceive of it also as a very subtle fluid in comparison with the third; but in comparison with the first there is need to attribute some size and shape to each of its parts and to imagine them as just about all round and joined together like grains of sand or dust. Thus, they cannot arrange themselves so well, nor so press against one another that there do not always remain around them many small intervals into which it is much easier for the first element to slide than for the parts of the second to change shape expressly in order to fill them. And so I am persuaded that this second element cannot be so pure anywhere in the world that there is not always some little matter of the first with it.

Beyond these two elements, I accept only a third, to wit, that of earth. Its parts I judge to be as much larger and to move as much less swiftly in comparison with those of the second as those of the second in comparison with those of the first. Indeed, I believe it is enough to conceive of it as one or more large masses, of which the parts have very little or no motion that might cause them to change position with respect to one another. [...]

From this extract, one reads a definition of ether which would be composed of two elements, the subtle air and the fire, the two closely overlapped and making a unique fluid which insinuates into matter and totally fill space, letting no place at all for vacuum. The constituents of these two mixed elements are compared to grains of sand with different thickness, such that their composition would be able to penetrate any interstice, and which would be animated by an "agitation" which would make them able to insinuate more easily in all directions (figures 1-2 and 1-3). Moreover, Descartes adds to these basic materials of his universe some "scrapings", produced by the restless friction of the previous between them and matter, as well as "fluted parts", which are kinds of twists capable of screwing themselves into the spaces let free by the spheres. Thanks to this cosmic Meccano of which he made all the parts, he explains everything: the movement of planets, light, sunspots, how magnets work, that is all the Universe, both macroscopic and microscopic. We must acknowledge that most of his "demonstrations" would be indefensible nowadays, and that, even in his time, almost all scientists, either friends or enemies, had something to contest about his statements. Statement is the right word to use in this case, because Descartes was quite self-assured, which caused him either the admiration of his disciples, or the jealousy and even hate of his sceptics, which was the result of his infamy. Despite of that, some of his discoveries were proved determining and used as references by his con-

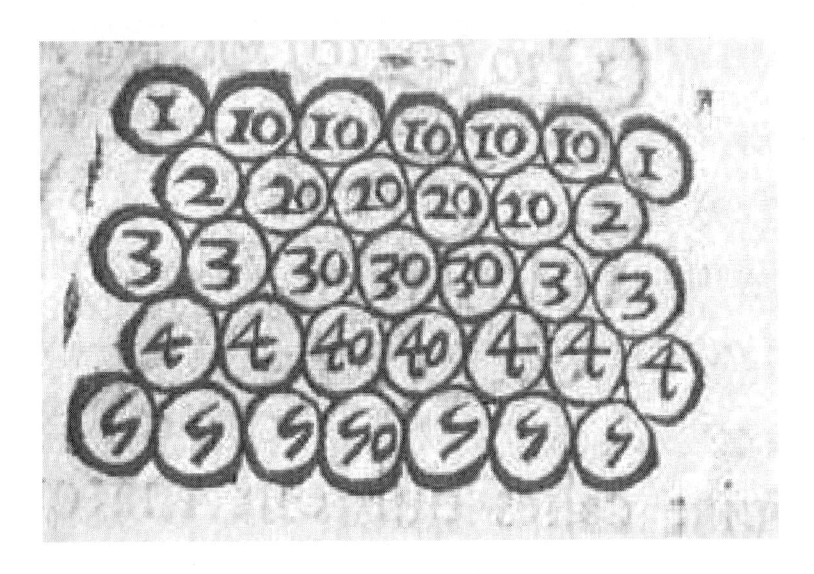

figure 1-2 : fine structure of ether by Descartes
first disposition

temporaries, as well as starting points for his followers. His work and his power of reasoning were incredible.

From the point of view that we are interested in, we'll especially retain his global vision of the Universe, where vacuum cannot exist, and which cannot be better summarized than in the titles of § 24 and the followings of the third part of the *Principles of Philosophy*, paragraphs which are the sign of an extraordinary premonition, because they are fully in accordance with not only the facts, but also with the most modern theories on ether, which will be exposed further[3]:

3 Translated from the Old French version

§ 24: "*That the Skies are liquid*".

§ 25: "*That they carry along with them all the bodies they contain*".

§ 26: "*That the Earth is at rest in its Sky, but it is nevertheless carried along by it*".

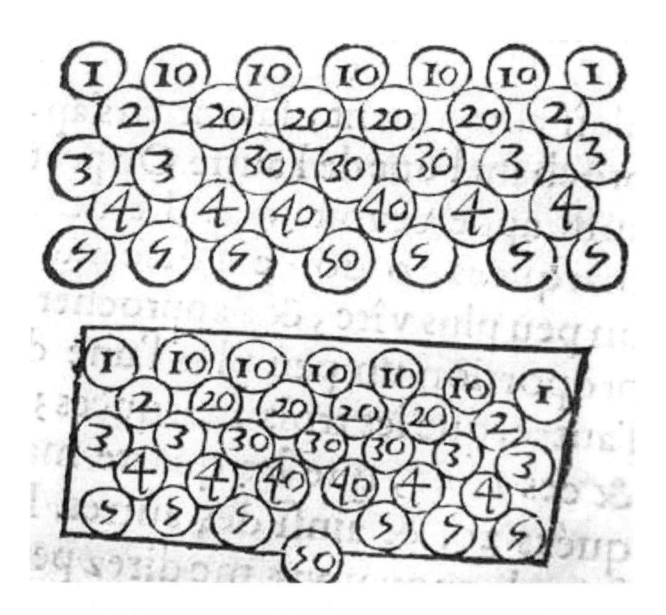

figure 1-3 : other disposition of ether balls

§ 27: "*That it is the same for all the other Planets*".

§ 28: "*That we cannot properly say that the Earth and the Planets move, though they are all carried along*".

§ 29: "*That even incorrectly speaking & following common usage, we must not attribute a motion to the Earth, but only to the other Planets*".

§ 30: "*That all the Planets are carried around the Sun by the Sky which contains them*".

§ 31: "*How they are carried along*".

§ 32: "*How they also are the sunspots that we can see on the surface of the Sun*".

§ 33: "*That the Earth also rotates about its own centre, and the Moon revolves around the Earth*".

§ 34: " *That the motions of the Skies are not perfectly circular*".

§ 35: "*That all the Planets are not always on the same plane*".

§ 36: "*And that each of them is not always equally distant from a same centre*".

This being granted and despite our admiration, we must agree that there is an enormous gap between the statement of the above titles, which all together constitute a clear summary of a powerful theory, and the justifications or explanations that Descartes tries to give in the content of these paragraphs. For a 21st-century reader, the Cartesian theory Descartes exposed in his development just seems as clear as mud. But let's be fair and realistic: in the 17th century, the level of knowledge in physics was such that it was not possible to link and verify things the way we can do it immediately nowadays, thanks to what we have been taught at school, and Descartes's approach is something huge anyway, though it includes indefensible theses today.

What we will remember of Descartes's work concerning the ether is that it is a fluid present everywhere, inside and outside the matter. It carries along the planets in a movement it originally creates, and each Planet possesses its own vortex which transmits and enforces its rotation. In regard to modern graphical processes, the illustration on figure 1-1 definitely looks like a prehistoric drawing, but it conveys, with the techniques of that time, the Cartesian vision of a "liquid Sky", where there are only invisible whirlpools with the Planets as their centres. In a more general way, it is the source and the explanation of all the physical phenomena where we can notice the action of forces of invisible origin, which more particularly concerns magnets and electrostatic charges. In this Universe, light is a pressure which propagates quasi instantaneously by the action of elementary spheres on one another. On this topic, some pretend that Descartes said that the speed in question was infinite, which was proved wrong at the same time by Römer after he observed the satellites of Jupiter. Actually he didn't assert that, as he had no intention to study the subject with precision without having serious arguments and adapted experimental means, but he only meant that the phenomenon is so fast that it can be considered instantaneous.

It's a pity that Descartes, who had the intuition of a powerful ether carrying along the planets, neither tried to define its physical parameters more precisely, nor succeeded to extend his model to the whole solar system, and to show that planetary whirlpools are secondary phenomena which participate in another phenomenon of the same kind, but more general and finally simpler. The fact that he used the term "subtle matter" led to the implication of a very small volumetric mass, which is in contradiction with the idea of a fluid capable of provoking the motion of enormous masses. This contradiction was perpetuated all along the history of sciences, from Newton to Einstein, including Fresnel and all the physicists who studied and demonstrated the laws of propagation of light, until 1970 and the Synergetic Theory by René-Louis Vallée who finally introduced the right hypothesis. Nevertheless, what Descartes achieved leading to the discovery of the ether was considerable, and can be seen as the basis of everything that would be discovered after him.

1-3 : Christiaan, Isaac, Gilles, Augustin and Co

Today, Christiaan Huygens is often considered the most brilliant spirit of the 17th century, in the domain of science. It was also true at that time, as shown by this passage of a letter Leibniz wrote to him in 1690 about one of his demonstrations in the domain of optics: "*...when I saw that the supposition of spherical waves helps you solve the problem of disdiaclastic refraction of the crystal of Iceland with the same facility, I moved from consideration to admiration.*". Descartes himself was dumbfounded when he read a communication of Huygens, who was then only 17, to father Mersenne, on the principle of equilibrium of funicular polygons. Newton would even call him "Summus Hugentus": no need to translate. These marks of respect must have been sincere, as they contrasted strongly with the usual ambiance of this elitist circle, where a tendency was to be at daggers drawn and always get at each others rather than pay compliments – something inconceivable today, isn't it?

The physics of Huygens is not as peremptory, as flamboyant as the one of Descartes. We can see in it someone who is more a physicist than a philosopher, who shows the desire to go to the bottom of things, but also someone modest who does not give an explanation if he is not sure of himself. Huygens was obviously an etherist, like everyone at that time, but he didn't try to explain everything with this philosophy as Descartes did. Huygens supported and even almost founded the undulatory theory of light. Indeed he can be considered as the precursor of this theory one century before Young and Fresnel definitely established its validity, so it is in this precise domain that he was led to better define its propagation medium and, consequently, to be in opposition with Newton's emission theory. Like his mentor Descartes, he considered the ether to be but deformable, hence a fluid, and he was brought to consider it as constituted of small solid balls in contact,, without trying to develop a more precise microscopic approach, which would have been totally useless for the study of light propagation. This way of representing the ether is somewhat natural and logical, and becomes evident as soon as we want to model, in the simplest way possible, an entity about which we only know that it is present everywhere, and nothing else: it is only evident to see it as an aggregate of minute elements in contact, like in a powder or a liquid, and only simple to imagine a ball to describe its main constitutive element.

This being granted, we must recall here that, in comparison with the conception of light by Descartes, Huygens had been given an advantage by the enormous discovery of Römer, whose observations on the occultations of the satellites of Jupiter allowed to show that light didn't propagate instantaneously, as Descartes believed it, but with a finite celerity that Cassini calculated and estimated. This first estimation was then refined by Delambre, who took into account those observations over 150 years, which gives a result that can't be ignored and the importance of which will be shown further.

But let's go back to Huygens. Contrary to Descartes, he did not have the ambition of explaining everything with the help of the subtle matter. For him, the Ether was principally the medium he needed, he couldn't do without, to attempt to define the very essence of light and understand how it makes its way in space, while taking care to always check that his thesis was in accordance with the newly discovered phenomena, like diffraction for example. One can find the essential of his idea in his *Treatise on Light*:

[…] *Now in applying this kind of movement to that which produces Light there is nothing to hinder us from estimating the particles of the ether to be of a substance as nearly approaching to perfect hardness and possessing a springiness as prompt as we choose. It is not necessary to examine here the causes of this hardness, or of that springiness, the consideration of which would lead us too far from our subject. I will say however, in passing that we may conceive that the particles of the ether, notwithstanding their smallness, are in turn composed of other parts and that their springiness consists in the very rapid movement of subtle matter which penetrates them from every side and constrains their structure to assume such a disposition as to give to this fluid matter the most overt and easy passage possible. This accords with the explanation which Mr Descartes gives for the spring, though I do not, like him, suppose the pores to be in the form of round hollow canals. And it must not be thought that in this there is anything absurd or impossible, it being on the contrary quite credible that it is this infinite series of different sizes of corpuscles, having different degrees of velocity, of which Nature makes use to produce so many marvellous effects.* […]

As far as the ether is concerned, this short extract sums up what is necessary for Huygens to legitimate the propagation of light in an elastic medium which he will not try to analyse more thoroughly: once this medium is correctly defined from the point of view of its estimated physical, or more exactly its logical characteristics, not in contradiction with his theory, he does not try to venture onto dangerous ground, but exclusively concentrates on his main goal, which is the methodical and experimental discovery of the still unknown laws of optics. The most important point is the care and the smart cautiousness he gives to the introduction of what is necessary for anyone who wants to consider light a vibration: the elasticity of the propagation medium. In order to account for the pliability of this medium, none of those who regarded its existence as an inescapable hypothesis could be satisfied with a simple assembly of small elementary spheres, the imbrication of which is such that they form a rigid solid when they are in maximum contact. For a possible deformation to occur, the balls need to roll on one another, change their position according to the neighbouring balls, so it must be admitted that at a given moment spaces are created everywhere, which allow them to move. That's the logical reason for which all the Etherists, in other words all the physicists of the 17th century and even after, supposed that, to justify the necessary elasticity as well as possible, in addition to the elementary spheres, there were either other particles, even smaller and in motion to fill the intervals, or permanent vibrations (an "agitation"), or maybe both. Without trying to go further in the study of a concept for which we can only make hypothesises, it means that we make available a material medium where a vibration can propagate in a legal and politically correct way.

Just like Descartes and Huygens, Isaac Newton also wrote a personal Treatise of Light. To engage himself in this way was obligatory for those who wanted to be regarded as great physicists, light and its different aspects being mystery number one in physics for everyone at that time. Like in the works of Descartes and Huygens, we can find in Newton's *Opticks: or, A treatise of the reflections, refractions, inflections and colours of light (1730)* - reissued in France by Gauthier-Villars in 1955 - a particular extract which explains, succinctly but clearly enough, his conception of the Ether:

Qu.18. If in two large tall cylindrical vessels of glass inverted, two small Thermometers be suspended so as not to touch the vessels, and the Air be drawn out of one of these vessels, and these vessels thus prepared be carried out of a cold place into a warm one: the Thermometer in vacuo *will grow warm as much, and almost as soon as the Thermometer which is not* in vacuo. *And when the vessels are carried back into the cold place, the Thermometer* in vacuo *will grow cold almost as soon as the other Thermometer. Is not the Heat of the warm Room conveyed through the Vacuum by the vibrations of a much subtiler Medium than Air, which after the Air was drawn out remained in the Va-*

cuum? And is not this Medium the same with that Medium by which light is refracted and reflected, and by whose Vibrations Light communicates Heat to Bodies, and is put into fits of easy Reflexion and easy Transmission? And do not the Vibrations of this Medium in hot Bodies contribute to the intenseness and duration of their Heat? And do not hot Bodies communicate their Heat to contiguous cold ones, by the Vibrations of this Medium propagated from them into the cold ones? And is not this Medium exceedingly more rare and subtile than the Air, and exceedingly more elastick and active? And doth it not readily pervade all Bodies? And is it not (by its elastick force) expanded through all the Heavens?

What strikes immediately, before analysing the content, is the excessively cautious style adopted by the author who suggests instead of asserting. In fact, this passage is part of an appendix to the Treatise, that Newton introduced as follows:

[…] *I shall conclude with proposing only some Queries, in order to a farther search to be made by others.*

We can also deduce that these questions were not of prime importance to him, and moreover, that he didn't probably want to get involved in quite a twisted matter in which Descartes and Huygens had already taken a few steps without really bringing sensational revelations about the nature of that famous “Medium”, which it is finally better to consider globally, while keeping a careful vagueness around its intimate constitution. As a result we shall not find small spheres in the Ether of Newton.

“Mon nom est Personne” (My Name is Nobody). That's the way Gilles Personne could have introduced himself, the man who is mostly known for having created a special kind of balance. Gilles Personne was born in a small village close to Beauvais called Roberval, and he decided to ennoble his name by adding the name of his village to his own, without knowing that his original name would never pass to posterity. Thanks to the editor Blanchard, Gilles Personne de Roberval was brought out of the shadows, because he was one of the most gifted and intuitive scientists of the 17th century. He spent his time learning and teaching and travelled a lot in France, which made him get quickly acquainted to all the theoretical-physicists of that period and to Father Mersenne and a few others in particular with whom he challenged the ideas of Descartes and the Aristotelian rationalists. Roberval was almost as rough-tempered as Descartes, and the relationships between the two men were never hearty. They often opposed, not only directly about personal theses, but also through other physicists. Thus Roberval joined Pascal (Etienne, not Blaise) to defend Fermat against Descartes and some others. Today Roberval is considered one of those who founded the French Academy of Sciences with Father Mersenne , or more exactly what was to become the Academy of Sciences. The "cordial relationships" and "mutual consideration" between Desacartes and Roberval can be summed up by two quotations :

-Descartes: "*... My meditations have lifted me over ordinary science enough to clearly and distinctly see that body and space are one and the same thing, while for you they are distinct by I don't know which intellectual blindness.*"

-Roberval: "*... A lot of friends and I have read your sublime meditations, but we have found absolutely nothing remarkable in them; nothing has appeared to us, except pure thoughts and vain sophisms.*"

What an atmosphere!

Apart from that, what is particularly interesting in the work of Roberval is precisely his opposition to Descartes about gravity and the structure of space - or vacuum, as you like. Before venturing a personal thesis, he first summarizes the three directions which his contemporary researchers-philosophers were moving towards[4]:

"...until *now, a question in the Schools was to know if gravity resided in the only weighing body; or if it was something common and reciprocal between this weighing body and the one towards which it is pushed, or if it was produced by the effort of a third element which pushes the weighing bodies.*

The authors of the first opinion think there is in the weighing body a quality which leads it downwards; those of the second opinion think there is a mutual and attractive quality between all the parts of a total body that makes them join together as much as

4 Translated from old French.

possible. And those of the third opinion usually have recourse to some very subtle element which moves very fast and which easily seeps between the parts of the other grosser bodies, so that, by pressing them, it pushes them downwards or upwards; and in this way they create either gravity or lightness."

This passage is both interesting and revealing because, thanks to the author's conciseness, honesty and teaching skills proper to the famous professor he was, it clearly shows the state of the reflections of the time on this major topic in physics, that is gravitation. Roberval, a man of great caution on this matter, contrary to Descartes, hesitated to tell his personal choice, even if he finally admitted he had a weakness for the second tendency: to his mind, it might consequently mean that any body, originally on the surface of the Earth, would weigh less if it were either closer to its centre or on the contrary farther from it, and that a neighbouring mass such as a mountain would disrupt its weighing. We know that all this is correct today, and Roberval's merit was all the greater for having deduced and identified it so smartly, through the reflections of his competing colleagues and his own. Moreover, his caution as an experimenter and a teacher led him to express his conclusions with quite an exceptional skill, which contrasts surprisingly with his temper and his social behaviour:

"*...I shall establish as he* (Archimedes) *did, my reasoning for mechanics, with making little effort in completely knowing the principles and the causes of gravity, saving myself for following the truth, if one day it shows up clearly and distinctly to my mind*".

In other words, he doesn't pronounce himself on the causality of gravity and scratches in front of a problem which seems to him so little prehensile, contrary to Huygens, for whom the third hypothesis is the right one: a body, released with no initial speed, falls because it is carried away by something which is already moving in the place where the body is, that is everywhere, since the phenomenon is present everywhere in the same way. And what we call something is an invisible fluid which permanently flows vertically from up to down and carries away the bodies, just like the water of a river carries away the sponge that we abandon in it. This way of seeing things, which didn't belonged only to Huygens, and which is almost four centuries old, is the only direct analogy, the only figure, the only simple explanation which has ever been given of gravity, and by generalisation of gravitation, but which obviously lays on the existence of the Ether as a postulate. But it is not enough: at the same time, when we come over to such a simple idea, it is then necessary to explain why the Ether would flow vertically towards the centre of the Earth in every place of it, and we must admit that the theses of Descartes on this topic make us go from the simplicity of the idea to extreme sophistication and confusion, difficult to be assimilated by everyone if we must analyse the mechanism in its details. Anyway, this theory establishes clearly the link between the Ether and gravitation, which is what matters here.

Without committing any injustice and considering the 17th-century scientists whose names have been put in history books, we could easily multiply by three or four the list of those who, in their physico-philosophical studies, focused on the problem of the existence of the Ether more or less time, either by mere scientific curiosity, or by the necessity of explaining the phenomena related to light or heat propagation for example, or even to do like everybody else and not to be outdone. It is not really useful because, eventually, the key ideas had always been the same and led to two recurrent themes that we find almost systematically in the scientific literature of that time: gravitation and light propagation, two major problems in physics, the first of which has not been solved yet. As for the second issue, it took another century to see Young and Fresnel break the Newtonian dictatorship and finally establish the irrefutable wave nature of light, without solving the remaining problems linked to the corpuscular appearance it shows in certain circumstances. But if the existence of the Ether is more or less explicitly implied in their reports, no one among them ever gave a new perspective in order to make its intimate nature and its physical constitution more precise: what remained was the concept of a quasi-imponderable fluid, as mysterious as the "subtle matter" of Descartes, and it would be hard to deny that, in the minds of his successors, that term put the perfidious idea of an extremely light fluid we don't know what to do with because of its inconsistency.

1-4 : From Descartes to Maxwell

The 18th century is called “the Age of the Enlightenment”. As a matter of fact, from a purely scientific point of view, the 17th century should be called that way, principally in France; in the 18th century, there was a certain stagnation of the ideas which made that France, considered the scientific and intellectual lighthouse of the world so far, was progressively caught up by the other dominating European countries before being left behind by them. Newton, lauded to the skies after the discovery of the gravitation laws, gradually supplanted Descartes as the reference of the rational thought, and the propagation of his ideas was greatly promoted by Voltaire and his friend Madame du Châtelet, who played a part in the accession of the Newtonians to the French Academy of Sciences. Then it took a century for the emission theory to be contested and to get rid of a dogmatic School which is reminiscent of today's relativistic movement, the one which has kept us in a dead end for more than a century now. It is the same process, the same gregarious defects which, three centuries apart, preside over this state of things, and the publications of Einstein on the Special Relativity (1905) and the General Relativity (1915), which are the present references in physics, are the equivalent of the *Principia* (1687) and *Opticks* (1704) by Newton. It is not that these books are not considerable, in the literary sense of the word, but mainly because it is human nature to constantly need leaders, to turn them quickly into gods (we say “genius”), and to create about them some kinds of religions which have all the same arbitrary, inflexible and unjust characteristics as the real ones, with sacred texts, priests, churches, believers, rites and also, unfortunately, fundamentalists. However, the Cartesians felt more and more uneasy about the criticisms from the Newtonians, all the more as fundamental discoveries like Römer's on light celerity, showing that it was not only finite but calculable, proved the falseness of certain assertions of the late master.

In this progressive ferment of physics, the hypotheses on the Ether were put to one side as the brains were occupied in studying the two domains where knowledge was exploding: mathematics and optics. Its existence was not contested, but the reflection on its constitution definitely took a secondary position, the expansion of the experimental methods having caused the necessity of knowing more to disappear. When science progresses, it always relegates the metaphysical reflection to a lower level of priority, compared to some new knowledge which suddenly flourishes and then the gregarious behaviour of man does the rest: every one follows, the way the hunting dog follows the hare.

The 18th century was nevertheless marked by a very important and determining event in the history of sciences, though people became aware of its importance only very progressively because of its staggering in time. This event is the merging of electricity as a scientific discipline, that is answerable and the subject of systematic study, both theoretical and experimental. Descartes had done his best to explain the basic electrostatic and magnetic phenomena by the way of his subtle matter, but his theses remained purely theoretical, debatable, and more and more controversial. The main point was to establish that the Ether existing everywhere, it would necessarily take part in all phenomena and all manifestations of facts related to physics. His successors having taken over the problems caused by invisible phenomena in a reasonable and organized way, it gave the Ether a new chance while opening a new independent way for physics: electromagnetism.

In fact, at the very beginning of the 19th century, optics was back on centre stage thanks to Thomas Young and Augustin Fresnel. The former was an intuitive gifted man, the latter a rigorous and methodical hard worker, but each of them made their contributions to a discipline which had been shelved after Newton's corpuscular theory of emission, the Einstein of that time, after a century of unconditional veneration for this man, a great figure but abusively revered by his peers and his epigones. But it is effectively by using the support of the Cartesian conception on the structure of the world that both Thomas Young and Augustin Fresnel - each one working in his own corner - had the feeling that an undulatory conception of light, propagating in an etherised medium, was a good way of dealing with the problem and the only argument able to provide a correct explanation of the new observations and acquisitions of science in this domain. Actually, the two scientists inadvertently shared the work, the first one by having the intuition of the phenomenon without trying to theorize it, the second - a faithful student of Lagrange, Monge and Poisson, his professors at the Ecole Polytechnique – by submitting the English great idea to the rigour of French mathematics, which for once did only what was required of them, and not more. So the Ether was rehabilitated, for more than a century, but it still remained enigmatic as far as its intimate nature and its physical characteristics were concerned. At least there it was, with all its mysteries but without being contested, neither by physicists, nor by philosophers.

The increasing knowledge in electricity and magnetism, and then the progressive discovery of the close interdependence between the two disciplines through the works of Oersted, Faraday and Ampere, which led to their unification and the creation of electromagnetism, once again launched the question of the Ether, since it was obvious that something was happening in that bloody "vacuum", but there was no clearer answer, quite the contrary, Nevertheless, the ownership of this new branch of physics by the mathematicians in the second half of the 19th century was to make it a totally distinct speciality, firstly completely independent from classical science dominated by mechanics, the main characteristic of which was the invisibility of the primary phenomena. In this respect, electromagnetism could already be compared to gravitation, as Coulomb showed it by establishing an attraction law comparable to Newton's, which would later on lead to closer and closer links, with some going as far as analogy. Mathematicians such as Poisson, Gauss, Weber, Henry or Lenz were at the origin of the notions of field and potential and all those bizarre new concepts which, while ignoring the etherised substance, however built a theoretical representation from it. In this way, the necessary number of new mathematical objects were invented, indispensable to theoreticians and meant to give to space a structure coherent enough to a develop a theory. As a particular example, vector notation would become of paramount importance in that process.

And then, during one of his many manipulations in his laboratory, towards the end of his life, Faraday noticed that a beam of light was diverted by a magnetic field. This discovery, only one among many others, actually gave birth to exceptional consequences: it showed that there was probably a family tie between optics and magnetism, the latter still separate from electricity, but less and less. It was then the arrival of an unconditional supporter of Faraday in the rich landscape of theoretical research that would mark the real start of what we now call electromagnetism, in the full meaning of the term: the Scotsman James-Clerk Maxwell.

1-5 : The Ether of Maxwell

The present book could have been called “The Physics of Descartes” instead of “The Universe of Maxwell”. The title meant to be a tribute to the man who made the greatest contribution to the knowledge of the Ether, and consequently to the structure of the Universe, and we must say that among our alumni, both names emerge with the greatest implication. Finally the choice fell on Maxwell because his work was less pretentious, more targeted and more complete than Descartes's and thus also more profitable and fecund to us. We can say that the physics of Descartes is the physics of the World, and that of Maxwell the physics of the modern world. So the title is a tribute to the one who genuinely launched today's physics, dominated by the industrial applications of electronics, which totally depends on electromagnetism.

Maxwell was a mathematician, but a mathematician respectful of the experimental method that he particularly admired in his older colleague Faraday for whom he was full of praise and whose conception of space, seen as the theatre of electric phenomena, struck his mind indelibly. Maxwell doesn't mention the Ether in his treatise on electricity, However its existence is implied, so slightly that it is almost invisible, but in fact it is totally present and necessary if we want to follow and well understand the genesis of his ideas. Faraday's lines and tubes of force, which are the beginning of the structure of the Ether through a certain interpretation of electric phenomena, constituted a representation of space, which replaced the contradictory subtle matter of Descartes and his contemporaries by something more consistent, more elaborate, while maintaining elements of its mystery. The powerful creative imagination of Maxwell transformed the intuitive notions of Faraday into a mathematical edifice, both rigorous and elegant, the beauty and coherence of which insolently left other theories in the shade – such as Relativity, to take a completely random example.

Maxwell mentions the Ether only in a few articles of his scientific papers. But there is also a trace of it in his *Elementary Treatise of Electricity*, issued two years earlier. In the foreword by William Garnett - the first Professor of Mathematics and Physics at University College, Nottingham - we can discover a detailed explanation of the unstated ideas of the author. If Maxwell is so cautious, it is because he feels somehow in the same situation as Descartes at the time of the publication of his book *The World*, with the Catholic Church instead of the scientific establishment for Maxwell, who considered his ideas the ridiculous figment of his imagination. This is a constant in the history of science, that each seminal discovery is regularly fought by those who should support and promote it instead: human nature, jealous, narrow-minded and half-witted, rages in all circles and at all times.

Faraday's discovery of the deflection of polarized light by an electromagnetic field relaunched the eternal question of the Ether, but only after Foucault, then Fizeau, measured the speed of light and thus allowed to settle once and for all between Newton's theory of emission and Young and Fresnel's ondulatory theory. Contrary to the Newtonian theory, the latter reckoned that light celerity might decrease when penetrating into a medium of higher refractive index. Foucault's rotating mirror was the only true way to make a direct comparison and to find a speed of 299,000 km/s in the air (315,000 for Fizeau who had probably been influenced by Römer's result), and 220 000 km/s in the water. The matter was closed, the corpuscular theory forgotten and the theory of Huygens finally admitted. But the other issue, the case of the Ether, was completely revived with a bigger mystery: it was a vibration all right, but a vibration of what? In what medium does this wave, which is disturbed by a magnetic field, propagate? And what is this magnetic field created in? What is its nature? Is there a second Ether which would be the support of electric and magnetic phenomena? In this ferment of questions which bothered the brainss of his contemporaries, Maxwell decided to tackle the problem.

After having carefully assimilated the results of Faraday, Max-

James-Clerk Maxwell (1831-1879)

well noticed in a short time that a simple division of two values (the units of charge), respectively relating to the electric and the magnetic fields, gave the celerity of light in the air as a result . The discovery of this extraordinary result definitely led him to the idea that all these phenomena were of similar natures, and that all of them could propagate in a unique and universal Ether. His calculations resulted in an equation of D'Alembert's type, similar to the one known for the propagation of sound waves, and which indicated that the electromagnetic field might propagate in the Ether with the same speed as light. But Maxwell was going too fast and too far to be followed by the rest of the scientific community, and when Faraday died, he was left alone against all, became the victim of misunderstanding and mockery, and withdrew from the world for a while. Nevertheless, he was called back to the direction of the Cavendish laboratory in Cambridge, where he wrote his *Treatise*, but died before Hertz experimentally proved the correctness of his calculations in 1833 and made way for the era of the radio waves at the same time. From that moment on, the existence of electromagnetic waves would no longer be contested, rapidly including not only light, which propagated at the same speed, but also "caloric waves", which became a category like the others and joined the family of the vibrations flowing through the Ether. One would then have thought that optics was going to become a crystal clear science, all the mysteries of which were now solved, but the discovery of the photon by the Einsteinians at the beginning of the 20th century, and the studies on the physiology of the eye in limit conditions, showed the opposite.

In any event, this extension and unification of electromagnetic

Michael Faraday (1791-1867)

waves were achieved after Maxwell died, amidst a general conviction that the Ether really existed, but without knowing more about it so far. It was necessarily the same fluid as Descartes's, the one which swirled around each planet, but which was also both legitimized by the rain of the recent discoveries, including Hertz's of course, and recognized as the necessary medium for ondulatory optics. However, despite that nearly consensual conviction, no new element could be added to an intimate portrait, still blurred and inconsistent.

Well, it is not true. The concept of lines of force imagined by Faraday had played a considerable role in the development of Maxwell's theory, but the latter, morally destroyed by the hostile reactions towards new ideas, didn't advertise his work for the reasons described above. It is really necessary to read *A Treatise on Electricity* with particular attention to find the few allusions to the etherised medium in it, but they are there, as solid foundations of the theory. However, even today, if these lines of force have been renamed "lines of field", they only serve to illustrate drawings of devices or electrical assemblies, without telling more on what they actually represent. For Faraday, as Maxwell explains it in the Scientific Papers (in the paragraph called "Action at a Distance"), the lines of force of the Ether are comparable to muscle fibres: they tense up while shortening when an effort is asked to the muscle. In the same way, the magnetic lines of force which, for example, represent the so-called field in the vicinity of a magnet, have a density, in other words a number of lines across a given perpendicular section, which varies with the value of the field and in the same direction: near the magnet, the field is strong and the lines are closer to each other, and they loosen when they are far from it. When the field is electric, it is represented in the same way, except that the lines of force go from a positively charged surface to a negatively charged one, starting from and arriving at right angle to these surfaces. But in the same article, Maxwell gives an additional description of the basic constituents of the Ether to this notion coming from Faraday, more refined and more detailed:

"*But the medium has other functions than the one which consists in permitting light to be propagated from human to human and from world to world, and also to highlight the absolute unity of the Universe scale. Its minute components must be the site of rotatory as well as vibratory movements, and their rotating axis make those magnetic lines of force which go on with unbroken continuity in spaces which no eye has ever been able to observe and which, through their action on our magnets and in a lan-*

guage still not cleared up, teach us what happens in an invisible world, by the minute and by the century".

In another note called "On Physical Lines of Force", Maxwell goes much further and dares to give a description of the Ether, actually not very different from those of Descartes or Huygens, but the progress achieved in electromagnetism allowed him to make reasonable hypotheses that his predecessors couldn't consider, by lack of experimental support. Nevertheless, they appeared as unbearably bold at that time. This note follows another one entitled "On Faraday Lines of Force", in which Maxwell begins with matching those "lines of force" to the magnetic field, the entity revealed when we sprinkle iron filings on a sheet of paper held over a magnet and then give gentle pats to see them move in the way we know. These notions having been introduced, Maxwell then shows that the magnetic "spectrum" - that is the patterns formed by the iron filings - can be identified to the lines of force which somehow characterise the portion of space that we call "field" and where these forces act. It is understood that this way of thinking takes its source in the sine qua non hypothesis of a fluid ether, subject to mechanical stresses and to various movements of different types, in particular swirling or linear, that he would describe later.

For Maxwell, the established part of the theory of magnetism is reduced to a combination of mathematical formulas, with no attempt at making a link with other branches of physics. This means limited possibilities of progress to a student, and what's more at the cost of substantial efforts of understanding. He drives the point home when he suggests the idea, still so obvious to some people, that progress in physics cannot be performed without a simultaneous and parallel development of the mathematical tool and intuitive understanding: if only one of the two branches of knowledge is developed, it can only lead to a lack of understanding.

Moreover, in the same way, some mathematically simple formulas, supposed to represent a physical law or phenomenon, may not bring any light on the mechanism of the investigated facts, or any explanation on their causality. A typical example cited by Maxwell is the gravitation law, but today we could add Einstein's theories without remorse. There are things that mathematicians cannot replace, like philosophical reflection (in the Cartesian sense of the term) or intuition (even if we believe with Marcel Boll[5] that the latter comes from knowledge), both being the age-old foundations of physics. To show to what extent mathematics need simple logics to really improve our knowledge, Maxwell borrows from William Thomson, aka Lord Kelvin, the following reflection on the analogies which can be noticed, with some attention, in two a priori independent branches of physics: "*There seems to be no relationship between gravitation and the propagation of heat, and yet we can find a $1/r^2$ law in both cases. Just replace "centre of gravity" by "heat source" and "potential" by "temperature" for the solution to a gravity problem to become the solution to a thermal problem*". Admitting the intervention of a medium in gravitation would make the two phenomena similar, easier to understand, and so it would mean progress. That was how Maxwell viewed it, while recognizing Faraday as the father of the reasoning, with his usual elegance.

But let us now move on to the most interesting paper, as far as the Ether is concerned: "On Physical Lines of Force". This work falls into four parts which deal with the theory of vortices applied to four different fields:

1- the theory of molecular vortices applied to magnetic phenomena
2- the theory of molecular vortices applied to electric currents
3- the theory of molecular vortices applied to statical electricity
4- the theory of molecular vortices applied to the action of magnetism on polarized light

Maxwell's declared objective is here, in his own words, "to bear the reflection on certain aspects of mechanical actions resulting from stresses and movements in a fluid, and then to compare them to various phenomena that are studied in electricity and magnetism". After that, he intends to re-examine these phenomena from a mechanical point of view only. There lies the major hypothesis that electromagnetic actions could be nothing but known mani-

5 French physicist and philosopher

festations in the classical domain of physics, but transposed in a partly invisible world, which even escapes from the totality of our usual senses, because such is the difference between electromagnetism and ordinary physics: the apparent obscurity of the deep causes. Besides, the same applies to gravitation, and possibly to all the phenomena where there are actions between bodies with $1/r^2$ laws. Maxwell's great Treatise is built on this philosophy, and it is not a pure mathematical edifice as it may seem when you first read it.

With this process, Maxwell progressively and rationally leads us to definitely consider the magnetic influence as the sign of pressures and stresses, acting in a certain medium, thus following the way opened by Faraday and William Thomson. Then he has to explain the origin of these various forces acting in this medium, and this is why he introduces his vortices, which are the intimate elements of the Ether and pile up along the lines of force and make them in fact. So what are the famous vortices of Maxwell? In spite of the numerous illustrations in “On Physical Lines of Force”, it is not so easy to put oneself in the Master's place and try to see what he saw. Nevertheless, given that Maxwell's equations have never proven wrong and henceforth constitute the unshakeable foundations of electromagnetism nowadays, it is necessary for us to try to imagine the Universe the way he must have seen it, at least if we want to understand a little the genesis of these fundamental relations which finally depend, it's good to insist, on hypotheses about the presumed medium where the studied phenomena take place.

Figure 1-4 shows the arrangement of vortices in a cut perpendicular to the lines of force. All of them turn in the same direction, and for it to be possible, Maxwell assumes that there is a layer of even smaller elements between them, comparable with the balls of a ball bearing, and with such consistencies and a shapes that there is no friction, and so no decrease of energy in thermal form. For a given vortex, the direction of rotation is clockwise when, following a line of force, one goes from a lower density area (weak field) to a higher density area where the lines are closer (strong field). We recognise here the premises of what would later become, at a macroscopic level, the so-called “Maxwell's corkscrew rule”.

Figures 1-5 and 1-6 are picked out from the same paper and are part of a bigger series, all placed on the same page. But the selected ones can be considered more representative, so there would be little point in showing more, even if all of them are interesting. It

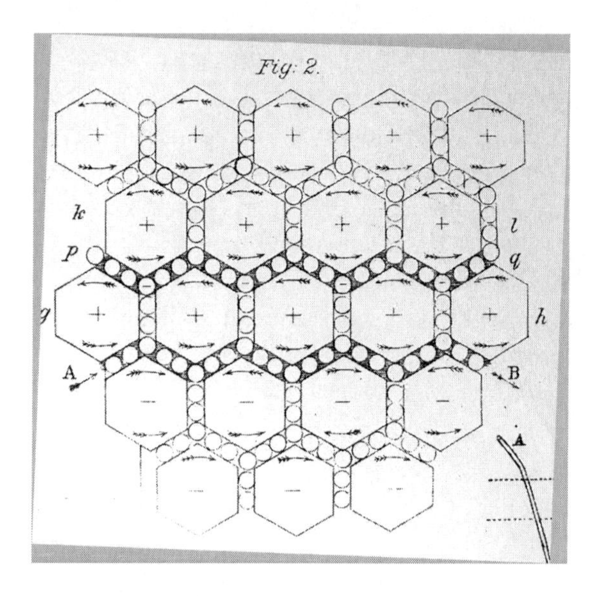

figure 1-4. The electric current AB perpendicularly crosses the lines of force seen in section (hexagons). The small balls inserted allow the vortices to rotate freely .

is recommended to everyone who wants to go deeper into the subject to consult the original paper, all the more so as all the Scientific Papers are a wealth of ideas, and may be the best example of this good old descriptive physics which doesn't exist any longer. The above sums up quite well Maxwell's representation of the Ether, though not exhaustively, and one must be fully aware of the fact that it enabled him to develop the concept of displacement currents subsequently, which ensured the completion of his “Equations”, considered a stroke of genius. However, this stroke of genius contains no sign of suddenness, as the expression usually implies; it is the outcome of a particular effort by Maxwell to go a bit further than his colleagues physicists in the discovery of the constitution of this medium- mysterious though indisputably real to him- where light propagates. So the Ether of Maxwell is constituted of very small vortices which pile up where there is a field, that is where it is possible to draw lines of force in one way

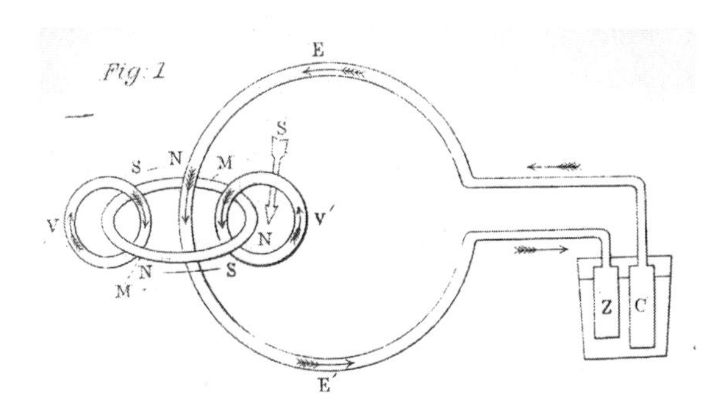

figure 1-5.
The ZC (Zinc-Copper) cell generates a current EE' which crosses a magnetic line of force around which Maxwell has drawn two vortices. Following the chosen circulation senses indicated by the arrows, these vortices will carry a particle up outside the line of force.

or another, and their rotation speed is proportional to the field intensity. But he didn't close his eyes to the frailty of the hypothesis he put forward, and still proceeded with the highest cautiousness:

Actually, we must now wonder what relationship can exist between these vortices and the electric currents, even though we don't know yet the nature of electricity, if this one is a substance, or two, or no substance at all, or how different it is from the matter, and what its links with the matter are.

Despite the cautiousness and the modesty that he shows in all his work, we can find in the same note on the lines of force extremely bold ideas which, collated with what we know about electromagnetism today, reveal the full extent of his intuition:

We have found that the revolution speed of each vortex must be proportional to the intensity of the magnetic force, and that the density of the matter of the vortex must be proportional to the aptitude of the medium for the magnetic induction.

There is something astounding in these lines, which are unfortunately never quoted in higher education: it was the first time that the question of the mass of the Ether was clearly raised, though not directly. Of course this is not its total mass, which is nonsense, but its specific mass. But it was even cleverer of him to see the implicit relationship between this mass and what he called the aptitude of the medium for magnetic induction, that we today know as magnetic permeability. It will be necessary to wait until the 1970s and René-Louis Vallée's *Synergetic Theory* to see a resurgence of this fundamental notion of mass, which cannot be evaded if the Ether is believed to carry away bodies.

So this was a rough presentation of the Ether of Maxwell. Though miserably summed up, it shows what essential role it played in a theory reinforced by thousands of experiments, and considered a steadfast foundation of modern science by everybody. And yet it has been taught as a mathematical edifice with such dryness – by not mentioning the primary hypotheses at all,

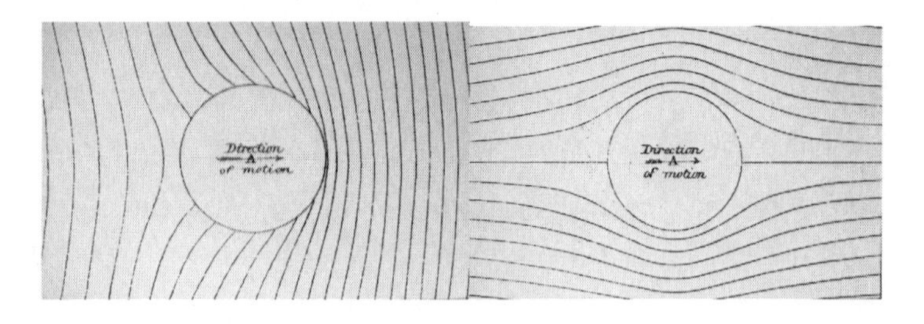

figure 1-6.
The central circle represents the section of a conductor which moves in a magnetic field, or reciprocally. Where the lines of force are compressed, the rotation speed of the vortices is higher and the field stronger.

long forgotten or completely ignored - that it discourages students to choose this option more and more. It is really a pity, all the more as the etheristic conceptions of electromagnetism pave the way to fabulous research and discoveries. Finally, it should be noted about mathematics and its tools that the use of curls by

Maxwell in his study on fields first intended the use of elementary vortices, that it was borrowed from Stokes in a general study on the dynamical theory of diffraction, and that it belongs to the usual theoretical arsenal of the fluid dynamics, which the studies on vortices belongs to and which will be the ultime weapon of any etheristic physics if it is properly redirected.

1-6 : The ether of Clémence Royer

Born in 1830 in Nantes, Clemence Royer was contemporary with Maxwell, but she didn't quote him and didn't seem to have ever had a passion for electromagnetism anyway. She was a very eclectic woman: she was a translator of Darwin as well as an acknowledged physicist, and was awarded the Legion of Honour[6]" for her work in this branch. This is the reason why a part of this book is dedicated to her, because the originality and the boldness of her theses on matter deserve a little attention, even if her theses have not been approved by today's specialists. Clémence Royer deserved all the more credit as, during the Napoleonic era, women were not very welcome in a male chauvinist scientific community where physicists and mathematicians were always bickering. From this point of view and considering the social environment of that time, we may see her as the Louise Michel[7] of physics, and a precursor of the feminist movements. However we are slightly surprised when we discover her conception of the Ether in her main publication on physics, "The Constitution of the World", where she builds in 800 pages an extraordinary theory which, it has to be admitted, is rather difficult to swallow today.

For Clémence Royer, the Universe is constituted of two entities: the Ether and matter - nothing really new. But the main difference with the previous theories, starting from Descartes's, is that the Ether doesn't penetrate matter any longer, because both entities are made up of the same atoms which together compose "cosmic matter". The principal idea is that these atoms are neither rigid nor not deformable, as suggested by the Epicurean time, but on

6 Légion d'Honneur: the highest honours awarded by the Republic of France
7 A French revolutionary

the contrary fluid and expansible, while at the same time not penetrable. If isolated, each of them would tend to form itself into a sphere and to occupy the whole space, but the presence of the other neighbouring atoms, which have the same behaviour, results in the uniform and total filling of the Universe. Its formability causes the individual spheres in groups, which is the general case, to be squashed together and finally take the shape of joined polyhedrons, filling the space completely. Consequently, there is no way for a fluid to flow between them; the Ether will then manifest in a constant pressure, against which the matter will permanently fight by developing an antagonistic pressure, maintaining in this way a spatial balance the geometrical limits of which will be related to the temperature of the considered body. Matter is seen as

Clémence Royer (1830-1902)

Figure 1-7 Molecular shapes of calcium

a set of degraded atoms which have therefore become measurable through an accidental process where the etheric atoms, originally imponderable, have lost energy to get inertia:

"If with the Ionian dynamists we consider the first elements of the cosmic substance, not like centres with an indefinite expansion force and all the absolute properties of perfect fluids, the relative size of these atoms will be, under the same pressures, pro-

portional to their quantity of substance, or to the sum of their expansive forces. They will be all the bigger as they will be more active. Their inertia or their mass will become an inverse function of their force and their volume, constantly modified itself by the variation of the pressures that they exert on one another, by virtue of their expansion force which makes them fight for each of them to appropriate its proportional part of space, without any vacuum between them.

In this respect only Descartes, by denying the existence of vacuum, would have been right against Leibniz who hesitated a lot on this subject.

figure 1-8 Propagation of colours.

As for gravity and inertia, far from being primary and essential properties, inherent to all atoms, they would be, on the contrary, acquired by only certain categories of them and for each of them proportional to their loss of substance or to the weakening of their expansionary force".

A little further:

"So the weightless Ether would be the virtual and primordial state of the cosmic substance. The weighing matter would be

made of ether atoms more or less weakened by a variable loss of their substance; their volume would be virtually in inverse reason of the cube of their inertia or of their mass, proportional to their weight. But this volume would be variable under certain conditions of pressure”. (La Constitution du Monde, p70/71)

From there, Clemence Royer unveils her conception of gravitation, which ensues from the previous hypotheses:

“*All the atoms, having a weight or not, pushing away one another when in contact, in accordance with their expansionary force, variable in intensity, proportionally to their volume, the weakest, the smallest, whose expansionary energy is lowered,* ***pushed*** *one against the others in direct relation with their inertia or their acquired mass, must seem to attract each others*”.

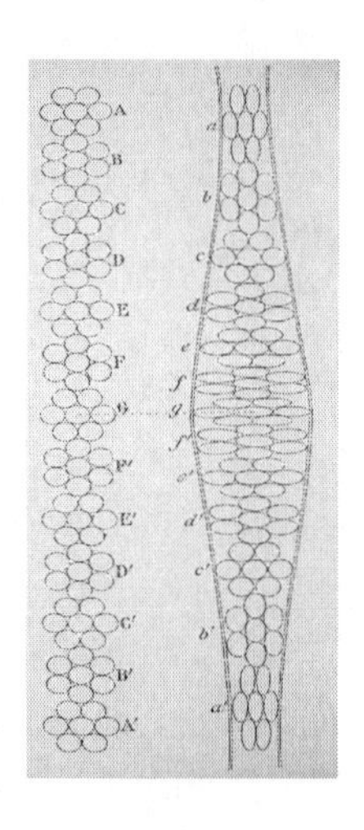

figure 1-9 Sound wave and deformation of atoms.

We can find here, expressed with force, the denying of the existence of attraction forces, which Newton himself was completely persuaded of, and their interpretation by static imbalances which push the bodies towards each others.

Bearing the mark of a continuous mix of justified critics to the official physics and too new ideas, the work of Clemence Royer is disconcerting. Certain conceptions, like this of the monad-atom with a variable volume, open the way to probably promising reflections, at the limit of metaphysics, but the break with classical thesis is too abrupt, too sharp, for not feeling a certain disease if we try to play the game and assimilate the successive demonstrations in chain swept along by the avalanche of her revolutionary hypothesis. As more as the scientist is very peremptory, self-assured, and that doubt seems not to habit her: it's evidently a “battling”, and her style singularly contrasts with the advised cautiousness of Maxwell. From this point of view, she is closer to Descartes than this last.

However, we notice that she frequently use of a conditional that we wonder how to interpret it: hesitation, lack of proves, embarrassment with a too narrow vocabulary for expressing so new ideas? But when we read Descartes or Einstein, is it so different?

On the other hand, when the question is to criticise the contestable foundations of the various physicians theories, and in particular the "systems of the World", Clemence Royer knows how to insist on weak points and show what she consider as ineptitudes with a strong logic, accessible to anybody. As witness, this passage which justify the existence of ether:

" *what we don't conceive at most, is the persistence of the movement of bodies in an absolute vacuum of any expanded substance and how it happens that, in this vacuum, all the celestial bodies, submitted to the gravity, which is introduced to us as a reciprocal attraction between masses, don't vertically fall the ones on the others.*

It's an else way, if any movement takes place in a completely full medium, full but fluid, plastic, elastic, weightless and without inertia, and specially not compressible by the fact of its incoercibility."

She probably goes too far in the enumeration of the properties of the etherised medium, those which justify her own theory, but not the others, but she uses a rational method which consists first in destroying what blocks her, for afterwards rebuild another edifice, more conform with the elementary logic, the one that everybody has got and which she constantly appeals to. Whatever it be, even if her work is nowadays completely unknown by students in physics, even if her constitution of the World doesn't stick any longer with the present givens, it's all very well that her critics would merit a new reading, so as to be used as a basis for a permanent examination of the soi-disant progreesion of knowledge.

1-7 : The ether of Tommasina

The Swiss Thomas Tommasina belongs to the group of the forgotten of physics, like Clemence Royer, but at a point that it is even not possible to find his biography on Internet. We sometimes wonder what sort of crimes certain people must have committed so as to be this way excommunicated and condamned to be perpetually forgotten, the worst sanction for a man of science. Well, man of science he was, as doctor ex science and honoured member of the National Institute of Geneva. Tommasina wrote in 1927 a book entitled "physics of gravitation and dynamic of the Universe", that he considered as his "scientific testimony". All inside is based on the postulation of the existence of ether but, to well show that is conception of the World is not an individual fantasy, but is in the continuity of a certain current of thought, he refers to other scientific philosophers of whom he has chosen the ideas as base of starting, and whom the citations he takes from them merit to be reproduced just like they are, so limpid and judicious they are.

1855-1935

First of all the one of the mathematician Joseph Bertrand, in an eulogy of Lamé:

"*Indexes too certain to let place to doubt reveal the existence of air. One sees it shaking the leaves of a tree; one ears it whistling in its branches; one understand that it resists the wings of birds and, stating that air exists, nobody is tempted to say:* ***in physics only****. No hand has touched ether, no eye has seen it, no scales have weight it. It is demonstrated, not shown; it however is as real as air, its existence is as certain: if I dared saying that it is more, I would be accused of exaggeration*".

And more, from the same:

"*Lamé, nevertheless, would have encouraged me. Whatever it be, all schools agree on this point. Fresnel has pushed the demonstration to the evidence; he has done more than convincing his challengers, he has reduced them to silence. The Universe is filled by ether, it is mare expanded, more universal and may be more active than weighing matter...*".

Then, this of Lamé himself:

"*How believe that this fluid, of which the intervention accords and conciliate up to the least details the facts relative to light, wouldn't intervene in the caloric phenomena? That, mixed with material molecules, it doesn't influence on elasticity? And that, present for the electric actions, it however doesn't play any role?*".

After having normally paid homage yo his predecessors, Tommasina engage himself with authority in his personal way, denying first the two fundamental mistakes, in his opinion, of the etherist conceptions, one concerning the driving of masses and the other the light celerity. To get out from the contradictions he has noticed, he will emit two master ideas of which the pertinence is still of actuality.

The first one is that, for himself as for Descartes, it's ether which drive planets, so it's no longer possible to attribute it its two usual characteristics so far: a zero mass and a status of immobility explaining the absence of action on the movement of an astral body, while making of it an absolute spatial reference.

After that, and may be it's there the strongest idea of his theory, the light celerity in such a medium cannot be a universal constant: "*...the pressure Maxwell-Bartoli undergoes a continuous damping as being by the isotropic mecanism of radiation propagation, and that, following this damping, the light celerity* **c** *is not a constant, like relativists suppose, but an average conveniently exact in the limits of our experiments and our observations.*"

These ideas are now springing again, after the non-acknowledged failure of the relativist school, and we'll see later that they probably have inspired the Synergetic Theory of René-Louis Vallée, who cannot have have missed reading Tommasina.

That's not all. There is in the approach of Tommasina on the problem of ether, in general, a third determining idea, extremely pedagogic, which wants to explain why it is so difficult for us to agree on this question, and which consists in more finely analyse the way we use our natural means to observe facts. If it is so painful to us to imagine the physic, real existence of a medium for light propagation, and more generally for all electromagnetic waves, it's because all our senses gather to tell us that there is nothing between an object that we are watching and our eye, apart from certain matters that we can easily identify: air, fog, a glass, etc...Starting from this fact, Tommasina will endeavour explaining to his reader why the impressions, the appearances, either visual, tactile, auditive or what we want, are not in contradiction with the existence of a matter filling up the whole Universe. It's so smartly done, so well presented, that we had rather integrally citing him than risking to deform his thought: every word counts.

"Relativists mathematicians have tried to get out of this difficulty by making intervene a purely symbolic entity to which one has not to attach, from them, a real substance. Recognizing that here it is only an illusive solution, and as besides we don't believe in the fathomless mystery supposed by Lord Kelvin, we have thoroughly examined the problem and admitted the necessity of taking into account its complexity, beginning by the study the modification that produces the vision by ether, before tackling these, astronomical, of the movement of stars.

We have wondered: what is seeing? For anyone, to see is to more or less clearly detect an object, following the distance it is. From his own experiences, every man knows that he stops seeing when an opaque screen is placed between him and the thing he is watching on. Every man has this way acquired the certainty that space is free between him and the thing he is seeing; or, if it is not completely empty, what is inside, let it be solid, liquid or gaseous, must be transparent. For example we know that water and glass, up to a certain thickness, are transparent, and that it is the same for air. This way of thinking is common to any child, and when one of these children, having finished his studies, has become a physics professor, he still keeps this way of thinking about the physical conditions which allow vision, either this of illuminated objects, or this of light-emitting ones. But however this is only illusion, as unreal as the march of sun, that every day of good weather our eyes can follow from rising to setting.

So why, when all people having got the most elementary necessary tuition accept to correct this wrong way of judgement, is it not the same for the phenomenon of vision? It's because, for one, it's enough to take into consideration the fact that Earth turns on itself in the opposite sense of the apparent march of Sun, and so that this march doesn't exist, and our visual illusion is only the result of the wrong supposition of our immobility; for the other, at the contrary, we must go inside the hypothetical physical mechanism which really constitutes what happens in the phenomenon of vision.

We are trying to explain that; but to make easier and clearer our explanation, we'll take the allowance of neglecting, for a while, the phenomenon of vision, to examine this of audition

Noises and sounds that our ear perceive arrive to us more or less clearly following their distance and their intensity, but we also know that it goes faster in solids than in liquids, and in these last better than in gases. We know that when we stick an ear onto a door we better ear what is said the other side. Nobody has problems to admit the special mecanism which allows sound transmission inside a solid body, interposed with no solution of continuity between our ear and the source of the sound or any noise.

Therefore, a quite identical mechanism leads vision, and our eye is constantly stuck to the solid ether of Lord Kelvin, it never can be removed. It's by and through ether that it receives the vibrations which make visible illuminated objects and luminous points, be they sparkles or stars.

But we add this notice that if we have got eyes to see the door on which we stick our ear, we have no organ to directly feel the presence of this solid ether, which is interposed between us and all surroundings, which even insinuate until the atoms of our organic stuffs.

From these few considerations, I think that the reader will not have any difficulty for correcting his habit, previously described, of considering the phenomenon of vision, in other words what is to see, and that he will be convinced that it is the filled space which allow him to see, because an empty space would constitute, at the contrary, absolute opacity, unable to transmit any vibration."

To be conscious of the existence of ether, to possibly try to convince others if it is necessary, the daily experience shows it is obligatory, indispensable, to do this preliminary dialectal work that uses Tommasina with so much mastering and subtlety.

We'll find in the present book, in the other chapters on ether, a development, or at least another presentation of the topic, but we must agree that the method of Tommasina is of a real elegance. His analysis is of an enormous weight. His book on the dynamics of Universe, from which are picked out the former citations, is of such a richness that it would a hundred times merit a reissue, be its logic, which has nothing to envy to these of Marcel Boll, still a reference in the category, is both powerful and pertinent. It's the absolute weapon to fight against the invasion of physics by relativists, whom more and more smoky thesis ends up discouraging normal people who tries to understand the content of scientific issues and remove them, step by step but surely, from the most beautiful intellectual activity it be: Science.

But let's get down again to ether, which is our subject. Once he has dismantled, through a few judiciously chosen examples, all the unlikelihood of relativist thesis and shown the necessity of the existence of a medium for EM waves propagation, Tommasina begins drawing what will be his own ether, and refers first to Maxwell, with whom he agrees on the fact that there is everywhere corpuscles which turn or vibrate. In other words, ether is active: it's a medium globally and locally motionless, but where all is moving, where all which seems stable is not at rest. What was vortex for Maxwell becomes "energon" for Tommasina, but this last completely admits the existence of the Faraday lines of force, and we may consider him, as a function of this parallelism between the two reasonings, as the prolonger of Maxwell's work for the particular subject of ether. So the energon is the ultimate particle, of which the movement is vibratory instead of rotatory, what makes the difference with the vortex and curiously makes it closer to Descartes ether and its "agitation" of elementary particles. The energon is then defined as a punctual electron having its own domain around it, domain of which the extension is a function of its energetic status, which itself is the result of the permanent action between ether and matter.

Another point, very important, of Tommasina theory, is the affirmation that ether has not a negligible specific mass, that it is not the infinitely light substance suggested by its qualifier of "subtle matter" given by Descartes, but at the contrary a fluid of an enormous density, able to give to astral bodies the movement they have got, to carry them away on their orbits or fix them on their positions.

The book of Tommasina, of which the exact title is "Physics of Gravitation and Dynamic of the Universe", is so rich that it is a pity to give only a few digests of it, when the complete reading is a rare happiness for whom likes anti-establishment physics. However it will be enough, to finish with this too short cursory glance, of a resume of the author ideas given by the French physicist Cornu during a lecture at the Cambridge University, in 1899:

"Mr Tommasina begins with establishing the following physical axioms:

*1- Any phenomenon goes on in **space and time.***

2- any phenomenon can only be produced by matter in movement.

3- No action can be transmitted between two bodies without a material intermediary.

4- The unlimited space, where Universe is evoluting, must be filled everywhere by moving matter.

5- Movement without matter is inconceivable.

Then Mr Tommasina passes to the development of his main hypothesis, which are the following:

1- Ether really exists, it is material, isotropic, symmetric, homogeneous and of a perfect elasticity, it is unlimited and fills all the space occupied by Universe.

2- Ether acts on the atoms of all bodies, with which it is always on contact by the vibratory movements of it cells (electrons) and by the pressure which exists between them. This last being the necessary and sufficient condition for energy transmission.

3- Universal gravitation and all physical and chemical phenomena are relevant to these two actions of ether, means radiating energy and etherical pressure.

4- The appearing phenomena of attraction and repulsion are always caused by pushing or carrying in a sense or the contrary sense, energy transmission being done by the communication of movements between ether and weighing atoms and reciprocally.

So is this, incompletely resumed, a theory of an exceptional quality, of which the prolongation by René-Louis Vallée and his Synergetic Theory will be detailed in the following chapter. The name of Tommasina is to-day completely ignored by universities people, just like Clemence Royer one's. Both have in common that they have refused the obligatory teaching of the official science to follow their personal ways and let speak their physicist instinct. But this is never done without irritating the Establishment, represented all upside by the Sciences Academy, overpowered bastion of the dogmatic thought and association of congealed-minded careerists, who in all times have considered themselves as the incorruptible and incontestable guardians of the acquired knowledge. Accordingly, the contesters cited upper were not specially timid and taught their revolutionary thesis with an unbearable vigour. Moreover, specially concerning Tommasina, this one was guilty of sacrilege and lese-majesty by daring openly criticize Einstein, the new idol of Theoretical Physics. Even now, practically one century after, we are still at the same point, practises and mentalities are any way the same. We must not forget that more than a hundred years were necessary to unbolt Newton from his pedestal. But cheer up: Einstein will follow, good riddance!

1-8 : The ether of Gustave le Bon

Edouard Branly and Gustave le Bon, knowing well each other, are two examples of eclectic minds working in several disciplines, as well in medicine as physics, this last being probably, even certainly, their common passion. The first one began studying physics before diverting lately to medicine, the second went the inverse way and made also his name known in anthropology and social sciences, but only his performance in physics will be considered here because, on one hand, he realised extraordinary experiments of which nobody tells to-day, and on the other hand he was a convinced etherist. His two major books in physics are “Evolution of Forces” and “Evolution of Matter”. What is interesting, in the reading of the works of this confidential type of scientists, is a particular style probably coming from a more large knowledge fan, and which is characterized by another logic, softer and closer to the reader than could be these of specialists in physics such as Poincaré or Duhem, for example, who has made of mathematics their main tool and whose style is in consequence. Those who compile the old physics books of licence level and issued at the beginning of the 20th century may testify on the particular clearness of those which are dedicated to students in biology, medicine, pharmacy or even letters, in comparison with these of the “rigid line” of the so-called exact sciences branch. Reading le Bon is enjoying physics like a romance, it's find again the pleasure of knowledge, the pleasure of learning and reflection. This is as more true that the care of avoiding an excess of mathematics, the intellectual effort that you must do to do it, often leads to more elegant demonstrations and to a presentation of the facts more susceptible, thanks to a simpler logic, of interesting the reader, be it a specialist in the branch or simple curious. Le Bon has this in common with Tommasina that he also wrote for a large public, contrarily to the “big scientists”, like Maxwell or Fresnel, who only wrote for the restricted sphere of the searchers of their level

Le Bon of course was etherist, but his approach of this subject is very different of the previous. It is first more intuitive, it is always strongly related to reality and doesn't progress without the care of explaining without modelling. He is, on this point of view, the "anti-Duhem". After that his scientific reports are often added of societal and human notices which translate his diversified culture and his antecedents, as well as, like all contesters in physics, his total distrust about official science. This being, he proposes, like Tommasina did, a digest of his master ideas that we can find at the beginning of "Evolution of Matter" and which constitutes a short of his conception of Universe:

"*From the experimental researches displayed in our diverse memoirs and which will be resumed in this book get out the following propositions:*

1- Matter, in times past considered as indestructible, slowly vanishes by the continuous dissociation of the atoms which constitute it.

2- The products of matter de-materialization constitute intermediary substances by their properties between weighing bodies and weightless ether, that is between two words that science had until now deeply separated.

3- Matter, previously seen as inert and only able to refund the energy that it has been given, is at the contrary a colossal tank of energy -intra-atomic energy- that it can spontaneously spend.

4- It's from the intra-atomic energy released for the dissociation of matter that result the majority of Universe forces, in particular electricity and solar heath.

5- Force and matter are two diverse forms of a same thing. Matter represent a relatively stable form of the intra-atomic energy. Heath, light, electricity, etc..., represent unstable forms of the same energy.

6- Dissociating atoms, or in other terms de-materializing matter, it's simply transform the condensed stable form of energy called matter into these unstable forms known under the names of electricity, light, heath, etc.

-7 So ordinary matter can be transmuted into diverse forms of energy, but it probably was at the beginning of things that energy could be condensed under the form of matter.

-8 The equilibriums of the colossal forces condensed in atoms give them a very big stability. However it's enough perturbing these equilibriums by an appropriate reactive as for that the desegregation of atoms begin. It's this way that certain light rays can easily dissociate the superficial parts of any body.

Gustave le Bon
1841-1931

-9 Light, electricity and most of the known forces coming from the de-materialization of matter, it follows that a body which radiates loses, by the simple fact of this radiation, a part of its mass; if it could radiate all its energy, it would completely vanish in ether.

-10 The evolution law cable to living beings is also available for primary elements. Chemical species no more than living species are invariable.

Like Tommasina and Maxwell, le Bon considers ether as being the siege of all physical phenomena. Like them, he supposes that it is constituted of, or it contains, particles of which one would well know the shape and the relative disposition, but which in final are in movement. He often use the term of "gyrostat", which bears in itself two notions: the fixed position in space and a very fast rotation on itself, which obligatorily confers it an axis of symmetry. In fact, vortices of Maxwell must be exactly the same thing that the gyrostats of le Bon, which by the way doesn't claim for this designation, that we can attribute to Tommasina as well as William Thomson. But le Bon doesn't really try to foresee the mystery, that he thinks too deep, of the intimate constitution of space, he takes interest in phenomena for the study of which the intervention of ether can bring a simpler or new explanation. There is indeed a phenomenal quantity of physical facts, known by everybody, of which the justifications given by science in scholar tuition let any one on his hunger: the kinetic theory of gases, which seems to be one of the most solid foundation of physics, is an example. Relativity, at the other end of the complexity scale, is another one.

Between the two, we have the choice: it's enough to pose a minimum of questions, what students scarcely do, and teachers no more. Figure 1-10 illustrates a master idea of le Bon on matter, that is its permanent interaction with ether. Here we see a person holding a sabre and vainly trying to cut a water jet which gets out from a pipe, after a fall of several hundreds meters. How a fluid, evidently without rigidity when it is on rest, can it get one as soon as we give it a speed? What the hell this speed can change in the intimate constitution of a liquid of which we know that its molecular cohesion is null? Has the rigidity of a solid something to see with a speed? And the speed of what?

One can at leisure increase this burst of questions, as many are they which come to lips as soon as we try to shell any quite odd phenomenon. This last is particularly interrogative and open the way to deep reflections on the behaviour and the constitution of matter, and le Bon gives us his ones in an article of "La Nature" (n° 1855, p17: the role of speed in phenomena), from which is picked he illustration. For him, "*the great constants of Universe are movement and resistance to movement. This resistance is inerty.*". The rigidity of the water jet acquired by speed is only one example, chosen for its evidence, among other macroscopic or microscopic phenomena, going from cyclone to the electron, where speed changes the nature of the forces at stake. But one cannot seriously ascertain this mutation of matter under the action of speed as an intrinsic evolution: there necessarily is something else, and an interaction with ether is an hypothesis which naturally goes to mind as soon as we admit the postulate of its existence. When we choose to go in this way, it's also necessary to suppose that ether has supplementary properties on which le Bon, by cautiousness, doesn't linger on, but that Tommasina has stated with voluntarism, and which will be developed in the second chapter to it (ether) consecrated. In particular, its specific mass. Though ether is the discrete but omnipresent fellow of le Bon, he finally tells few about it, like if its existence was something natural, the same as for air that we breathe and which definitely has for us the same vital importance. His aim, what he wanted to elucidate and which was the motive of a ten years research, partly with Branly, was a new approach of matter, which was previously considered as absolutely stable but which, for him, continuously destroying, despite not visibly, because the evolution is so slow that one cannot notice it. One of his formulas is, to paraphtase Lavoisier: "*Nothing is created, all is being lost*". We may be against, but it is well sent, no?

Forty years before the invention of the atomic bomb, the intuitions of le Bon have something exceptional, as witnesses this article untitled "de-materialization of matter", issued in the n_o 1699 of the weekly La Nature:

" *In a suite of researches of which the first were published in 1897, I have succeeded in proving that the dissociation of matter, radio-activity as we say to-day, is a universal phenomenon which can be observed with all the bodies and which goes on, either spontaneously, or artificially, in a lot of circumstances...*

...so called radio-active substances, like uranium and radium, only present at a high degree a property that any body has at a casual degree.

... matter is, contrarily to all ancient conceptions, a giganteous tank of force in a state of extreme condensation. It is to this force that I have given the name of ***intra-atomic energy***.".

One often attributes to Einstein, to-day and by those who knows only superficially the small history of physics and the chronology of discoveries, this knowledge of the energetic properties of matter. We see, through this occasion, when making acquainted with works of forgotten scientists, that reading the books of the past can reveal many surprises, and that the paternity is sometimes, even often, very difficult to determine. It also happens, consequently, that homages and distinctions could be distributed with fantasy and injustice. That's life.

Apart from that, and it is the most weightly, works of le Bon are, like these by people evoked in the former paragraphs, completely ignored by most of present physicists. Typical example: in year 2000 and some, a horrific shout escaped from the anguished breasts of the guardians of the standard kilogram; the mass of this one had decreased! After a first moment of panic, they looked for determining the causes of the phenomenon, confirmed after that the possibility of a measuring mistake was eliminated, first reason proposed but quickly abandoned after checking. One then attended, by the experts of fundamental physics, a torrent of stupendous theories of which the most amazing was, very seriously advanced, that the gravitation field had possibly varied in the vicinity of the Pavillon des Poids et Mesures[8]. Those who have read the experimental reports of le Bon on metals evaporation have killed themselves laughing, so evident is the right answer. But in fact this is not funny at all: this particularly edifying story shows to all of us to what sort of stupidities can lead, on one hand the lacunas of tuition and the selective lack of memory for our ancients, of whom many have passed their whole life trying to make science progress, and on the other hand the judgement deformations to which mathematical physics leads, more generally helped by the present scientific teaching methods and the programs which go with.

Le Bon was contemporary with the Curie, Henri Becquerel, Poincaré, Einstein, Brillouin, Tommasina and all the other members of what was called at a moment the "Copenhagen School", hyper-snob short to design a group of scientists, congealed for the posterity on the mug-scope photo of a Solvay congress in the beginning of the 20th century. In other terms, each of them was aware of what were doing the others, of what they were thinking, of the main orientations of their works, so that we may wonder, through reading them to-day, in what way they have been each other influenced, and if some of them have not simply swiped a

8 French location for standards

few ideas by the others. The rules of ethic, honesty and good education in the sphere of the most advanced research, have always been at the level of buccaneering. Le Bon cites, on this subject, in the preface of "Matter Evolution", a pertinent description by the American philosopher William James:

"*Every new doctrine crosses three status. It is attacked because*

figure 1-10

Rigidity of a high pressure water jet.

declared absurd; then it is agreed that it is true but evident and irrelevant. Finally one recognizes its effective importance and the challengers then claim for having the honour of its discovery."

Has le Bon inspired Einstein, that's something we'll never know, in return the clarity of expression of the first breaks with the constantly ambiguous style of the second. It's however this last, the man who has engaged physics in a blind way for more than a century, that history has chosen for immortality. In his fight for the promotion of ether, great promise notion which will be the coming big scientific rise, it will be fair to greet le Bon for his courageous participation, and to whom we'll let the closing word of this paragraph:

"*It seems that physicists should have seen since a long time, that is far before the recent discoveries, that matter and ether, intimately bound, exchange their energies don't constitute at all separate worlds. Matter constantly emits light or heath vibrations and can absorb of them. Till to the absolute zero it constantly radiates, which means projecting etheric vibrations. The movements of matter propagate to ether and these of ether to matter, there wouldn't be neither light nor heath without this propagation. Ether and matter are the same thing under different forms and they cannot be separated. If we had not started from this narrow mind that light and heath are weightless agents because they seem not to add anything to the weights of bodies, the distinction between matter and ether, to which scientists attach a so large importance, would have vanished since long.*

Of course ether is a mysterious agent that we cannot isolate, but its reality asserts itself since no phenomenon could be explained without. Its existence is as certain as the very matter. We cannot isolate it, but it is impossible to say that we cannot neither see it nor touch. It's, at the contrary, the substance that we see and touch the most often..."

1-9 : The ether of Lakhovsky

We now are tackling a quite particular category of scientists, and if Georges Lakhovsky is mentioned here, it's as one says after "ripe reflection", and also after a long hesitation. Indeed, despite of the fact that the value of his person was recognized by his contemporaries of the beginning of the 20th century, because it's not anyone who may be issued by Gauthier-Villars, Douin or Alcan and prefaced by professor D'Arsonval, the part of his works which is more relevant to mysticism than to science makes that we have to be cautious before eulogizing and to well separate the wheat from the chaff. By the way his case is neither unique nor exceptional, the astronomer Flammarion also has dared going into the dark zones of spiritualism, and he was not excommunicated for that. And there are many others, known or unknown, who have been adventuring in regions that cautious physicists carefully thwart if they want to keep immaculate their reputations of scientists. Should have been able to witness on this ostracism Yves Rocard, eminent physicist, laboratory manager at the ENS[9], executed by his colleagues for having tried to scientifically tackle the problem of water diviners. These reservations being made, the conception of ether of Lakhovsky have the particular interest of having as a frame the medical domain where, despite having a formation of physicist engineer, he accomplished a unique performance in the treatment of cancerous tumours, that could have made him definitely honoured but which took end tragically. Lakhovsky belongs to these individuals both illuminated and pragmatic which, upheld by a faultless willingness and by the total conviction of being right, go to the end of their ideas. Helped by his friend Nikola Tesla, who provided him with the necessary advises and elements to realise his equipment for the medical treatment of tumours, he very early had the intuition that space was crossed by an infinity of electromagnetic waves which probably might play a prime role in the creation and the maintenance of animal life. This reminds us something, already evoked in the previous paragraphs about Maxwell and Tommasina who, let's repeat again, were his contemporaries and the theories and works of whom he was very probably aware. The first one is by the way cited in several of his books.

The Universe of Lakhovsky is exactly the same that the one of Tommasina. But where this last tries to build his physics, the former wants to find the source of all our mortal beings diseases and also the explanation of the great mystery of existence. What physicists call ether, Lakhovsky gives the name of "*Universion*", contraction of the two words "univers" and "ion", and he gives of

9 Famous French High School

it this description in the first part of his book "the Great Problem":

"*In my precedent books, I tried to establish the existence of a weightless medium, infinitely subtle, that I have called Univer-sion, which would fill up the whole Universe and would be present, in the immensity of intersidereal spaces as well as the molecular and atomic interstices of the most subtle bodies.*

This medium, which by certain of its properties remains the old ether of physicists -and that we presently call cosmic rays- would be the ideal pro-matter from which derive all the known sub-stances. It's it which the same way regulates the course of stars across the celestial regions, as well as the movements of the infin-itely small particles -ions, electrons, protons, etc...- which are shaking inside the atoms, in vortices of an extraordinary speed and which constitute the immense tank of all the energies of which the effect are felt everywhere in the Universe.

So the Universion, as I have shown, is the vehicle of waves and radiations of all sorts which cross space in all directions and which cannot be transmitted from a point to another without its intervening. Consequently, it also explains light propagation or radio waves in vacuum, as well as the passage of electricity in the denser bodies like metals."

Unfortunately, Lakhovsky is then engulfed in a metaphysic the-ory where he mixes all the irrational themes which are the basis of the human thought myths and which would want to end in the di-vine knowledge of all. Thanks to scientific arguing or considered as, which constitute the tactics of all those who propose and up-hold esoteric thesis, he wants to embark us in a serial of pseudo-demonstrations of the existence of soul and immortality. So we'll only remember here, in this walk inside the etherist thesis, the few upper lines, which are largely enough to give a relatively clear image of what represented vacuum for Lakhovsky.

It doesn't the less remain that, apart from this digression in the mystery of the next world, that we reject as dangerous, Lak-hovsky was a scientist, and that his general score in the medical domain, that he had chosen by idealism, is of interest when it is only considered under this angle. It is assumed that his friendship with Tesla has not been for nothing in his knowledge and mastering of electromagnetism and radio, but Lakhovsky had in supplement this curiosity and this sense of analogy on which we have slightly insisted in the introduction, and which make great physicists, or better great discoverers, because one can be one without the other. Being totally filled with his idea of an active ether, constantly crossed like this of Tommasina and, later, this of Vallée, by a double infinity -in direction and frequency- of electromagnetic waves, Lakhovsky has tried to know what this reality could have as consequences on living beings, whatever they could be: microbes, plants, animals and mankind, all has been prospected. And the results are stupefying.

His first systemic studies applied on vineyards and more specially on the influence of sunspots on wine quality. He made evident, this way, a correlation between years of good wine and eruption cycles. He also tackled, in the same order of ideas, the relationship of the nature of ground and the quantity of cancers, region by region or even arrondissement by arrondissement in the case of Paris. The observation of the behaviour of pigeons, which seemed to loose their orientation sense near the high power radio transmitters, comforted the feeling he already had that living cells were a sort of receivers for EM waves, and that these ones could have profitable or not profitable effects, following their frequencies and their powers, on different things like health of people or growing of plants. From that came naturally the idea that the fact of submitting an organism in disease to "good" EM waves could fix it by providing the damaged cells with the vibratory energy adapted to their situation. Lakhovsky decided then to make, being 53 and with the help of his friend Tesla, a wave transmitter at a frequency of about 150 MHz, device that we would call to-day a VHF power transmitter. Once the equipment constructed, it was neces-

Georges Lakhovsky

1869-1942

sary to check the validity of the principle, what was done first on geraniums, submissive subjects unable to protest in case of maltreatment, to which he began to graft tumours, and then to submit them to his transmitter. And it was totally successful: plants made ill re-found health, and even went developing more than witness examples. It just remained then to try on man. One year after, from 1924 to 1928, with professor Gosset, clinic manager at Salpétrière Hospital[10], he tackles condemned ill people and brought about the recovery of cases reputed hopeless. Successes come in stack and begin to raise jealousies, But Lakhovsky insists and improve his care appliance by inventing the"multiple

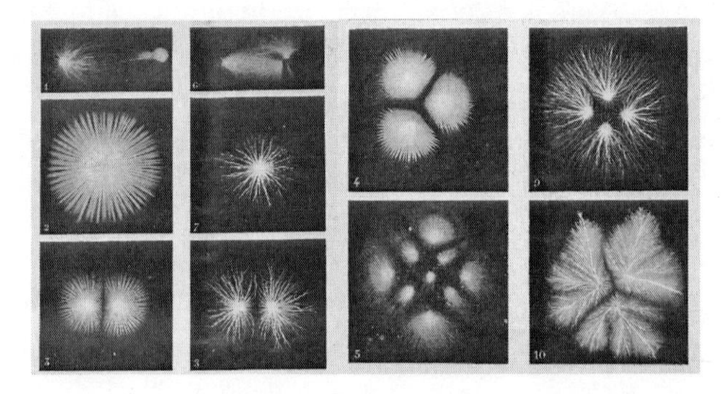

La Nature no 1699. Figures obtained by direct contact on a sensitive plate of metallic points connected to a HT generator, and showing lines of force.

Figure 1-11

wavelength oscillator", a multi frequency transmitter providing to two large band antennas, between which was placed the organism to cure, bursts of power controlled EM pulses.

Fleeing Europa in war, he migrates in 1941 to USA, where after having proved again the efficiency of his technique with hiw new equipment, he dies in suspect circumstances the following year: after having been knocked down by a car, he is taken against his will to a hospital where there is a sightseeing with two men, and he is found dead. Has he been executed by hospital lobbies, which could have seen with a critical eye a too cheap and too fast treat-

10 Famous old hospital in Paris

ment against cancer? Was he the victim of a Nazi revenge? Nobody will ever know. But as a matter of fact, all his equipments and archives disappeared the same day, and since cancerous people is no longer cured by waves in American and French hospitals.

Figures 1-11 and 1-12 illustrate the representation that had Lakhovsky of cellular phenomena. He thought that cells and micro-

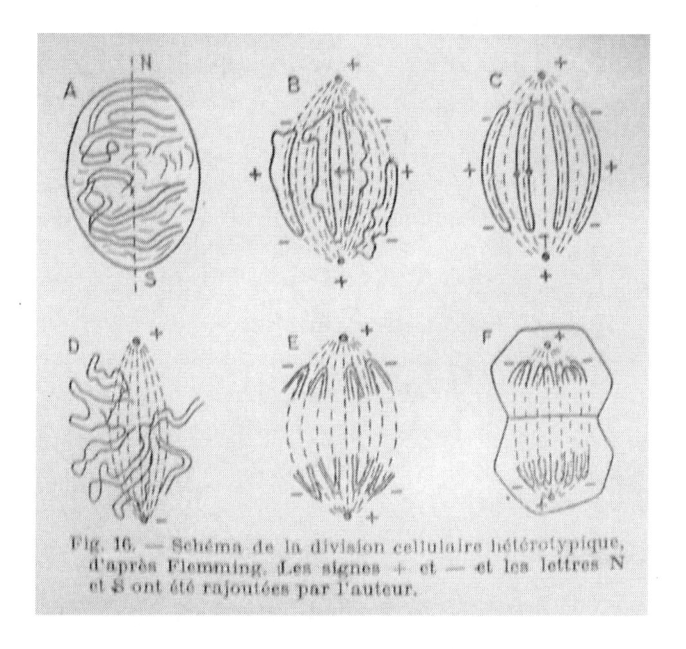

figure 1-12
Cellular division and lines of Faraday

organisms were built like receivers including oscillating circuits, at the ordinary and electrical meaning of the term, that is made of inductances and capacities. So put he in evidence, in the microscopic structures, filaments which had the shape of a coil. Capacities being only a geometric characteristic of vicinity with, either a ground plane, or between the turns of a coil, there was not any problem in this way for finding them. In order to complete the list of the electrical materials necessary for his theory, he measured the conductivities of the different elements to well show that there was in the cellular components all a scale of values, going from the good conductor to the good isolator, these values being besides susceptible of big changes in case of brutal and too fast raising of the temperature. Ether being supposed constantly crossed by EM waves, there obviously was a link to do with the cellular life, and this is the basis of all his observations, of all his experiments and of all his results in the treatment of tumours, and in the explanation of a considerable number of biologic phenomena, never explored before under this angle.

Figure 1-12, particularly, is there to appeal us on questioning analogies between the lines of force of Faraday and Maxwell and the geometrical arrangements taken by the cellular constituents seen at microscope by Fleming during their multiplying phases, and simply annotated to introduce into the pictures classical EM marks, susceptible of reinforcing these analogies to make them more evident. Be Lakhovsky right or wrong in his interpretation of facts, be he a genius or simply an illuminate served by a lucky chance, that's to be appreciated by everybody. Indeed his he an example more of an individual about whom nothing can be told but right, but who, for reasons that he didn't known himself, has been rubbed from the collective memory.

1-10 : conclusion of chapter 1

Reminders on the short story of ether, in the preceding paragraphs, are done to refresh some memories, or may be discover for others, the avatars of what appears as the common point between physics, metaphysics, philosophy and theoretic: all the basic reflection on vacuum and Universe constitution. All the physicists and all the thinkers in general have worked on this question, and all of them have renounced digging a notion that some, like Huygens or Fresnel, have used as a reasonable but not affordable work hypothesis, without really trying to clear its mystery. As a matter of fact, all the history of physics had followed the same laws than History in general, laws that we can resume in one only: might is right. In the domain of science, the stronger has always been the Establishment, Anglicism which quite vaguely but finally clearly enough for anyone points out the whole of official science instances: all upside Science Academy, far nebulae which doesn't communicate with the plebs but claims to be the guardian of knowledge, and just below all the careerists, with impressing diplomas, who has made their duns in the system by obstinately climbing the stairs, generally licking boots of hierarchic superiors and stamping colleagues mugs and who nowadays have got the decisional posts. This is a resume, perhaps too stiff but so true, of human condition, and which is available in each microcosm of our vanguard society, including, as is right and proper, in the scientific world. In the previously treated examples in this chapter, all the cited scientists have had to deal, for a moment or for ever, with this status of fact. Half of them are now hushed up, ignored in all the documents of the official archives which are the stuff of tuition. They have been punished for their intolerable emancipation mind, and condemned to the supreme sentence in this corporation: forgetfulness.

In what concerns ether, there is obviously scandal. To-day, at this beginning of the 21th century, a growing number of faculty professors feel more and more awkward when they have to inculcate to their students notions that they hardly ingest themselves, but they are paid to do it, so they do. Nevertheless, during some confidences done in conversations out of the academical frame, we can see that there is somewhat a discomfort. Relativists have got an enormous weight in the science society. In scientific reviews, in faculty courses, even in secondary teaching, it's normal to insist as much as possible on the “full” accordance with Einstein's theory of such fact, such theorem, such discovery which would bring a supplementary proof, forgetting that this theory has been so much modified that Einstein himself couldn't recognize it. First of all, no explanation with: it would be too intricate. Secondly, it is often forgotten that a theory cannot be proved. To be in accordance with facts can reinforce its well-funded statement, or eliminate a challenger theory, but all we can prove about a theory is its faultiness, and in this case it must be either modified or suppressed. Theories which are too easily according, with too much easiness, while being incomprehensible for almost all the population, are suspect.

Ether, in which we live in such a daily and ordinary way that it is impossible to get conscious of its reality, so it is part of us, cannot be detected by any of our natural senses. Since Descartes named it "subtle matter", it carries the incoercible idea of a so light fluid that it is impossible to attribute it a specific mass. But things become quickly complicated if we accord it the power of carrying away planets, or conversely. How explain an ether-matter interaction if there is mass only on one side? An article of the French geologist and physicist Louis de Launay (La Nature n_o 1852, nov 1908, p 390), not yet, when he wrote it, a member of the Science Academy but worldly reputed scientist, perfectly lights us on the wrong reasoning used in any epoch, even by very smart people:

"*Ether is all different of matter. In a first approximation, one must see it as a whole, a homogeneous* ***plenum****, everywhere similar to itself...*

... Ether, contrarily to what happens for matter, doesn't move (or does it in very special conditions of whirling); it is eminently capable to manifest stresses, status of efforts, vibrations, waves, at a point that it could be purely and simply defined "what is waving", or still "a medium, in which something periodic propagates without the need of seeing it as a movement". This ether, the notion of which one must get accustomed step by step, must be seen as a fluid, with first a negligible mass, since it doesn't slower the movement of stars in a sensible way, and however of an enormous elasticity which allows light to propagate with a speed of 300 000 km per second...

And in the same article we read this, following:

"*...Moreover, ether has a density higher than this of all the known elements and a rigidity higher than this of steel...*"

It isn't worth insisting on these contradictions which have contributed, partly by their constant repetition, the abandon of the half-theoretical half-philosophic theoretical researches on the nature of space, till Einstein who, exasperated by the lack of new and serious propositions from his colleagues, and in despair over the possibility of going farer in this way, cut the Gordian knot and initialized a new physics from which ether was excluded.

But there is something more. Apart from all the reasonings tried till now and which have not led to anything because wrong-funded, it exists another cause, at least as important, which explains the difficulty in the progression of ideas on ether.

An eminent professor of the Paris VI University (Pierre et Marie Curie), teacher-searcher and laboratory manager at the end of his career, told one of his friends, during one of the last lecture he was attending: "*me, the day of my retirement, I take my bag, I piss off and I don't want to ear again a word on physics*". This is of a big sadness but also well describes the moral exhaustion in which are being some valuable teachers, may be these ones in priority, who have chosen this profession by vocation and who have got the vague feeling, after forty years of good and loyal service, of having been swindled and taught insanities to their students, enforced and against their wishes. Others realize that young people are no longer able, in the third cycle of scientific Faculties, to do something else than solving equations, without trying to really understand the phenomena they are working on. But are they guilty? Is it not the method they are invited to follow? In fact, if we read the books of the beginning of the 20th century, we can see that this kind of critics were already expressed by certain professors like Bouasse[11], who castigated the National Education authorities just like Descartes did against the "Scholars", as he named them in his time.

Tuition of electromagnetism is to-day a mathematics course. Throwing out ether by relativists was the first cause. But these last, however, now convinced that its existence is necessary, but on another hand not ready to admit that they constantly had lost their way through a wrong-funded logic, preferred to re-create a new one of their own, so complicated that it poses more problems than it helps solving ones, and isolates them in their ivory tower from a large public, unable they are to explain their works with ordinary words. Good luck to those who want to try the cords theory, for example, since it's the fashion, but there are a lot of other examples which will be developed farer.

They are not afraid by contradictions, of which they have both habit and skill. But a big problem is that, having the power, they

11 Professor at Toulouse University in 1930

enforce them to the others. Luckily, no dictatorship is total, and when we closer examine the relativist wall, we can see that there are crevices in it. The Bureau des Longitudes[12] sends to issuing, every five years, a physics encyclopedia, very well made, in five volumes, and where, in the volume treating of electromagnetism in the 1981 edition, page 243, we can read this which is really edifying:

"*Vacuum is a physical medium, capable of propagating electromagnetic actions. It is characterized by three physical constants:*

1- Vacuum permittivity, $\varepsilon_0 = 8{,}854187.10^{-12}$ *Farad/meter*

2- Vacuum permeability, $\mu_0 = 1{,}256637.10^{-6}$ *Henry/meter*

3- Light celerity in vacuum, $c_0 = 299\ 792\ 458$ *m/s*

The Bureau des Longitudes is a very official institution, an instance to be compared, for its seriousness, to the French Science Academy, or the Ecole des Mines, or Polytechnique[13]. And what do we read in this encyclopedia? That vacuum is not vacuum? That it is a medium? That we can define this medium using three fundamental constants, which have been determined with a precision of at least six decimals? What a cheek!

Restricted Relativity tells us that there is no ether and that light celerity is a universal constant, which is wrong as we'll establish it in the following chapters. But the definition given by the Bureau des Longitudes says that ether exists, and as scientists know that permittivity can vary, there is indeed no reason for light celerity to be a constant, even in a vacuum which, we now are aware thanks to the Bureau de Longitudes, is not a vacuum.

This is one of the causes of the discomfort of certain professors, who have difficulties in teaching to their students things to which they don't believe themselves. There are others, of course, which are inherent to the very job, about which it 'll never be enough to say that it is exhausting, but when, after having added the final

12 French Physical Institution

13 Two French Highschools

touch to their course, lovely prepared after unpaid work hours, they pass to the true reviewing, taking for a moment the place of those who will listen to, they cannot after analysis miss noticing all the fundamental incoherences, due to relativist ukases and first to the negation of ether, attitude which commands all the hypothesis in physics. This is why some of them don't want to ear speaking physics when they retire. This is why they feel embarrassed when students ask certain puzzling questions. This is why, nowadays, there is a whisper, a simmer of rebellion in this profession, obviously disciplined, but which is fed up swallowing anything.

The notion of ether needs to be re-visited to be again of actuality, to offer to scientists a new interest, susceptible of make them mind this fundamental notion, but with new arguments. We cannot do that to-day like Descartes did, with the background of his time: science has advanced, world has changed with technology, electromagnetism has led to inventions which have changed man. Our knowledge has any way progressed, too. But there are also many snags which should make mind: apart that we have the right of not believe in Relativity or Big Bang, the confirmed fiascos such as controlled fusion or those, foreseeable, of the big colliding torus of Geneva, show that we don't master matter as the beginning of nuclear industry had shown us the hope. The way is long, and not straight either.

Einstein theories close us since 1905 in a blind way from which we cannot get out without going backwards. And going backwards means read again the genius ideas of Tommasina and other outlaws of physics, to sieve them in the light of what new we have learnt since their time, and launch for once in the right direction, simply agreeing that light and all other EM waves propagate in a well existing medium, and not in nothing, which would have no sense.

Besides, the concept of ether has never been completely abandoned, except by relativists who, all the same, begin being turncoats so much their theory becomes abstruse and unbearable. In the course on accumulators, written by the students, given at Supelec in school year 1931-1932, we can read the following in chapter 1, about the constitution of electrolytes:

"To understand this analogy, we may suppose that a gas is solved in the medium ether."

So we can see that, 25 years after the promulgation of the Restricted Relativity, the most famous French High school in electricity was resisting and prized keeping the notion, may be as an homage to Maxwell the Ancient, of what constitutes the soul and substance of his invisible universe

The author of this book, electronics engineer now retired, has made published by Hermes[14], in 2007, his first technical book, intended for the narrow club of engineers and technicians specialized in radio-communications, and treating the arid subject of high-selectivity HF filtering. Having made of ether, during his professional career, an effective tool for work, which allowed him to design extremely practical mechanical models of resonant cavities, as well as other objects typical of his profession, the temptation came to him writing a few words on the topic as a foreword. However he was anxious, doing that, to be directly and for ever listed in the category of dangerous eccentrics, and see refused the access to an issue to which he was very tied up. That's the reason for which he preferred, by precaution, ask first his collection manager, Faculty professor, his opinion on the pertinence of the thing. Contrarily to his fears, and at his huge release, he was surprised by an unreservedly approbation and even an encouragement to take part in the reaction to what he called "burying one's head in the sand". He defined this way the laxity status, in middle and superior tuition, which consists in teaching, sometimes with regret, things in which hardly believe, and so pointing all those who refuse to admit that it is necessary to visit again this basic problem of the existence of a propagation medium for EM radio-waves.

This anecdote and the various former notices show that some possibility of seeing physics move is not lost, and that something is going on in the substratum of the scientific world. Hope of seeing a new birth of powerful etheric theories is not dead. But the battle will be long and painful. The social sphere of Research and Tuition is an enormous machine which has, by construction, a

14French science editor

fantastic inertia. New ideas find there what sea waves meet when they shock a cliff or a dyke: a hermetic rampart which cannot be broken except by abrasion, with the help of time. Scientific research is perhaps what is the most difficult, in our countries, to move or change its course. There are both too many interests and too many habits to be fight, and principally this quasi-certainty that science can do nothing but progress, that in any case it cannot misleads, that there are too many searchers going forwards so as all of them could mistake, that if it was true it would be known, and so on, and so on...

Man is a gregarious animal, like lamb. When there is a leader, everyone follows him, and when Panurge[15] intervenes, it's the catastrophe. Contesters, even when they are right, generally are splitted off in the conflict, and the value of their arguments is possibly and exceptionally admitted when they are dead or out of play, and when their ideas has been recuperated. See le Bon. The scientific social sphere is not spared by this universal social rule which well settles hight and quality of human values. Ether will not reappear from the oubliettes before an individual will experimentally prove its existence, by suddenly presenting to the community an invention which will make evident a new property, only explainable by its presence. Moreover, it will be necessary that this invention has a huge resounding, with social implications fit to make change their mind the most refractory relativists and sceptics, like for example give men a free and endless source of energy, or master gravitation, or something enormous in this vein. May be that, in this case, the battle will be won.

Why “may be”? Because the Establishment doesn't admit, doesn't stand competition. Great discoveries cannot be conceived out of the restricted sphere of the official research, this which goes on in state organisms, of which the organisation is made by professionals of management. And just below these specialists of financial balance are high level careerists, having got loads of qualifications and international awards, and the majority of whom have forgotten, along their curses, the scientific vocation, on behalf of social success and well paid honours. These peoples are

15Mythic hero who threw a lamb into sea to make the others follow

ready to all for preserving the control they have on orientations and internal and external rules of the Research.

The forgotten scientists of now and of whom it was question in this first chapter have meticulously been eliminated from the history of science. It's out of question that a member of the scientific community, whatever his level, his merit, his hierarchical ^position, ventures out of the highway guarded by relativists. Besides, this highway will possibly, one day, change its name and its orientation, when research will have too long stayed for that protestations overtake the self-satisfaction and the lying promises of an imminent progress which never comes.

Waiting for that, any one who thinks he has found something important in the scientific domain have to immediately warn the uppermost authorities, which will do the necessary in order that the invention doesn’t leave their hands, and also that it doesn't lead on a scramble and so doesn't harm the general harmony. The contrary would be the proof of their incompetence and could have disastrous consequences on their reputation, their budgets and their internal order: horror! The one who overcomes this implicit rule is devoted to the road roller lethal torture, every time it's the struggle between the iron pot against the earthenware pot with the end that we know. All the scientists who has been cited before in this chapter have been more or less the victims of this system, which unfortunately is universal. But the highest award goes to another man, who could have been at the end of this shortened list, but his performance has been such -he died in 2007- that we have decided to dedicate him a whole chapter. It's question of the French engineer René-Louis Vallée and his field theory that he named “Synergetic Theory”.

Chapter 2 : The Vallée brief

2-1 : The man and his course

René-Louis Vallée, electronics engineer graduate but also, probably, one of the greatest French physicists of the post-war period, completes for now the non-exhaustive list of the first chapter, which pays homage to scientists wounded or dead at the science front -it's an image-, victims of the incomprehension of their colleagues and the rigidity of the institutions.

René-Louis Vallée
1926-2007

He was born in 1926 in the town of Constantine, in a still French Algeria, but carried a scientific course of study in France, where he was graduate at Supelec in 1951. Having had in this high school eminent professors such as Louis de Broglie, who had in charge the tuition of undulatory mechanics, he got at their contact the virus of physics and began to have a look to the great theories, in particular Relativity. In that period already, he showed a quite exceptional characteristic touch among physicists, by learning German in order not to be betrayed by a translation which, despite being made by acknowledged specialists, like Solovine in the precise case of Einstein theory, always brings a suspicion in the study of any document. Learning foreigner languages so as to better study scientific archives is neither the fact of everybody, nor in his grasp. Practising English and German at the same level, that is fluently, he got the necessary weapons for the compilation, in the original language, of the most important works of our various predecessors and masters in the matter of physics, necessary pre-work for any serious study. This rigour was finally going to play a nasty trick, because he was often reproached of being obstinate at the point of not agreeing on the opposite arguments. Specially, he never accepted to bend in front of the authority when he felt he was right, and he didn't understand quickly enough that economical interests always win against any discovery, with greater reason when this last is important.

Whatever it be, Vallée quickly revealed himself as an excellent mathematician, having got the gift, like J-C Maxwell, Henri Poincaré or Pierre Duhem, of perfectly mastering modelling. To model, in physics, is in short to put into equation the phenomenon to be studied, by the mean of something which is often a mechanical equivalent, but which sometimes can be purely mathematical. Contrarily to what it could be thought, this operation is very difficult to be correctly realised. It's even the big weak point of the theoretical physics, because a bad modelling, even if it can lead to such aberrant consequences that evidence commands to go backwards soon, can also, in another approach, leads to something *a-priori* coherent, in accordance with the known facts, but which finally will generate a non-available theory. Damages are generally repairable, but time has been lost for ever, and history of sciences is unfortunately full of this kind of adventures. The last one is called Relativity, and we'll never talk too badly about it.

When, in 1959, Vallée arrived at the CEA, this institution was managed by Jean Debiesse, Companion of the Liberation and placed there by the Research Minister of that time, Francis Perrin. Debiesse quite rapidly noticed the Vallée's talents for mathematics and physics. This last used to tell his students, far later, that his boss would have suggested to him that the photon, behaving like a wave guide, it would be profitable for him to follow this track for a new study of this particle, still mysterious and with a contradictory behaviour, sometimes wave and sometimes matter. It could be from that suggestion that the Synergetis Theory was born. In fact,

Debiesse was neither the first, nor the only one, in the CEA, who had seen the potential of Vallée, as this last was able to work not only in computer logics, having passed a Master in Communication Engineering, but also in fundamental physics, as well as in electronics. From 1959 to 1976, either in Saclay or in Fontenay-aux-Roses, he was in a way the universal man, in the good mean-

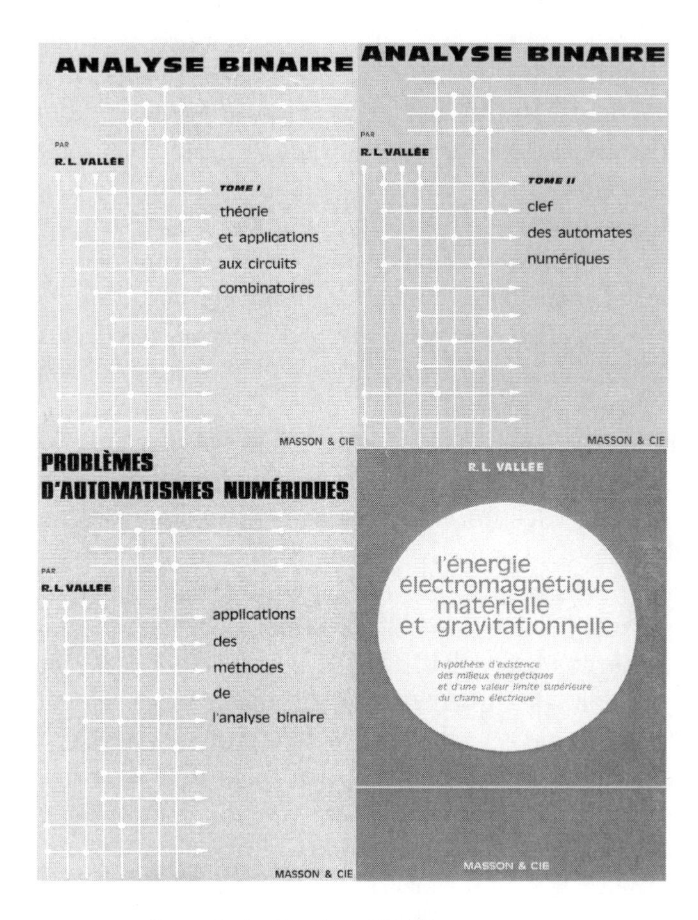

figure 2-1 : Books of Vallée

ing of the term, the one who was consulted, the one the advice of whom was asked when there was some hypothesis or calculation to be checked, and more simply someone that any colleague liked to discuss with, when ideas were in shortage.

In 1966, he gave in Saclay a lecture on binary analysis, a mathematical tool for electronic logics he had created during the former years. He was at that time working in the team of Debraine, a specialist in electronic automates, helped by the engineers Chicheportiche and Vergès, on the problems of high speed logics, just after the integrated circuits were born and put on the market, only a few years ago. At that time, and still like now, the CEA needed, among others and specially for the equipments of particles count, to possess the "nec plus ultra" in the domain of logic components. But it's not enough having got the best materials: you also must take the most of them, and for that have a tool able to optimize, for a given task, the number of elementary functions, and to reduce it to the strict minimal. Doing that decreases as much as possible the signal transit time, in other words its response time. Since there, hardware engineers could only use, to determine and draw the schemas of interconnection circuits between gates and functions, in the study and the design of automates, the Boole algebra. It was the unique tool at their disposal, a tool which gave good results but which was not able, since its special construction out of classical mathematics, the exhaustiveness of the simplification possibilities for the optimization of the circuits. Vallée then invented and perfected, with the help of the small team of specialists, a new algebra called "binary", completely integrated to the new edifice which had been introduced in 1956 into the superior mathematics courses under the name of "modern algebra" or still "theory of the ensembles". By using this new concept, he gave to the binary functions a "ring" structure, already known in, and so re-integrated them in the field of "normal" mathematics, while developing operative modes proper to this particular ensemble. The members of Debraine's team got acquaintance with the new tool to the detriment of Boole algebra, that they progressively abandoned, and during the lecture they brought their witness-ship as satisfied users. This didn't prevent, at the end, obstinate supporters of Boole algebra to contest the utility of another new algebra and to intervene with a surprising fierceness, which led to an animated and interesting verbal battles, during which Vallée proved the pugnacity and the lecturer talent which had made a good part of his reputation. All this proves, if necessary, that there is always someone that you worry when you propose something new, whatever the domain, the place and the hour. From Vallée's point of view, it was there only the lesser evil, since it was a simple improvement which concerned only a few people, only in the frame of the CEA, or at the limit a small number of private enterprises, interested by novelties, and for the rest without political or financial consequence. The situation was completely different in the other domain where he had decided to put his nose, the domain of nuclear physics.

2-2 : The discovery

While the electronics engineers of the Debraine's team were playing the fools with the binary analysis, Vallée was working in the direction suggested to him by Debiesse, these of the photon and its behaviour similarity with this of a waveguide. Deliberately getting rid of the fundamental hypothesis of the Restricted Relativity, which sets that light celerity is a constant, he recuperated at once, this way, all the mathematical arsenal bound to the refound derivatives of c_0 and began to test what happens to Maxwell equations when you get free from the universal constants principle. It's besides very curious that, among the numerous opponents of Einstein, not one tried to do this attempt, will it be only to see what happens. But for that you must have the necessary aptitudes, and it's not the case for anyone. Moreover, like all those who refuse to believe that light celerity is a constant, Vallée was an etherist, but with the philosophy of Tommmasina, which means that he not only considered that ether is crossed by an infinity of EM waves, but that it is the interlacing of these waves that constitutes the texture, the matter of the ether. Besides and cautiously, he never used the word “ether”, because he had already noticed, since a long time, during the discussions on this topic, that the only fact of using it generally led to the end of the speech. So he gave it the name of “diffuse medium”, just to give it a new virginity and thwart the condescending jeerings of relativists mules. The fact remains that, after two years of lonesome work and trituration of the Maxwell equations, manipulated using the Heaviside formalism, he was obliged to admit that these last were irremediably leading to the same relation, of a strange simplicity, but the signification of which was of a considerable importance:

$$\boxed{\vec{\gamma} = -\overrightarrow{grad}\, c^2} \quad (1)$$

and which can be read this way: in every point of the EM waves propagation space, the acceleration (or gravity) is equal to the gradient of c^2.

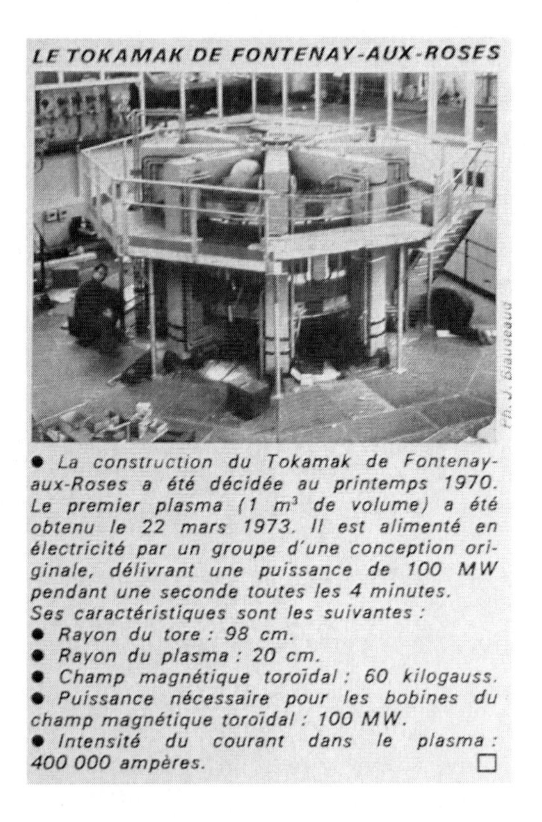
LE TOKAMAK DE FONTENAY-AUX-ROSES

● La construction du Tokamak de Fontenay-aux-Roses a été décidée au printemps 1970. Le premier plasma (1 m³ de volume) a été obtenu le 22 mars 1973. Il est alimenté en électricité par un groupe d'une conception originale, délivrant une puissance de 100 MW pendant une seconde toutes les 4 minutes.
Ses caractéristiques sont les suivantes :
- Rayon du tore : 98 cm.
- Rayon du plasma : 20 cm.
- Champ magnétique toroïdal : 60 kilogauss.
- Puissance nécessaire pour les bobines du champ magnétique toroïdal : 100 MW.
- Intensité du courant dans le plasma : 400 000 ampères.

figure 2-2 Science et Vie n_o 703

The first time this result appeared under the eyes of Vallée, he remained incredulous for a long while: there was, in this simple formula, the expression of the physical fact that acceleration, in every place in space, was bound to to light celerity! In other words and more precisely, it meant that the gravitation field was the derivative of a potential, numerically equal to the square of light celerity and that he named at once “gravitation potential”. It clearly means that light celerity is lower in a gravitation field than in free inter-sidereal space, and that the higher the gravity is somewhere, the lower is that speed. For two years, two years of doubt and incredulity, he checked tens of times his calculations and that formula which always was appearing before him, tirelessly, faithfully, like a domestic pet, and which was opening a large window onto a previously obscure world. He was not really ready to believe all this, so enormous was the discovery, but the calculation appeared solid and resisting any cross-checking. So was created, year after year, overlapped in the multiple paths where the unquenchable curiosity of Vallée was exerting, what was becoming the “Synergetic Theory”. A few after the issuing of the three volumes of “binary analysis” (Masson editor), the book entitled “ l’Energie Electromagnétique Matérielle et Gravitationnelle” was published.

But besides these purely theoretical researches, Vallée was also very closely interested in the experiments going on in the TFR (Tokamak of Fontenay-aux Roses), and this equipment, design to make a magnetic confining of deuterium or hydrogen plasma in the aim of producing a controlled nuclear fusion, was going to inspire him a program of manipulations directly coming from his theory. That is how, during year 1974, he proposed to the team working on the Tokamak an experiment planned to check a practical consequence of Synergetics, and consisting in the emission of slow neutrons in plasma, in precise conditions of magnetic flux and of direction of the secondary flux. At his big surprise, instead of preparing a long prior theoretical justification, he discovered that the physicists with whom he was working were already aware of the phenomenon, that they had seen several times before but that, by lack of explanation, they had only mentioned it in their reports as a parasite event. This attitude is very typical of a led research, like the nuclear one, where there is an official goal which has allowed to obtain important credits, never enough for searchers but considerable for the tax-payer, and from which, due to the importance in time and means of the device, it is out of question to divert: if an experimental result is not in accordance with what is foreseen, it is immediately classified as a manipulation or measurement error, or some fable is invented not to cause paying attention to. Then Vallée, who perfectly knew what he was speaking about, explained them "his" concept of the phenomenon classified "odd" and proposed to go farer by endeavouring another experiment of his: introduce in the fusion torus -it's the literal translation of Tokamak, Tok = current and Mak for magnetic-, with the hydrogen and the deuterium, a supplementary element, also light and in gaseous phase, but with an atomic number slightly higher, in the occurrence nitrogen 15, to see if there was production, as foreseen by his theory, of a radio-active isobar emitting a β radiation, that is to say energetic electrons. The men in charge, not yet recuperated by the staff but quite enthusiastic, agreed for preparing the trick. A few years after, Vallée told that this way to his students:

"*First electric shock, nothing happens. Second shock, parameters are slightly modified, still nothing special. Third shock: 5 millions francs damage!*"

The vacuum chamber at the exit of the torus had been literally disintegrated by a flow of energetic particles, not expected by the technicians but planned in a smuggling theory, coming from a not concerned engineer who has put the trouble in a State Research program!

2-3 : The quarrel

We could think, a priori, that such a discovery should have immediately initiate, if not enthusiasm, at least interest and consideration from the upper spheres, and that the discoverer be the target of particular praise. Instead of praise, Vallée found himself in front of a college of men and women completely panicked by this thunderclap in the atemporal humdrum of the CEA: discoveries are beloved, but they must be expected, appearing as the results of a research plan elaborated a long time ago and perfectly organised, and coming from above, not besides. It must not be possible that high level persons in charge, those who decide, appear like phoney searchers who have constantly mistaken in the orientation they have given for the use of public money, hypothesis happily very improbable, that goes without saying (except that...). But may be it was not there the most serious: it's that Vallée, incidentally, had also explained to the physicists of the Tokamak why this last, designed to increase plasma temperature through an increase of the efficient section, will never do it this way, because the effect of that increase was, at the contrary, a decrease of this section. We'll notice at this occasion that there are no more Tokamak, neither in Orsay nor in Fontenay-aux-Roses, which would tend to prove that Vallée was right, instead of which the famous -or smoky- controlled fusion should now be at the industrial stage since a long time. In short, Vallée had just demonstrated that the nuclear French Plan was travelling since the beginning in a blind way, and that it was actually bound to fail. It was there an unforgivable fault of him, and the staff began taking coercive measures against the incautious troublemaker, while it was obliged to develop a large screen of smoke on what had happened in Fontenay-aux-Roses.

In the 1974 bulletin of technical and scientific activities of the CEA, diffused in the public domain, we can read this haunting official description of the phenomenon provoked by Vallée:

"*The study of the current collected on the observation plan and its X-ray radiation show that it is question of high transverse energy electrons which are carried towards the wall at the favour of a weak undulation of the magnetic field. The existence domain of this phenomenon suggests that these electrons are created by an instability which takes its source in the uncoupled electronic beam. The radiation braking spectrum of the uncoupled electrons, measured during a low density flash, has given a 6 MeV energy. Besides, the current carried by these electrons can reach 0,15 % of the main current: TFR works then like a high intensity betatron.*".

This is exactly the processing way, when you manage a big enterprise, to definitely bury a case that you don't control any more and which is for you an unsolvable problem of authority and competence: you make a black pudding juice report, something confuse enough and about which you know that, by construction, it will not be understood by anyone, and then you lock the grid.

So, in function of all this commotion, to finally put an end to a situation blocked for questions coming from persons having an authoritative and obstinate behaviour, happened what might happen.

In front of the posture of the staff, Vallée rapidly understood the situation. He realised that his case was over and that his days in the CEA were numbered since, instead of being, if not congratulated, at least encouraged, he was firmly advised, at the contrary, that he should have interest in making himself small and discrete, and over all to only do what he was asked to do. But Vallée was all except a lamb, and he decided that it was not because there was a coalition of rigids, at the top of the CEA, who had decided to throw a hermetic veil on an invention which would break their status of order givers, that the nation was going to be deprived of a definite solution to the energy crisis, and moreover a French solution. Then he really began to kick over the traces.

His first dissident action was to communicate to the Science Academy the results of his theoretical works. It's to be known, indeed, that all the documents relative to the Synergetic Theory are there, or were there, and that nobody to-day, in the large public sphere, had eared of them. Then, or may be before, we don't remember well, Vallée decided to publicize his works. To do that, he entrusted his confidences to the review "Science et Vie", and more specially to the journalist Renaud de la Taille, who at once was enthusiastic, both for the man and for the theory, and who was going to open for him, for two years, the columns of the periodic, before being himself the target of unbearable pressures. The first article on the Synergetic Theory took place in the number 677 of februar 1974, p 15, under the title "*a French discovers: vacuum is a form of energy*". It was to be followed by three others, until the last in 1976:

- n 688, january 1975, p 32: "*a French physicist: nuclear fusion is twice impossible*"
- n 698, november 1975, p 52: "*Who will dare refute the Synergetic Theory?*"

- n 700, january 1976, p 45: "*Synergetics is still waiting for serious contradictors*".

We can even add to this list the box seen in the n 703, of April 1976, inserted in an article consecrated to the fusion, p 45, where we can see that the situation of Vallée and the orientation of researches in the CEA have no longer a chance of changing:

"*Higher temperatures would have been reached without the losses foreseen and explained by Synergetics*".

This box, which was like a swansong, was even not signed by Renaud de la Taille, who had been recommended not to keep busy with this case, but by one of his colleagues who had for a moment taken over from him, before being himself called to order.

Vallée also tried, at the very moment he was going to send his communication to the Sciences Academy, to first ask for the advice of his old master and professor Louis de Broglie, perpetual secretary of this institution, knowing he was receptive to his ideas. He then discovered to what point some scientists, even renowned, but from whom there is a possibility that they guiltily divert from the straight line of the "politically correct", were invigilated: the letter was intercepted by the team around de Broglie, who never got it.

The pressures, which had passed from the level of the CEA staff to this of the Research Ministry, acted not only on Vallée, but also on all people or organisms able to help him. The editor Masson was obliged to stop issuing his books, they are practically unfindable to-day and there is only the figure 2-1 of this chapter that proves they have really existed.

To finish with this long episode on the CEA, Vallée was convoked in November 1976 at the personnel office. He was proposed a cushy promotion at the DAM (Direction des Applicaions Militaires of the CEA), with the obligation not to make waves any longer and forget the Tokamak story. These conditions from his employer being for him unthinkable -shutting up? What more?-, he received his termination letter on the first of December. The method, initiated by Athenians to get rid of people classified dangerous for their society, and which consists in rotting their life until they cannot fight any longer and go away from themselves, is called ostracism. It's a practice which has become universal in our democracies, which forbids violent actions against trouble-makers.

2-4 : The exile and the resistance

Very fortunately, Vallée had not just enemies. Moreover, his diploma in electronic logics and his well-known performance in this domain with the Binary Analysis, and also well-placed friendships, allowed him to find an employment as a professor at the CIEFOP, continuing education organisation of the Thomson-CSF company. So the social and material survival were saved, but any way the nuclear lobby had made him out of the possibility of being harmful. It's at that moment that the destiny, which loves those who fight, came to the camp of the outlaw. It happened that the engineer Vergès, previously cited as a user of the Binary Analysis, friend of Vallée and active member of the logicians team, married the daughter of an old man, specialised in the communication, in the "Seguela"[1] meaning of the term, previously professor at the School of War and having his own company specialised in dialectics tuition. Being aware of the Vallée case, in all its details, by the way of his son-in-law, he decided that this unjustly treated man merited being helped. Georges Sauge, that was the name of the providential man, began by making better acquaintance with the individual concerned, which achieved trusting in him, and decided to found with him a new society, of which the aim would be spraying and diffusing as largely as possible the Vallée's theory, allowing this last to go back to fight with new weapons. So was created the SEPED, acronym for Société d'Etude et de Promotion pour l'Energie Diffuse.

From that moment, Vallée and Sauge were going to cross France to make the Synergetic Theory known by those capable of being interested and understand it, and above all propagate it: students, professors, engineers, simple curious of a good level, let's say any brain and any good willingness present for tackling the subject. The fact is that, at each programmed meeting, the two men filled the lecture halls, and that it was the occasion to see and

1 French specialist of political communicaion

measure to what point the public opinion is, not only receptive to novelty in the matter of science, but asking for.

Must Vallée be considered as a simple engineer or also as a professor, as some people call him? It seems that he was very early attracted by public contact and a certain form of professorship since, having got his engineer degree in 1951, he wrote his first publication on the Maxwell equations in 1956, in the revew “l'Onde Electrique”. This was indeed only a beginning, and other articles were issued under the aegis of the CEA, and among them, already, one dedicated in 1962 to the hypothesis that light celerity could not be a universal constant, and even not a constant at all. Vallée has really taught electromagnetism and Relativity at the National Institute of nuclear Sciences and Techniques of Saclay, thus he can be considered as a fully qualified professor. He is not a usurper. Then came the long episode we know, during which we must noted that the CEA staff, who hadn't yet taken the right measure of the Vallée's thesis, had not opposed to the publication of his books by Masson, including this treating of Synergetics. It's still during he was in CEA that he sent, under sealed orders, without the prime advice of de Broglie for the previously exposed reasons, the essentials of his theoretical discoveries to the Sciences Academy. These last, which communicates few, and about whom a large public often wonders what exactly is its utility -and sure not to the promotion of new ideas-, appears in the occurrence like a bottomless well were all that worries is buried. Let's hope they are competent!

This parenthesis being done, let's go back to the subject. Vallée and Sauge, encouraged in their crusade by the supporters of the CEA, Supelec and all the research institutions where people had been convinced by the oratory performance of the Master, spent the essential of their time spraying as much as possible the new ideas in physics.

We may say without lying that everywhere they passed, either in engineering schools or public halls, they created both enthusiasm and a lot of questions, always the same:

- *“Why were you sacked from the CEA?”*
- *“Why did your experiments stop?”*
- *“Why don't you try to find a buyer for your discovery?”*

- *"Why is nobody interested in your case?"*
- *"Why was the Tokamak put in pieces?"*.......

Hum...yes, by the way, why were the Tokamaks put in pieces?

When R-L Vallée left the CEA, he left behind him a state of perplexity that the staff and the national instances of the nuclear research couldn't obliterate with a simple memorandum. What had happened in the Tokamak of Fontenay-aux-Roses was really something extraordinary and should have normally initiate a series of decisions leading to get a deeper insight on the necessary knowledge to get for the comprehension of the phenomenon, and then, possibly, to define a new line of action. Nevertheless nothing was done for that, on one hand because it's quite impossible to replace someone having the great competence of Vallée, and on the other hand for other more casual reasons already evoked and which can be resumed in the principle that the authority cannot be wrong, never.

It was nonetheless decided to drive, with all the necessary discretion, a supplementary campaign of experiment but, as their was no more guide and as the observed phenomena were mastered by nobody, the initial program was modified in proportions which had not been planned at the beginning. First, protecting walls were built around the torus, in case, while a remote control system was installed, so as to do the manipulations from far and so protect the experiment staff against, firstly, a device now considered as non predictable, and secondly the possible consequences of something that had become a potential danger. Controlled fusion still refusing to show the tip of its nose, and Tokamaks taking more and more the aspect of a giant cactus, it was finally decided to suppress them, purely and simply. It is also good to know that Vallée, before going away from the CEA, had warned the borough council of Fontenay-aux Roses that this same CEA was experimenting on these devices without exactly knowing were he was going to, and taking risks which staked the security of the inhabitants. We can easily imagine the panic falling on a mayor and his borough councillors when arrives this kind of revelation: what a shambles! We also imagine, as well, the round trips between the town-hall and the CEA, the immediate meetings and the desperate efforts of the State organism managers who, while still more cursing Vallée, mobilised all they possessed as high-placed relationships and persuasion means to calm down the game, doing his utmost to make so as there would be no advertising on a so embarrassing affair, finally correctly mastered.

While Sauge and Vallée were going on in their information campaign, persuaded that their restlessness, in the good meaning of the word, will at the end emote some representative of the public powers or some sympathising Nobel Prize, taking consciousness and knowledge of the theory, study groups were organised here and there: in Saclay, Fontenay-aux-Roses, in the CIEFOP, at the SEPED which was now installed in Paris, and in several places of the Ile-de-France[2]. This diaspora of synergeticians had meetings, generally in the evening, after work, like conspirational groups, to analyse the work of Vallée in the smallest details, with a very precise goal: to do again all the calculations of the theory after the Master, so as first to check by oneself its validity, and then to be able to argue against opponents or learners, and so bring one's stone to the propagation of the new theory. They were engineers, students, simple curious having the passion of physics and novelties, and also a few discreet emissaries of big organisms implicated in commerce and energy, like the Oil Institute or EDF[3], come here for evaluation. At the beginning, halls or rooms were full, but after some time the audience decreased to be reduced to a small nucleus of regulars, but the remainder was enough to maintain action. This is in fact a process which was not specific to the synergetics courses. Those who followed, for example, the lecture cycles of Pierre-Gilles de Gênes at the Collège de France could notice the same phenomenon: at the beginning of the course, the attenders are enthusiastic for the discovery. The Amphitheatre Marguerite de Navarre, the biggest of the college, is too small to contain all the people. At the end, it remains only the small group of supporters, let's the quarter of those of the beginning, but more solid. Whatever it be, there are still a few hundreds people, in France and in the world, who have spent between 100 and 200 hours of study in a synergetic group or at the direct contact with Vallée, who know in the detail his work and are able

2 Surroundings of Paris

3 Electricité de France

to demonstrate the main results of his book. As an indicative, a DEA or a DESS represent a course of about 500 hours, not including an industrial session of 4 months and a memory in the form of a mini thesis.

It's now the occasion and the moment to give some practical precisions on the necessary means to get to be able to tackle the Synergetic Theory, by comparison with the Relativity. We say “the” Relativity to simplify, knowing well that the Restricted -or- wether Special- and the General Relativity are not compatible, from the preliminary hypothesis of the first one -light celerity as a universal constant, no propagation medium-. Relativity, as it is proposed to-day, is not accessible to the average man, despite of the fact that it is a sacrilege not to bend in front and pretend to understand it, for example when you have to publish a high level scientific article in relation with the Cosmos and the Universe. Synergetics, at the contrary, even if it is not accessible to anyone either, we must be fair on this point, concerns a far larger public: we can settle the maximal level of the mathematical tools to be possessed, to launch oneself into its precise study, to a good mastering of vectorial calculation, added to a good knowledge of theoretical electromagnetism, which is the fact of engineers, the professors in mathematics and physics, and the 2d and 3d cycle students in electricity. It sure remains something relatively elitist, but nothing to see with the level of high spheres searchers, who ramble in multidimensional spaces like we do in the Jardin des Plantes[4]. Moreover, a theory using the ether has always the advantage to give a way towards simple mechanical analogies, which make very easy the comprehension and the memorisation of the results. Lord Kelvin and all the English School of the 18th century physics, with in particular Maxwell, thought that a theory which had no representation or mechanical model was suspect. Perhaps this philosophy could be promoted again?

While Sauge and Vallée were going from meeting to meeting in the surroundings of Paris, the supporters were besieging the media, well conscious that they were the owners of the key to the public audience: an apparition on the TV or a few minutes at a national radio makes more than ten meetings anywhere. It's the

---

4 Garden of the Museum of Paris

reason for which the synergeticians used of the opportunity of all the forums on the energy crisis to try, by the channel of the "questions by the listeners", to ask what were the continuations given to the experiments on the Tokamaks and the Vallée's theory. But there also, a coercive action coming from the ministry of the Research and Industry built a filtering dam between the studios and the listeners, and a lead covering felt again on the small active world of the supporters of the new physics. The battle was unequal, lost by advance, but it is necessary to take part to be aware of realities.

It's really difficult to believe in this story, in a time when transparency, as we say, should be applied everywhere. Still a fable! For what concerns synergeticians, a true anecdote related by one of them, Gilles Brunebarbe, in the bulletin "Synergetics", allows to be more conscious of what can happen when the power, under its multiple forms, has decided to stuff a case. There was in 1979, on the waves of France-Inter[5], a daily broadcasting called "the telephone rings", during which the public was invited to directly ask questions to an influential personality, in the most diverse subjects. On February 26, the personality of the day was André Giraud, Minister of the Industry and so representing the highest supervision authority in the matter of energy: he was the man that the supporters of Vallée tried to reach by all the possible ways, to inform him of their existence, of their opinion, and may be have with him a public dialogue. However, the rules of the game settled by France-Inter were to eliminate both the non-desired people, and also the possibility to ask the evil questions. To ensure this goal, it was obligatory, before being on air, to declare with precision the chosen theme, which was transmitted to both the guest and the presenter, who so were able, in case of potential danger, to keep silence and ignore the question. Being aware of this system, after having tested it several times before, Brunebarbe decided to make no bones to lying and pass the barrage. He invented a hackneyed question, casual enough to give to the minister an occasion of placing his stock reply. Miracle! The question was drawn, and the antenna was opened to the conspirator, who was eager to throw to the minister's face all a serial of questions

5 French national radio

learnt by heart: “why is Synergetics systematically forgotten when you talk about new energies, why a certain number of physicists think that space is energetic and that we can capture it anywhere, in big quantity and directly in electrical form, as proved in Fontenay-aux-Roses by the Tokamak experiments, why is it forbidden to these physicists to express themselves in the media, etc...”. the minister, stuck in the studio and well obliged to answer, said this:

“*I think that if the synergetic energy demonstrates its existence and its capability of being captured, The minister of Industry I am will be enchanted to join his efforts to the promotion of this new energy. I don’t want to enlarge on the Vallée case, it's a problem of person, but the problem of synergetic energy is that physicists are at least divided about its existence. That's all we can say today and those who are in charge of these experiments fully contest that it would be made evident.*”

Morality: the minister was perfectly aware of the case, but being only a high level manager, moreover not seasoned in physics, he was obliged to trust in the management staff of Saclay to have a personal opinion on the Vallée case. After the presenter Claude Guillaumin expressed his anger to Brunebarbe for having made a fool of him, the hole in the water closed again and synergeticians, thoroughly annoyed to see their last resort fail, came back to their studies with a furled and low morale.

All what had come before is all the more not understandable that all the great Presidents of the 5th Republic, from de Gaulle to Chirac, have built their electoral speeches on an insisting willingness to promote by all the possible means the research and the new ideas. Even Sarkozy did so. This proves that competence and voluntarism are not where they should be, in particular in the tutelary ministries. There was a hope, in 1981, after the election of a left wing President, that things were going to lastly change, so true it is that there are a left wing science and a right wing science, each of them having the mentalities of respectively left and right wing. Alas, nothing more happened, and the second following anecdote on this subject is particularly edifying:

In 1983, a senior engineer of the microwave links division of Thomson-CSF, also synergetician, feeling targeted by the coming redundancy tumbrels, decided to profit of the dispositions of the continuous learning to pass a specialized DESS in the microwaves speciality at the Pierre et Marie Curie University. This new degree had just been created by the doctor ex science Arlette Fourrier, searcher-teacher at the tower 12, bastion of the far infrared specialists. So the young 46 years old student followed, like his 25 years colleagues, a formation which takes end by an obligatory session in an enterprise. In the occurrence, it was at the "radar" division of TH-CSF, in the town of Bagneux. It happened that the engineer shared, the while of his work placement, the office of Claude Jeanlin, "conseiller général" and mayor of Evry, who was half-time employed there to keep his rights to the retirement. During a rambling conversation, the two men came to speak about Vallée and his misfortune, and the political man was extremely interested by this extraordinary story that he had never eared before. He proposed then to the trainee to put him in relation with the Research Ministry, separated from the Industry Ministry and managed in that time by Jean-Pierre Chevènement, to try a new attempt for informing the high spheres, now in the hands of a left wing presidency and so, *a priori* and generally, more open to changes and to listening people. It's good to precise that the two "scientific" advisers of the minister were Catoire and Lorino, both socialist mayors of towns in the suburbs of Paris. The engineer, happily surprised by this wink of the destiny, informed at once Georges Sauge of this unexpected possibility, for the synergetics campaign, to access to the ear of the minister, but surprisingly this didn't start any enthusiasm. The old man, tired out by all these years of sterile battles besides Vallée, moreover annihilated by the recent death of his wife, simply wished him good luck after having predicted a snag for his coming action, even supported by a political relation, and to show both his comprehension and his exhaustion, gave him several copies of the book of Vallée, reissued by the SEPED, to give to anybody who wants. Slightly shaken by this unexpected reaction, the engineer still persisted and sent to Mr Lorino a mail including all the necessary documents for him or others to be aware of the events, and have a sufficient knowledge both of Vallée's theory and about what happened in Fontenat-aux-Roses. Followed a round trip correspondence for one year, at the end of which Fabius replaced Chevènement at the head of the Ministry, with the same advisers. The documents passed from the Ministry of Industry, self-declared incompetent, to the Ministry of Research, which definitely buried it. Sauge had well seen the strike.

A few years before, the friends of Vallée had worn out themselves in this kind of steps, aiming to break a silence wall that nobody understood, except the initialisers. But in that time the government was a right wing one, and the rigidity of the superior instances could possibly be explained by an attitude that we generally lend to this half of the French political opinion, at least by the other half. But Georges Sauge had understood that it was far worst, and that national interests have no party, except this of business, from where the warning he had addressed to the little engineer of the suburbs who wanted to howl with wolves. The result of this terrible case is probably that France has let pass a determining discovery in the domain of physics, discovery which will spring again a next day and which could have given to our small country a role already played in the past, this of the intellectual and scientific lighthouse of the World. There is also a morality: no discovery can ever come into existence without the agreement of the financial world, either directly or through the political channel. Nobody may put in danger the economy, nobody has the right of making waves, everyone must obey the chiefs, beck and call. Any individual who thinks he has found something interesting for the collective must immediately make aware the authority and wait for the orders. That's right; but we also are in the country of the Revolution, and those who have built this nation and left their mark in the History are all of them rebels with strong personalities. And this people is the nightmare of mandarins.

Vallée was of that race. His fight has not failed, simply he'll never see his victory, and unfortunately it is very probable that he'll never receive, either, the homage he merits. However, we find on Internet a lot of sites which talk about him and which show that scientists continue to study his theory. Some of them make experiments, others calculate, but success will only come when a high-placed and competent representative of the public powers will be prepared to take back the document and will propose to give it a budget and a specific program. In this period of energy crisis, where all the industrials of the branch try to promote their more or less matched products, it would sure be a good thing.

2-5 : The synergetics and the media

All the French physicists are, at different degrees, aware of the Synergetic Theory. To be aware doesn't mean to know it, because for that it is necessary to study it, which asks for interest, good willingness and availability: it's not so easy to assemble all these conditions. Among the mass media, only the revue Science et Vie

PHYSIQUE

UN FRANÇAIS DÉCOUVRE: LE VIDE EST UNE FORME D'ÉNERGIE

Les théories unitaires de la physique avaient échoué car l'univers n'est pas relativiste. Un ingénieur de Saclay, R.L. Vallée, montre que le monde est fait d'énergie même le vide intersidéral

figure 2-3 Science et Vie n_0 677

has really taken an interest in the discovery of Vallée, contrarily to the other scientific revue La Recherche, which has only published one article in 1976, and even it was to shoot it down in flames. We'll see that again further. At Science et Vie, it's the reporter Renaud de la Taille which had the initiative, the courage and the perseverance to support Vallée and to provide the public with the only detailed popularisation of Synergetics. The list of his articles in relation with the case is already indicated p100, the first is illustrated in the figure 2-3 hereabove, only the last is not his. He has passed long hours with the physicist to assimilate the master ideas of the theory and translate them in a language accessible to the large public, in other words the usual customers of the revue since 1913. So it is, to pay homage to him, his version and his vision of things which will be resumed the best possible, with the invitation for the reader to consult the archives of the revue to get the complete text.

1974 February, first article.

Firstly, it is a question to make the lambda reader implicated, from what a very short calling in mind of the evolution of ideas in physics: the notion of energy, at first bound to movement, then the discovery that mechanical energy and energy are identical, then the equivalence with heath, the most degraded form of energy, the one in which any other can be spontaneously transformed. Never the contrary, conversely: to transform heath into mechanical energy, for instance, we need a machine working according to the rules expressed by Carnot, that is to say between a hot source and a cold source, with a terrible efficiency ratio. All that constitutes the physics of the 19th century, let's say the classical mechanics. Further, the research in the invisible phenomena of electromagnetism and atomistic, followed by the nuclear, and also the reflections started by the new acquisitions of the "far" astronomy, if we may say, gave rise to obligatory specialisations for searchers, who were not able any longer to have a complete view on a physics which had become carrying contradictions, and any way too complicated. This splitting of the formations, of the university courses, of the industrial branches, of the theories, pushed scientists towards more and more mathematized domains, each of them needing ultra-specialised tools, of which the utility was not necessarily evident, but which digged a gap between branches of the knowledge previously close, not to say welded one to the other. To-day, it's no more possible to be generalist in physics, except if you are a minister paid to say no matter what, but this is a very particular case. The study of newly discovered phenomena and the longer and longer duration of formations having lead to slice up the knowledge, we found ourselves provided, and it is the present state of our physics, with an extraordinary catalogue of different forces of the Nature, as of numerous particles in a constantly increasing quantity, of concurrent theories, of so many costly and impressing equipments, etc, etc...., all this to remain at the same point in what concerns the search of the miraculous energy, this which is free, inexhaustible, under our conk, and of which only Vallée and a fistful of contesters had the intuition.

After these necessary callings in mind, after having castigated Relativity and Quantum Physics, Renauld de la Taille comes to the heart of the matter, which is to put his reader in condition to have a representation of the World seen by Vallée, while making him conscious of the fragility of the theoreticians affirmations. For that, a few pictures and some notices of a biblical simplicity make at once evident the defects of the relativist physics. At first, we have to notice that the mass energy given by Einstein's formula, $E = mc^2$, is available in this form in each location of space. Therefore, if we take a mass m at a certain altitude to put it higher, this change can only be done by providing a certain mechanical energy. Consequently, the mass has got in this location change a supplement of potential energy, that the relativist formula superbly ignores. So Vallée will use another formula, $S = mc^2$, where S represents the **sum** of all the countable energies which have concerned the history of the mass m. It's the Synergy, instead of the simple mass energy, which has given to the complete theory the qualifying of "Synergetics". This is of a range which goes well farer than the simple replacement of a letter by another. We must imagine, indeed, what can represent the mechanical energy when the trajectory is done in a gravity field like this of the sun: it's all that you want, except negligible. Besides, if we see that other forces, of different natures, intervene in the energetic history of the given mass, they will be taken into account and be included in the "S". This is a concrete example, and there are a thousand of others, of relativist absurdities corrected by Synergetics.

Despite of the fact that Vallée, in front of the rejecting reactions of some people, has decided not to use the word "ether" any longer, his "diffuse medium", that he uses instead, is exactly the same. Simply, and this is perhaps a justification for changing the vocabulary, that medium has now got well precise properties, which were lacking to the former. First of all, it is active, in the sense that it is constantly crossed by a double infinity -in direction and in frequency- of electromagnetic waves, the ensemble of which, for Vallée or the same for Tommasina, even constitutes the actual matter. So this matter doesn't correspond to the usual notion attached to the objects that we can see or touch, it's a special substance which has particular properties, among which some are still mysterious, and that several physicists, like Wheeler, tried to describe with much embarrass as a sort of magical jelly, at times soft and at times rigid, in fact taking as many forms as necessary to justify his puzzling behaviour. It is well evident that we'll never know the complete truth, since it will always be, for us, invisible and untouchable, but we know that it is a perfect fluid, since it is able to transmit light without any distance limitation and, as the other main characteristic with its active nature, that it has a very high specific mass, any way largely above these of the heaviest known elements, submitted like the lightest to the same gravitational laws. There is in fact only one alternative for the ether: either it is very light, if we think in a theory that it is the reason for not opposing a resistance to the movement of bodies, or it is at the contrary very heavy if, in another theory, we want it to be the cause of the gravity forces and the movement of planets.

The statement of this fundamental property of being the siege of a ceaseless propagation of EM waves immediately leads to the desire of knowing more about these waves, what is their distribution, their origin, their intensity, and what practical consequences it could be possible to take out from this hypothesis which, as we'll see in the further chapter, is perfectly justified by a new interpretation of the elementary physical laws, re-visited in function of the existence of a medium. The figure 2-4, coming from the book of Vallée, shows the spectral distribution as he has imagined it, knowing that the frequency order has been verified. It postulates that there is in the space a continuous energetic spectrum which presents peaks, *id est* maxima, which correspond to the so-called Compton frequencies. They are the vibration frequencies of elementary particles, each of them being deducible from the two main, these of the proton and the electron, by the way of the simple operation that knows by heart any electronician involved in intermodulation problems: if two frequencies F_1 and F_2 are mixed in a non-linear system, they generate secondary frequencies on the following form:

$$F = mF_1 \pm nF_2 \text{ (m et n entiers positifs)}$$

By giving diverse values to m and n, we effectively find again, starting from the Compton frequencies of the proton and the neutron, the frequencies of the other constitutive elements of the matter, first this of the electron (fig 2-4), but also these of other less stable particles, foreseen by the Synergetics and then experiment-

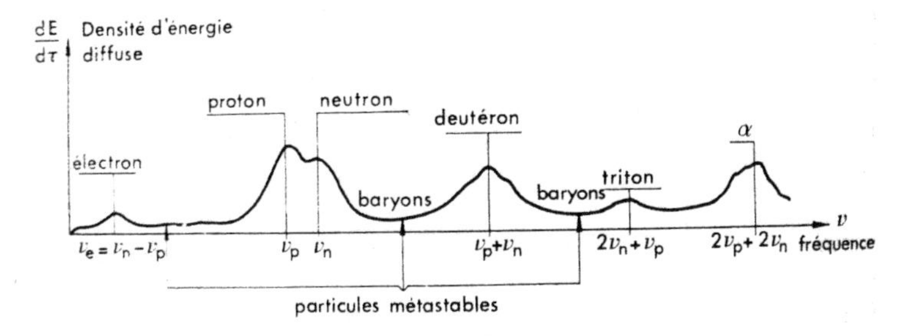

figure 2-4 Spectral distribution of EM waves.

ally discovered after that Vallée had mentioned them. It's the case of the so-called "charm" particles, predicted in 1970 before their existence was experimentally confirmed. As to the medium non-linearity necessary to the creation of intermodulations, it's effective only in well particular conditions where the ether, ordinarily linear when it only carries the EM waves, is locally submitted to electric fields which reach a limit value, defined by Vallée as this which corresponds to the creation of matter by an electromagnetic cavitation of the diffuse medium, basic phenomenon on which we'll come back further. We arrive there to an imagery to which only the hypothesis of an ether, whatever it be, can give access to. The ether of Vallée, mixing of mass and energy, opens the way to a richness of reflection, consequences and mechanical analogies which is at a hundred miles of the complexity, or even impossibility of any concrete representation, of the Generalized Relativity and its smoky theoretical objects. We suddenly find ourselves in a living universe, almost familiar and where, the same way that are created bubbles in the backwashes of a ship helix, a sufficient agitation of the diffuse medium will give birth to this other kind of bubbles that constitutes elementary particles, and which will tell us the unexpected and uncredible truth: the soi-disant vacuum is the mass, and it's only in the heart of matter that we can find vacuum.

Vallée formulates his "materialization law" this way:

"*If it happens, in an isotropic stationary inertia medium, that during electromagnetic events the energy is concentrated in places where the electric field can reach the limit value* ε_d,, *the properties of space, in this places limited to elementary volumes* $\Delta\tau$, *are then modified so that the divergence of the electric field take a non-zero value, in order to forbid overtaking the limit value. It exists then, at least, two microscopic contiguous limited volumes* $\Delta\tau_0$ *and* $\Delta\tau_1$, *constituting the* $\Delta\tau$ *zone, in which the defined integral of the electric induction divergence gives respectively the quantified values +q and -q, with* $q = 1{,}60.10^{-19}$ *Coulomb.*"

We must agree that this is a bit intricate under this form, and that we must have learned the basic definitions of Synergetics to make the link, but Vallée practises theoretical physics in the line of Poincaré-Duhem-Brillouin, and his preferred language is mathematical, with all the consequences it leads to, good or not. Indeed, the translation in imaged terms is simpler, it says in ordinary language that, when the electric field has reached, in a certain place, a predetermined value that cannot be overtaken, a couple of electric particles is created, one positive and the other negative, both coming from the energetic and structural transformation into matter of a small part of the ether. This new matter will be represented, like by Gustave le Bon, as sorts of gyrostats, except that Vallée goes off the beaten path of generalities and give them more precise characteristics, quantified, as well as a geometric structure which explains, in particular, the spin moment of the electron. We can then see, with not too much surprise, that the new theory joins the older ones, not only this of le Bon but also these of Maxwell and Tommasina, on the particular point of the representation that we can imagine for the constituting elements of matter, as soon as we agree on the existence of an ether.

With the ether by Vallée, gravitation becomes again a clear phenomenon, understandable by everyone, pretty far from the elitist flights of fancy of the Generalised Relativity. You first get rid of the attraction forces concept, to which even Newton didn't believe in, to replace it, the ether being crossed by EM waves, by the radiation pressure that goes with. We find again, there, a phenomenon and a principle well known in acoustics, either aerial or submarine, where the experimentation is easy and allow relatively practicable measurements. In what concerns EM waves, it's a bit more intricate, but Maxwell, Bartoli and Lebedev have left their names in this study field. Later on, to pay them a legitimate homage, we'll name “MBL” the electromagnetic radiation pressure, whatever it be, luminous or any else. This MBL pressure exists each time an electromagnetic wave finds an obstacle on its way, and exerts on this obstacle a force which tends to push it in the same direction. In the case of a body placed in the ether, and that we'll suppose spherical to be simpler, the pressure acts homogeneously on its surface and has no other effect than compress it, or let's say ensure its cohesion. There is no other movement. Contrarily, if another body is in the vicinity, this last partially intercepts the radiations coming from the direction where it is, and the first body receives less pressure from that side, and reciprocally. As a result, the two bodies are pushed one towards the other, which gives the illusion that they are attracting one another. And this is the key to the attraction forces! Newton had perfectly suspected the phenomenon but, having not made the right hypothesis on the ether, and very embarrassed with the problem, he nevertheless succeeded in expressing the great gravitation law that everyone knows, pretending that attraction forces existed while keeping in himself an enormous feeling of dissatisfaction, perfectly expressed in a letter to the reverend Bentley. Figures 2-5 and 2-6 illustrate, naively but clearly, the idea to be made of gravitation in the universe of Vallée or Tommasina, on earth for the first one and in interstellar space for the second. Later on, we'll go still further

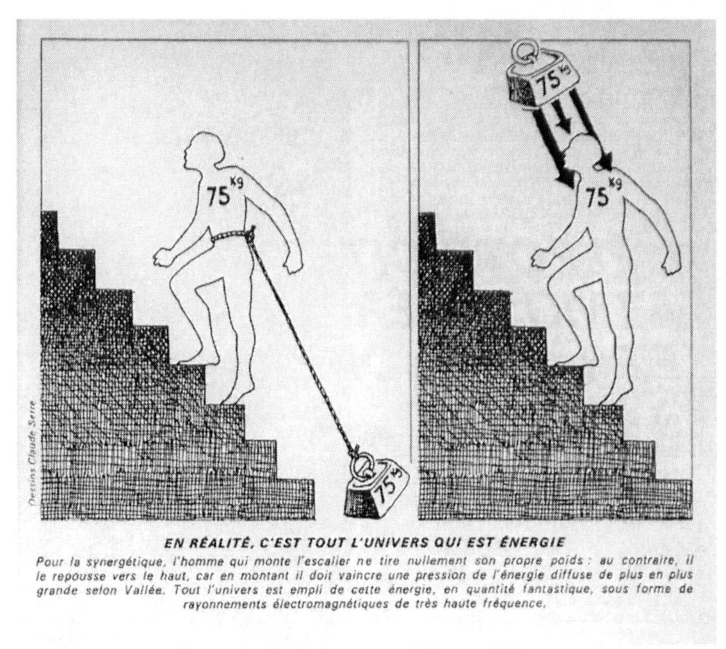

figure 2-5 Science et Vie n_0 677

in formally denying the existence of attraction forces, relegating them to the category of mathematical tools of theoretical physics, where they are useful.

1975 January, second article.

This time the target is changed. The previous article was written to initiate the Science et Vie reader to the Synergetic Theory, the next is explaining how the tax-payer money is thrown through the window and why nuclear fusion will never be realised. It shouldn't be overlooked, indeed, that the most important part of Vallée's efforts was to give us a fantastic alternative to the disastrous experimental chain of the Tokamaks, and to allow the CEA to orientate its researches in a new way, announcing a definite progress in the quest of clean energy. But the open war was now declared between the management and the trouble-maker, inspired or not. However, this last was still keeping the hope, wrongly, that intelligence, reason and comprehension would

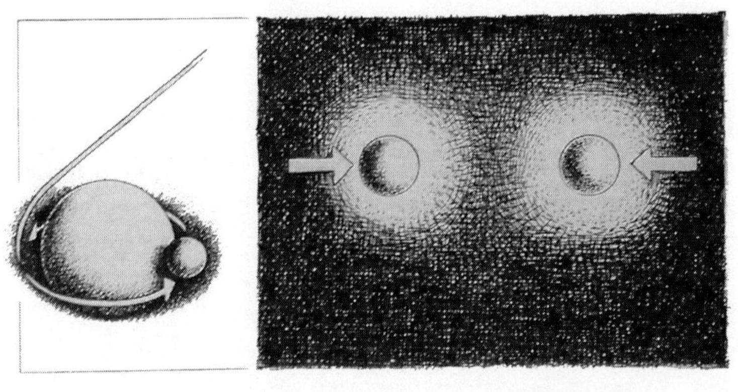

L'ATTRACTION UNIVERSELLE NE SERAIT QUE « LE POIDS DES CIEUX »

Newton avait constaté que deux masses quelconques s'attirent. La relativité y voyait une propriété géométrique de l'espace temps, la masse creusant une cuvette dans laquelle vient tourbillonner tout autre corps. Pour la synergétique, l'attraction est un mirage : en réalité, c'est l'énergie diffuse de l'univers qui pousse deux masses l'une vers l'autre.

17

figure 2-6 Science et Vie n_0 677

eventually win. Renaud de la Taille shared this opinion, and he kept supporting, without ulterior motive, the efforts of his mentor, having the conviction, by the fact of opening him the columns of the famous popular review, to take part in a more than exciting venture. So is it without any doubt that he joined up to process consisting in explaining why it was impossible to realise the fusion in the Tokamak, which needs a short calling in mind on the way it works.

The idea of the nuclear fusion came from the observation of Sun. In light of what we already know about it to-day, it is considered as a permanent, or at least of long-duration, source of energy, and of which the motor is the fusion of light elements (hydrogen, deuterium, etc...), of which the spectral analysis reveals us both the presence and the abundance. So our reference star is seen by the physicists of the atom as a huge plasma boiler of which the temperature overtakes a hundred million degrees and where, given this extreme condition, are going on energetic reactions which feed its radiation. The aim of the "fusionists" is thus to try to reproduce in a machine the solar conditions and to inject light elements, the fusion of these is accompanied with the loss of a part of the total mass, which is turned into energy. This machine is the Tokamak. The plasma is created by an enormous electric flash which increases the rarefied gas temperature up to twenty million degrees. It's not enough to generate a fusion, but largely sufficient to burn the torus envelop which constitutes the essential of the device. This one is consequently equipped, around the flash path, with coils where pass a strong current which generates a magnetic field, this last confining the energy and so protecting the walls, as well as concentrating the plasma and so increase its temperature.

It's what is hoped, at least. In reality, Vallée has shown that the magnetic confining didn't increase the cross section of the plasma, but at the contrary lowered it. Thus the fusion is impossible with that sort of equipment. Only the experiments he had proposed were able to exploit it, and we presently know how enthusiastically they were welcome and what followed. 40 years later, since the first article of "La Recherche" was published back in 1970, we are obliged to see that the facts are in favour of the Synergetics: just like the Arlesian girl[6], we are still waiting for the fusion, this fabulous promise which only moves further away, but the tone of the publications on the subject has changed very much. From the confidence of the first days, fed by a legitimate but too optimistic hope, we have passed to a so excessive modesty that it more and more takes the appearance of an avowal of impotence. This doesn't prevent continuing to insist in this channels, with bigger and more and more costly devices, but the index of confidence, like we say at Meteo-France, has drastically decreased, and

6 Someone you are waiting for and who never comes.

to-day we wonder if the controlled fusion problem is not of the same nature that the controlled explosion of a dynamite stick or a mix of air and gasoline, chain reaction that we can only use by the way of thermal engines.

75 November, third article.

That year, the synergetic fight goes on still more actively, especially since that the emulators of Vallée announce several experiments which seem to prove something, if not the very theory. A Belgian student, Eric d'Hoker, claims having realised a mounting,

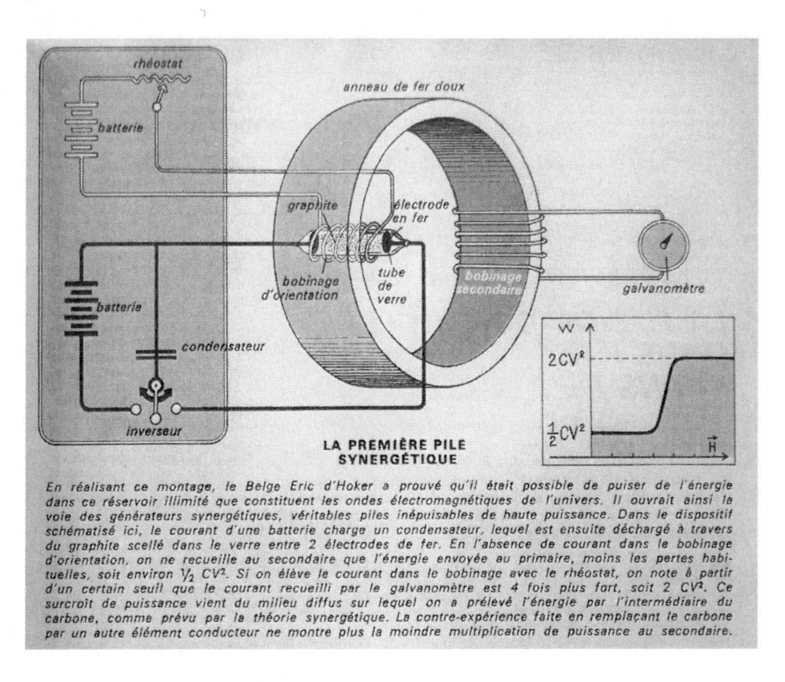

figure 2-7 Science et Vie n_0 698

not very sophisticated, where superposing an electric field and a magnetic field in a graphite bar, recuperated in an ordinary cell, would have led to the transmutation of carbon 12 into bore 12, with immediate return to the carbon while an emission of one electron for each transmuted atom, which, like in the Tokamak, would have caused the production of a supplementary energy three or four times upper to the very bias of the device. It seems that Vallée, who had said that this production of short period radio-active isobar could only be done in gaseous phases, was very suspicious on this experiment and its author. But thirty years after, his student Jean-Louis Naudin, about whom anyone can find, on Internet, all the information on himself and his experiments, seems to have built a device which furiously resembles to the former one. Who is to be believed? The TFR[7] case, finally, seems to be far more serious. However, nothing moves in CEA. It's difficult to think that there would have been no reaction, even negative, inside or outside the big French institution of atomic energy, but it's so: whatever could be the importance and the value of an invention like Synergetics, the most earthy human problems always overcome the general interest. We'd rather not touch an organisation and a program which have too many political and financial implications, instead of looking at a new and promising way, but which would oblige some "enarques"[8] to agree that they have failed. Inconceivable, isn't it?

And yet, it was so easy and simple, while respecting the program for studying fusion, to let Vallée work and allow him to have an experiment from time to time, possibly allocating him a small crew, without spoiling the main budget which, as shows the results of the study on controlled fusion after forty years of unfruitful efforts, is wasted money. This would have allowed a progression of CEA in a royal path where France would have taken a clear lead over other countries and given it back the role of an intellectual and scientific lighthouse that it already had in the course of its history. Moreover, this would have meant for everybody the end of the run to oil fields, gas pockets, uranium mines, to greet the final advent of the "all electric".

But that didn't happen. Would there be, by chance, some particular interests above the general ones? Hardly thinkable! But we must be conscious of the fact that, in every organisation, the more we climb in the hierarchy and the more individuals forget the basic knowledge, to go towards management and careerist rivalries: no much place for competence any longer. This last remains the

7 Tokamak de Fontenay-aux-Roses
8 Graduates of the Ecole Nationale d'Administration

property of the poor technicians of the labs and their engineers, who promote their personal value before any ambition and refuse the addiction to the social ascension. Pushing this argument to extreme, it appears that the less competent, technically speaking, is the President of the Republique, slightly before the Sciences Minister and his advisers. This about whom we are thinking of right now was going to distinguishing himself few after the Vallée case, when he gave the start signal and opened the credits to the study of “sniffing planes”, these ultra-sophisticated devices which should find oil fields only by overfly them, thanks to an embarked electronics of which the study has made the happy days of a certain number of enterprises. Just the money spent for this master bullshit, which had made some true scientists explode laughing, could have made Synergetics definitely progress. But no: there is an enormous gap between what a minister has called the “down world”, in short the one who works and creates, and this of our managers, who decide what is the more profitable for them, in the short term, which means for the duration of their mandates.

May be we are going a bit too far from the technical domain when we are thrashing this way the nomenklatura of our leaders, but unfortunately all that has just been said is a major key to understand why the progress, in general, is not faster, and we cannot be convinced of that without having passed decades studying the social life, in particular of its industrial aspect, which is not only technological. In brief, the Vallée case perfectly illustrates what happens when someone tries to swim against the current. Even if you are aware of the importance of something that you are bringing to the mankind, if this last doesn't want of it, you should neither insist nor try to fight, alone against all, if you don't want to be stuffed by the incoercible pressure coming from higher. How many Vallée have understood that in time, we don't know. There are probably a surprising number of them, but most have not the course, the importance of ideas, the tenacity, in the good meaning of the term, which are at the origin of the synergetic venture, which is only slumbering.

It's necessary to insist on a point, which is suggested in the title of the article “who will dare contesting Synergetics”. It's indeed a constituted theory, which means that it had been published, with a mathematical support, in a book the access of which, when it was issued, was possible for everybody. We fully agree that it's now an unfindable book, for the reasons here-upper developed, but at one moment hundreds of copies have been in competent hands and under competent eyes, capable to assimilate the contents. Moreover, there was inside the CEA a non negligible number of engineers who, not only were interested by the thesis of Vallée, but also gave him some support, some prolongation, by finding in their personal experiments and reflections complementary proves. Do we see well, consequently, the fantastic inertia of the scientific community in front of new ideas? It's almost incredible that high level searchers, at least simply competent in the domain of atom, had not joined up to help one of their ours to express. But however, we have seen that it's this way that things have happened. Any individual willingness to help the dissident has been killed in the egg, any association has been unnerved, all the low-key processes have been used to destroy Vallée and stuff the Tokamak brief.

In this country -France-, but also in many others, it seems that there is a barrier at a certain level of social hierarchy. It is approximately located in the sphere of the management and share the individuals in two groups: those who control and those who obey; those who think and those who work; those who want to climb the higher possible in the social scales -it was the profession of faith of the CGPME on one of their tract, in1980-, and those who simply wants to have a correct situation in the exercise of the job they have chosen. Our biggest problem, which seems unsolvable so long-running it is, is that our promotion system doesn't automatically send to the management of the Country the most competent, but often the most covetous, those who are less scrupling, who have no consideration for others, who miss this so important feeling of the honest man and that we call humanism.

We are now quite far from science and techniques, isn't it? Not at all. What happened to Vallée is the perfect application of these principles, as simple as despicable.

1976 January, forth article.

This last article from Renaud de la Taille doesn't namely cite Vallée, but is more dedicated to his active disciples who experiment, with the restricted means of the amateurs, and who try to design simple equipments working under the basic physical principles coming from the hypothesis that a diffuse energy exists. By this way they put in evidence an apparent fault in the rule of the energy conservation, by showing at moments energetic ratios greater than one. After the experiments of Eric d'Hoker, we now get acquaintance with another synergetician-handyman called Michel Meyer, a student again, who besides doesn't claim taking part to the synergetic movement, but places the realisation of his electronic mountings in the frame of a personal action, in the classical backdrop of the Bohr atom. His idea is, using a variable frequency generator, to excite copper atoms at such a frequency that it makes the external electron orbit resonate, these which is the most remote from the nucleus. He uses a squared signal, able to give sufficient high-range harmonics, among which he hopes finding one which will make a peripheral electron enter in resonance, until the orbit will be so perturbed that it opens up and the electron is ejected. It thus happens that there is an emission of free electrons in the copper beam, which is placed in the electromagnetic field of a coil, and so appears an extra current detected by an ampere-meter. Meyer said he had noted a multiplicative coefficient of 30 or 40.

If we credit his experiments, we are then in front of another process, simpler than this used by Vallée in the TFR, but which uses the same principle: perturb a judiciously chosen atom so as to oblige a peripheral electron to leave its orbital system, orientate it with the others so created by either an electric or a magnetic field, or both, following the method or the variant used. So it appears that, in some places, a certain number of independent searchers have obtained interesting results, and the majority of them only explainable by the existence of the diffuse medium. Even if we know that the history of sciences is marked by cheating and falsifications, it's difficult to think that all these experimenters are liars or forgers. The briefs of energy and space are indeed thicker and more solid than it appears.

Another very interesting point of this article is the box just besides it, page 45 of this number 700 issue of "Science et Vie". It is entitled "Synergetics is still waiting for serious contesters and contains a paragraph which says much on the mentality of systematic detractors of the theory. It is so significant and representative of the action led against Vallée that it merits to be integrally cited:

"*right now, experiments similar to this of E. d'Hocker are being realised in several laboratories. Let's precise that they are conducted under the critical eye of some authorities -in particular representatives of Sciences Faculty-, who have declared that they will not fail to give all the necessary advertising in case of negative result; and in the case of success, they any way will explain the result in the frame of classical physics. A very beautiful example of scientific spirit and intellectual honesty...*".

The following gives some examples of the consternate arguments, trivial and stupid, of some contesters who, it is fair to recognise their effort, has accepted to write a mail to thrash these pseudo-physicists and their forged manipulations. And this leads us to a quite particular article issued in the number 69 -July- August, 1976-, of the great informative scientific review "La Recherche", under the title "the synergetic theory of Mr Vallée". Sheltered inside this famous reference monthly publication, they were two of them to destroy someone who was then fighting for his survival in the CEA, and who had a scientific seasoning far greater than theirs. One of the two, supposed to be a laboratory assistant, describes an experiment that he would have driven in the Research Unit in physics of Paris 7 University, and of which the schema lets think that it is an attempt of copying the manipulation of d'Hocker, robbed in Science et Vie. The operating details, for which Vallée was not a witness and had not been consulted, show that the operator has not got a deep knowledge of the basic principles, since he uses graphite powder solidified by araldite, when the β emission can only be realised in gaseous medium. Despite of that, it seems that this man only did what he was asked to do, that is to forge the most seriously possible an experiment programmed to fail and be followed by a failing report. He will be half forgiven as someone being only a disciplined element, obeying the orders.

The case of the other, this who signed the article and for whom we'll do not make any unmerited advertising, is completely different. Synergeticians will never forgive him the processes he used, unworthy of someone claiming to be a great scientist. What he did is stabbing, knocking under the belt a colleague already grappling with an avenging mob. To be against a theory is something nor-

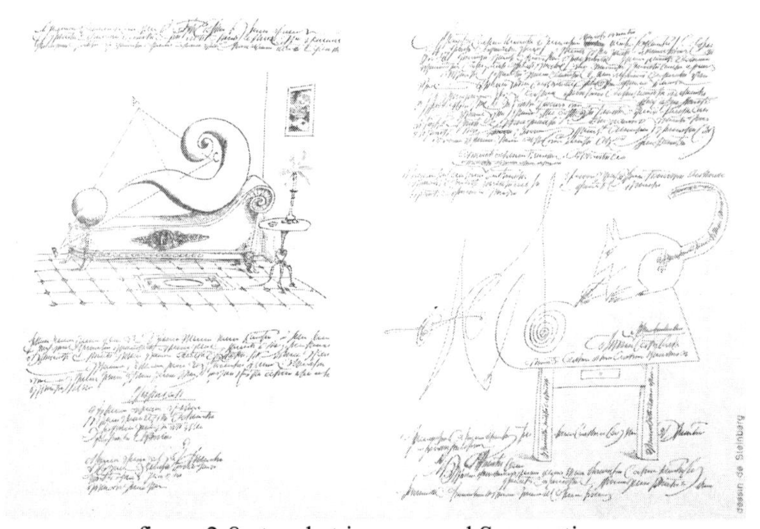

figure 2-8 : to what is compared Synergetics
in the n_0 69 of La Recherche

mal and current, and the history of sciences has given us hundreds of examples. It's generally the occasion for heated and possibly exciting debates. The first rule is to directly discuss with the contester, with correction. The second rule, in fact more important than the first one, is to well know the subject and have taken pains in studying the thesis to be criticized. It's precisely on this point that the article becomes straight ignoble: the signatory compares the Synergetic Theory to the “graphics of Steinberg”, semblance of writing that we can take as such when you see it from far, but that just happens to be revealed as insignificant drawings as soon as we get closer (figure 2-8). When someone who has studied Synergetics see that, he may wonder how a so high-level review like La Recherche has been capable to let pass such a flow of ineptitudes and snorting expressions, without any serious arguments.

"...His propositions didn't receive any echo in the physics laboratories..."

"...Criticize the official science on behalf of his own criteria, want, better than it, follow its norms and so accept the ones and the others, is to be more royalist than the Queen and get condemned to derisory...

"...So about M. Vallée, of whom the writings resemble physics like calligraphy resemble these graphics of Steinberg which, looking from far like a true conventional writing, reveal them as insignificant drawings. All the same can one read to see the difference, and in front of the same proceedings applied to Chinese writing, we'll remain mute. It's this way that the public announcements of M. Vallée, taken here as the simple example of a more general process, strongly resemble the usual popularisation articles. The ideas he uses are moreover all borrowed from the normal arsenal of physics: limit electric field,..."

"...It is not question, so far, to refute step by step the thesis of M. Vallée: insofar as there is no more, there, than a pseudo-theoretical speech and not a formalized and predictive theory, such an enterprise would be useless...".

If he would only have a glance on the book of Vallée, the man who has written these lines would also have noticed that one of the master new ideas of the theory is, precisely, this notion of limit field that we don't see elsewhere, because it's typical of an electromagnetic theory acting in a medium, whatever this medium.

But it is useless to let us be carried away by anger, there are not here reasoned and funded reflections. All the rest is in keeping. It's just a settling of accounts, an attempt more, despicable and unjust like the others, to discredit Vallée to the eyes of public, to reduce him to silence whatever the means to be used: nothing daunts. The executioner will continue gratifying the readers with his pompous epistemic-philosophical notes, small pretentious dungs that nobody reads, and destroy anyone of his wish with impunity: he is sheltered, untouchable in an editorial stronghold which, in this precise case and exceptionally, didn't give to Vallée the right of reply.

Two other articles made furious Vallée and his students. The first one, in the chronology, is in n 6, vol 241, of the review Scientific American, 1979 December, under the title "The decay of the vacuum". Instead of decay they'd rather take the word "miracle" to describe what happens to vacuum, following the authors. This article, signed by the three American searchers Fulcher, Rafelski and Klein, is one of the numerous examples of plagiarism of Synergetics and an attempt to appropriate the ideas of Vallée:

"*Vacuum is usually defined as a status of absence: we say that there is vacuum in a certain part of space when there is nothing. In quantum theories, which describes physics of elementary particles, vacuum becomes something more complicated. Even in space without matter, this last can spontaneously appear, in function of fluctuations of vacuum. For example, an electron and a positron can be created from nothing...*"

Follow 9 pages which will not be resumed here, but which contain extraordinary notions like this of "loaded vacuum" or this other of "virtual particles", to qualify those which have such a short life, despite of having been foreseen by the three genius, that it is impossible to detect them. And especially, we learn that this decay in question, and the creation of these ghost particles, cannot be realised without the presence of an intense electric field! The guys didn't dare the term "disruptive field", but it was very close to. Let's pass on the infamy.

The second targeted article was issued in the number 130, page 223, of La Recherche (1982 Februar), under the pen of Serge Haroche, at that time working at Normale Sup[9] before further entering the Collège de France. We read this:

"*Quantum electrodynamics, theory which describes electromagnetic phenomena at the atoms and particles scale, shows that even in the absence of any radiation emitted by sources, all happens as if space was occupied by random electric waves, with a very large frequency spectrum and propagating in all directions. We say that this is electromagnetic fluctuations of vacuum, the existence of which is intimately bound to the very concept of the photon.*"

We recognize there, without the least possible ambiguity, the definition of space given by Vallée in his book on the gravitational energy which, let's remind, was issued in 1971. Serge Haroche, very cultured person, could not ignore it.

To put an end to the different written opinions, expressed by the various interveners, it just remains to mention some scientific authors, wanting like Vallée to leave beaten tracks and affirm their dissatisfaction for the official science. Among them we find H. Hugolin, present in the biblio for his theory on inertia and gravitation, and who practically his the only one having drawn up a comparative balance of all the explanation attempts of these phenomena by his counterparts. And among them we can find, at the end of chapter IV, Synergetics, indeed resumed in a probably too short way to give a really precise idea, but enough to want to know more. In any case, Hugolin shows his difference in an honest manner:

"*Taking into account our ideas on inertia, on gravitation and on the medium which fills space, we don't agree on the preceding results, which want to explain the gravity phenomena through electromagnetic ones.*"

We'll come back on the scope of this comment, extremely important in the point of view of sheer logic, because each of these two men has a personal feeling of a theoretical physics which

9 Highest professors school in France.

make them blind when they want to interpret it, but the deep philosophy of which is much closer one to the other than it seems, what only mathematics cannot detect.

2-6 : The theory

It's not necessary to perfectly know the demonstrations of the Synergetic Theory to seize its substance, the notices done before are enough for drawing its outlines. But it's good to answer the attacks of sceptics, who say that it is not a constituted theory and who only trust in the power of formulas and equations. It's also worth giving some necessary keys for those who will have the courage of launching themselves in a meticulous verification, while probably a discovery, of the work of Vallée. We must indeed admit that his book is dramatically condensed for an average student or engineer, and that 138 pages on a so huge subject cannot be something else than a sort of aide-mémoire for advised scientists. In other words, the book is not sufficient in itself, 500 pages at least would have been necessary to make it really didactic. Those who don't appreciate calculations can easily pass to the following paragraph, there's something for everybody.

The starting hypothesis of the theory are the followings:

1- Light and EM waves propagate in a medium called "diffuse" and which is constituted, like the ether of Tommasina, by the whole of these same vibrations. Let's already say right now that this concept on the physical nature of this medium will be contested in chapter 3. But it remains that there is a propagation medium, any way.

2- Light celerity is not a universal constant, it's even not a constant at all. If we ingest the explanations generally given about black holes, these fierce ogres crouching in the bottom of infinite skies and which oppose light to leave them, so enormous is their mass, we may believe that it must vary from zero to 300 000 km/s. It is the exact reflection of the gravitation field in any place of the space and is materialized by the relation (1) already given page 96. We may call this relation "fundamental formula of the gravitation", it's very representative of Synergetics and gives a remarkable example of all this theory can bring as novelties in the so congested way of the fundamental physics. There are others, but may be it's the most spectacular in Vallée's theory, because it is the first which had brusquely emerged from his stubborn calculations, the first to show him that he was on the right way. One cannot test the validity of Synergetics without trying to do all the calculation by oneself, it's the reason for which the enthusiasts will find hereunder the indispensable elements to dare the venture. There are not all the lines, but the main steps are given and might allow a student having the level of a Master in microwaves, for example, to understand.

There is a first key to tackle these calculations, which is writing the Maxwell equations in the symbolic system of Heaviside, and it is necessary to first define this symbolism.

Maxwell 's equations in a charge less space are classically written as follow:

$$\boxed{rotE = -\mu \frac{\partial H}{\partial t}} \quad (2) \qquad \boxed{divE = 0} \quad (3)$$

$$\boxed{rotH = \varepsilon \frac{\partial E}{\partial t}} \quad (4) \qquad \boxed{divH = 0} \quad (5)$$

Under this static form, electric and magnetic fields are distinct, when in free propagation we have an electromagnetic field where the two vectors, perpendicular one to the other and to the propagation direction, have similar roles. So it is interesting to find a global form for the equations, and reduce them to a minimum without privileging one. To do that we begin by multiplying the first line by $\sqrt{\varepsilon}$, and then the second by $j\sqrt{\mu}$. Adding member to member it comes:

$$rot\,(\sqrt{\varepsilon}E + j\sqrt{\mu}H) = j\sqrt{\varepsilon\mu}\,(\sqrt{\varepsilon}\frac{\partial E}{\partial t} + j\sqrt{\mu}\frac{\partial H}{\partial t})$$

and $\quad div\,(\sqrt{\varepsilon}E + j\sqrt{\mu}H) = 0$

then we pose :

$$\sqrt{\varepsilon}E + j\sqrt{\mu}H = Q \quad \text{and} \quad j\frac{t}{\sqrt{\varepsilon\mu}} = T$$

from where the Maxwell-Heaviside's equations :

$$\text{rot Q + dQ/ dt} = 0 \quad (6) \qquad div\, Q = 0 \quad (7)$$

Equations (6) and (7) are thus a condensing of equations (2) to (5), where we consider a global electromagnetic field in which the two components, electric and magnetic, are not taken separately but constitute an indivisible whole which will allow simpler manipulations.

We then need, in function of the Vallée's hypothesis, to characterize the space in which the EM waves are propagating. By analogy with gases, we can define a stationary energy medium, that is a medium where the quantity of movement is null in an elementary volume, which we can write $\overline{\iiint \rho.v.d\tau_0} = 0$, and we'll say in the same way that the local sum of the electromagnetic pulses $D_0 \wedge B_0$ has an average value equal to zero.

The line by line demonstration of the simple formula of the gravitation acceleration as a function of light celerity needs about ten pages of calculation, principally in vectorial analysis, which doesn't fascinate many people. We'll only give here the essential indications for those who have a big mathematical appetite and want to try that interesting exercise. In the aim of lighting the writing, the arrows of the vectors like E , H or Q will be omitted. We'll moreover use the intrinsic definition of the gradient, the divergence and curl:

$$gradU = \sum_{i=1}^{n} \frac{\partial U}{\partial s_i}.grad\, s_i$$

$$divV = \sum_{i=1}^{n} grad\, s_i . \frac{\partial V}{\partial s_i}$$

$$rotV = \sum_{i=1}^{n} grad\, s_i \wedge \frac{\partial V}{\partial s_i}$$

Space being constituted of EM waves, we'll call s_i the corresponding wave surfaces, which will be defined by the relation $s_i(x,y,z,T)=Cte)$, and which verify the Maxwell-Heaviside equations. We then can write these under the form:

$$\sum_{i=1}^{n}\left(grad\, s_i \wedge \frac{\partial Q}{\partial s_i}+\frac{\partial Q}{\partial s_i}\frac{\partial s_i}{\partial T}\right)=0$$

and $$\sum_{i=1}^{n} grad\, s_i . \frac{\partial Q}{\partial s_i}=0$$

By conveniently manipulating the basic here-above equations, we deduce the following relations:

$$\left(\frac{\partial Q}{\partial s_i}\right)^2 . \frac{\partial s_i}{\partial T}=0 \qquad (8)$$

and $$\left[(grad\, s_i)^2+\left(\frac{\partial s_i}{\partial T}\right)^2\right]\frac{\partial Q}{\partial s_i}=0 \qquad (9)$$

x,y and z are related to a coordinates system for which the propagation medium is a stationary inertia one. Each point of a surface s_i moves in this system on a distance dOM following the normal defined by $grad\, s_i$.

Using the unit vector $u_i=\frac{1}{\sqrt{\varepsilon\mu}}.\frac{grad\, s_i}{\left(\frac{\partial s_i}{\partial t}\right)}$, we find the relation:

$\left|\frac{dOM}{dt}\right|=\frac{1}{\sqrt{\varepsilon\mu}}$, which shows that the wave surfaces moves with the celerity of the light. After, we use the fact that a solution of

Maxwell-Heaviside equations takes the form $Q=rot\left(rot\,A-\frac{\partial A}{\partial t}\right)$, and that the vector A has the form:

$A(r,T)=\frac{1}{r}\left[V\,(r+jT)+V'\,(r-jT)\right]+grad U(r)$ to finally arrive to the relations:

$$\gamma+grad V=0\text{, with } V=c^2$$

$$\text{and } div\,\gamma+\frac{1}{V}\cdot\frac{\partial^2 V}{\partial t^2}=0$$

When the field is static, it remains:

$$\boxed{\begin{array}{l}\gamma=-grad\,c^2\\ div\,\gamma=0\end{array}}$$

Most of the detractors of Synergetics, if not all, have never endeavoured verifying these calculations, of which we give here only the principal steps, and also the few formulas which allow the brave candidates to know if they have got wrong in their progression. A few hundreds people, in the years 80, had joined up in study groups and remade Vallée's calculations. All of them were actually convinced of the legitimacy of the theory: as soon as we get rid of the fallacious hypothesis that the light celerity is a constant, Maxwell equations, correctly manipulated, inexorably lead to the here-above results. Conscious that the necessary level in mathematics to break doubts was, nevertheless, a bit too high for the average supporters, Vallée applied himself to find other means, may be less general, less universal, but more accessible than those which are used in the "dynamic" demonstration given just before. As a matter of fact, it exists tens of them, and for example the following "static" demonstration is a small jewel of smartness.

We still start from the same principle, it's the hypothesis of Synergetics, that the light celerity is not a constant, and consequently that ε, in particular, depends on the space coordinates x,y and z, μ varying few and being considered as keeping equal to μ_0. In other words, light celerity in any place can be written $c=\frac{1}{\sqrt{\varepsilon\mu_0}}$, with a maximum value $c_0=\frac{1}{\sqrt{\varepsilon_0\mu_0}}$, far from any mass. Let be a sphere having a capacity $C=4\pi\varepsilon a$, bearing a charge q. The corresponding energy is $\frac{1}{2}\frac{q^2}{C}=\frac{q^2}{8\pi\varepsilon a}$. Let's apply the virtual work:

$$\frac{\partial W}{\partial x}dx+\frac{\partial W}{\partial y}dy+\frac{\partial W}{\partial z}dz+d\mathrm{T}=0$$, that we may

write $gradW.dl+F.dl=0$, from which $F=-gradW$.

So W depends on x,y,z only by the intermediary of ε, and we may write: $F=-\frac{q^2}{8\pi a}grad\frac{1}{\varepsilon}$. Let be W_0 the value of W_0 which correspond to ε_0. By multiplying up and down by $\varepsilon_0\mu_0$ we have:

$$F=-\frac{q^2\,\varepsilon_0\mu_0}{8\pi a\varepsilon_0}grad\frac{1}{\varepsilon\mu_0}$$

that is:

$$F=-\varepsilon_0\mu_0 W_0\, grad\frac{1}{\varepsilon\mu_0}$$

but:

$$-\varepsilon_0\mu_0 W_0=\frac{W_0}{c_0^2}=m_0$$

thus:

$$F=-m_0\, grad\, c^2=m_0\gamma$$

and finally:

$$\boxed{\gamma=-grad\, c^2}$$

The addicts of mathematics having had their oxygen intake, let's pass to rational physics and let's make working together our intuition and our sense of logics. We cannot now deny any longer the evidence: each time we'll get rid of this restrictive hypothesis of the Restricted Relativity that light celerity is a constant, each time we'll do the effort of revisiting the acquired knowledge and take profit of the recuperation of the missing mathematical tools, this gravitation formula will unceasingly come back, obstinately, tirelessly, through multiple ways, to scoff at the sceptics and tickle the neurones of amateur physicists. It is not contestable that throwing away the relativist restraints open extraordinary prospects in what concerns the great philosophical questions, which unavoidably emerge from science discoveries, and in particular on the structure of the Universe and the various "World Systems". What does indeed mean that formula of which we have here above given two demonstration methods?

First, that gravity and light celerity are intimately bound, without knowing *a priori* their causal relationship. We cannot affirm, indeed, that one is the consequence of the other. All inclines to believe that they are in reality two concomitant aspects of a phenomenon that we usually call gravitation, without this word dissipates in anyway the mystery and the intimate mechanism. And then, that light celerity in the inter-sidereal medium, far from any mass and so of any attraction, has an average maximum value which only can decrease when we go into a gravitation field, probably even to zero in the case of a black hole, if we believe the description and the interpretation of these astronomic curiosities, naughty beasts nestled in the fathomless depth of the infinite skies, like gigantic ant-lions hungry for matter. By the way, precisely concerning black holes and their capacity to block the light, classifying gravitation in the weak interactions, like modern physicists do, seems to be a perfectly preposterous idea. But this is another story.

The politically correct attitude to have, in front of this new presentation of the physical facts, is to check if there is accordance with observation, and first with Michelson and Morley experiment, of which the so-called "negative" result led Einstein to think that light celerity could be considered as never varying. From what its premature and fast election to the rank of universal constant: why not, while we are at it! Useless is to come back to the cutting down of the mathematical tool involved this unwarranted hypothesis. It's easy to notice that no calculation proposed in the previous pages can exist in Relativity. But if the light speed varies, why didn't the upper cited experiment put it in evidence?

There are two explanations, non excluding one another. The first one is that if the ether carries away the planets, there is no relative motion between them, and consequently it is impossible to detect it, as it doesn't exist. The ether moving with Earth, or more exactly driving Earth, should we say, it is normal that the light propagates in all directions with the same speed.

But we are not obliged to agree with this thesis, which nevertheless will be largely developed in the following chapters, and it is then convenient, in second recourse, to precise the orders of magnitude, to see if the variation $\frac{\Delta c}{c}$ is detectable or not. The calculation is simple: let be M the mass of Earth, and m this of a body placed on its surface, and so submitted, if we let it fall, to an acceleration $\gamma = 9{,}81 m/s^2$. This body has thus got an energy mc^2, noted W_0 in Relativity and S in Synergetics, with $S = W_0 + \sum W_n$. $\sum W_n = \Delta W_0$ represent all the energies, different of the mass energy, acquired by m ,and that we'll suppose limited to the mechanical work of a fictitious travel in the gravity field, between a far point where the gravity is null to the point where it is now, and which represents the acquisition of its potential energy. There are two paths to do that, the first going to a point so remote from Earth that we could consider it having a zero gravity, the second starting from the centre of Earth, where gravity is also rigorously null. In the first case, gravity varies exponentially, in the second linearly. If we calculate the mechanical work in the first case, we have:

$$\Delta W_0 = \int_R^\infty G\frac{Mm}{R^2}dR = G\frac{Mm}{R} = m\gamma R$$

G is the gravitation constant: $6{,}67.10^{-11}$ in MKSA.

Besides, if we call c_0 the light celerity in the inter-sidereal space and c its value at the surface of Earth, we'll have:

$$\Delta W_0 = mc_0^2 - mc^2 = m(c_0^2 - c^2) \cong 2mc_0\Delta c_0$$

From what: $\Delta c = \frac{\gamma R}{2c}$. Making $\gamma = 9{,}81 m/s^2$ and $R = 6{,}335.10^6 m$, we finally find for Earth about $0{,}1 m/s$, and for Sun $300 m/s$. We can see, consequently, that the variation of

c_0 which accounts for the terrestrial gravity is so small that it is not measurable. Therefore there is no discordance with the experiment, we simply may be surprised of the smallness of a variation which means so much. Vallée used of a sally to qualify this curiosity: "Nothing *is less constant which varies few."*. He wanted to say, by this way, that decreeing something to be a constant because you don't see it varying is always incautious and premature, and he was thinking of course to the Relativity and its prime hypothesis. Believe then that constants don't exist, mathematically speaking, is a reasoning which naturally comes to follow and easily insinuate in a criticizer mind. At the light of the outlooks open by what is written before, we hardly dare imagining to what theoretical discoveries could lead the clear and total abandon of the notion of absolute constant, while keeping the possibility of using the better-named "quasi-constants". So is there a good chance that what we call "constants" should be some quantities which, globally, afford this name, but which are indeed eminently variable and so accompanied by their multiple derivatives.

This being agreed, we may do two notices. First, the discrepancy of the light speed between the free space and the Earth surface seems to be, following the calculations of Römer, refreshed by Delambre, and those of Bradley, well above Vallée's results. Second, that this quantity, following the same calculations, apparently only depends on the radius of the considered planet, and not on the specific mass. This is very strange: it is difficult to believe, indeed, that two planets with identical volumes but with different densities, and so with different gravities, could have the same action on the light speed. Besides, those who use to work on gravitation have already noticed that, in a certain number of calculations based on Newton law, for example this of the attraction force inside a filled sphere, we see unforeseen simplifications appear, which make phenomena so simple that they become suspect. There's something wrong, and we'll come back on that later.

2-7 : The Vallée's photon

It's not possible to resume in a few pages a book *-L'Energie Electromagnétique...-* which is already dramatically condensed, and for the reading of which it is recommended to keep pencil in hand. Vallée did his utmost to list there all the outs supposed to be the attainments of the Relativity and of the classical physics, specially in the atomic and gravitational domains, but giving also his synergetic version of the phenomena and showing to the reader that his theory answers too, with generally better efficaciousness, to the great questions that Einstein had pretended having solved or discovered, and the others which have no explanation at all in Relativity. Among them, the photon.

Since the works of Fresnel and Young have annihilated the theory of the emission by Newton, although it had been considered as a dogma for more than a century, the corpuscular aspect of light has reappeared, with in particular the quantum theory, but not only. As a matter of fact, though the various experiments of interferences cannot leave the least doubt on the wave nature of the light, this last continues to show to the physicists a double face.

In the threshold vision, for example, when the light received by a target is so weak that we can count the photons, these have effectively the aspect of small balls of energy, with a straight trajectory and resembling projectiles. Relativity, the over-towering Relativity, has never solved the problem and left to the quantum theory the task of treating it in its small particular world. Synergetics, contrarily, is the only generalized theory which explains this dual aspect of a same phenomenon and, moreover, quantifies it. The model of the photon by Vallée is, we are not in pain to say it, of an elegance and an intelligence which should have lead him to the Nobel prize. It's the only success among all the attempts tried in this domain, it's to-day the only presentation of the light behaviour which makes compatible the two aspects, which previously seemed contradictory. And despite of that, nobody accepts to study this splendid model, clear, transparent, so coherent that it cannot be wrong.

When a body, let's take for example a ball of cast-iron, is heated, it begins to emit infra-red radiations, periodic shaking of ether which propagates in all directions and constitute one of the heat energy dissipation means in the non-material space. If we insist heating, after a certain time the colour changes, the emission becomes visible, but it's still a wave form. Then, if the temperature still increases, the photons appear, bringing with us what makes the difference between them and sheer radiation: a radial rectilinear propagation and an impact energy, different of the radiation pressure because quantified. We can remember -beginning of the paragraph- the orientations that Debiesse had proposed to Vallée to build his first axis of research: the photon, in several aspects of its behaviour, leads to think of a wave-guide. We then must have a precise idea of what could be this wave-guide, if it is

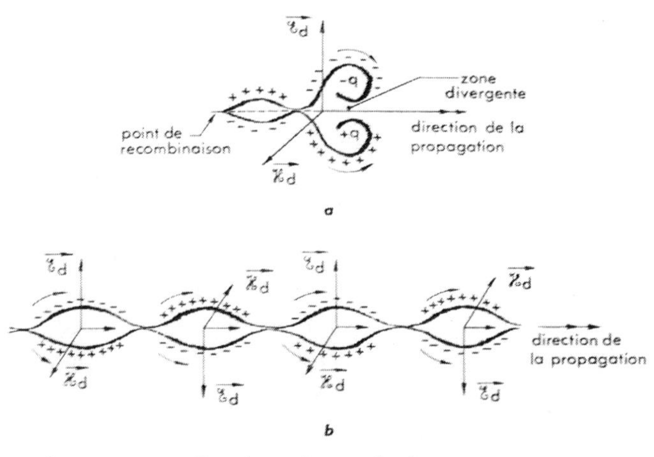

FIG. 4. — *Images de photons telles que la loi de matérialisation permet de les imaginer.*
a) Aspect probable des zones divergentes d'un photon isolé (photon γ)
b) Aspect probable d'un train de photons (photons de lumière visible).

figure 2-9 : The photon by Vallée

real or a simple equivalent, and if it isreal, what is the mechanism of its formation and its structure.

The electromagnetic ether of Vallée, the "diffuse medium", has exactly the same use than this of the Generalized Relativity: it's the propagation medium for EM waves, necessary support the existence of which was judged indispensable by Einstein himself. This being said, the comparison stops there. The second mentioned is a sheer mathematical object, peopled with other sheer mathematical objects, like cords or any other invention which could be necessary to the coherence and the ad-equation of a fictive ensemble with the real world. This of Vallée is defined from physical hypothesis inspired by Tommasina, that is considered as a medium of an electromagnetic nature, identical to this of the individual waves which cross it permanently. In this medium we consider, by hypothesis, that the electric field cannot exceed a limit value, called disruptive field. An attempt of exceeding it, when the corresponding energy density is reached, is traduced by the apparition of matter, under the fugitive form of an electron-positron couple, globally neutral, which are instantaneously recombined to return to a full vibratory state, and to recombine again farer on a cyclical basis. This apparition of symmetrical divergent zones, illustrated by figure 2-9, seems to be common with a sort of cavitation of the propagation medium, and there we are close to a fully mechanical interpretation of an electromagnetic phenomenon, interpretation which will be revived and developed in the next chapters.

So the photon is not exactly a particle, it is a periodic phenomenon which intervenes in the propagation of an EM wave, when its source reaches a certain level of electric field, called divergent because its divergence, in the classical mathematical meaning, is no more null. This implicates the creation of electric charges. In these conditions the energy, luminous or else, because it is not a question of frequency, propagates periodically and changes along its progression in the space: sometimes wave, sometimes matter, it's like a transitional phenomenon where there is an alternation between two possible aspects.

May be an analogy, in a very different domain, could better illustrate this character of threshold phenomenon. The geologist-speleologist-vulcanologist Haroun Tazieff had noticed, during one of his courses in the Himalaya, a curiosity of the Nature: there was in a high-altitude torrent, a spring, in which was progressing, at the same speed and in the centre of the main flow, a secondary one, made of a coloured liquid, normally mixing with water, but not at this place. This “river in the river” had its own existence, while sharing the same path that the main source. Tazieff didn't see again something identical and supposed that, at the place where he had seen the phenomenon, and only there, were gathered all the necessary conditions for it to exist. It was indeed, like the photon, a "fringe phenomenon", where matter hesitates between two status without taking a decision, and chooses to let one cohabit with the other. Energy, in the same way, has got two faces which are called movement and matter, which generally are well distinct but which accept that we could pass from one to the other under certain conditions, and that's one of the meanings of the formula $W = mc^2$. So is there, the same way, a "fringe phenomenon" called photon, in which energy "hesitates" between the two possible forms -moving mass and wave- and alternately changes from one to the other.

So are dissipated all the mysteries of the two schools which have always been in war along centuries, the corpuscular school and the wave school, about which the Quantum Theory had tried to calm the conflict by the way of a new modelling, but without bringing a convenient solution in the domain of the comprehension. The photon by Vallée is the only theory which settles this phenomenon in a form which makes compatible two partial con-

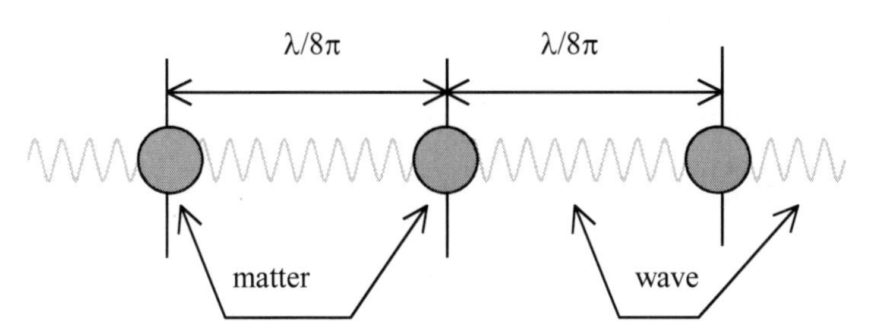

Figure 2-10 The phenomenon « photon »

ceptions, excluding one another until now. It's a model of such a clearness, a smartness and an aestheticism that all the physicists should have run to learn it, but no. What consideration was shown to the author and his thesis? None. We are really stupefied, in this country, in front of the passivity, the slowness, the dogmatism and the mediocrity of the scientific authorities. When someone possesses such an idea, when moreover the discoverer doesn't ask money for making it the benefit of the community, it seems that the least would be to create a committee for its evaluation and the checking of its merits, even for destroying it case of solid arguments against. But still no, not even that. At the contrary, it's scorn, oubliettes, forgetting, except if somewhere, it's another possibility, some people in charge of the energetic lobbies could have seen there a serious threat for their lucrative industries and consulted each other to find the best mean to get rid of the trouble-maker. Only God knows.

The photon seen by Vallée, to come back to the heart of the problem, is not a well defined particle, it's a global phenomenon,

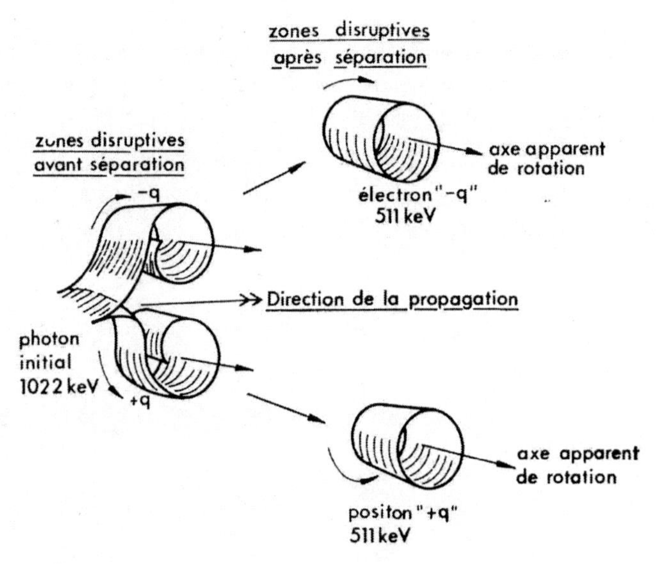

figure 2-11 Formation of pairs by Vallée.

a particular, fringing, propagation mode, during which wave is materialized at regular intervals, that we can determine by calculation.

When the energy density becomes locally high enough, there is no more recombination of the materialized zones, and then happens the creation of elementary particles, stable and permanent, electron and positron first (figure 2-11). So is revealed another mystery of the wave mechanics, this of the escort wave, the existence of which having been only supposed by Louis de Broglie, who had not imagined the dualism of the Vallée's model, but which nevertheless had propelled him to the Nobel Prize. This concept of the escort wave of the photon and of the electron, which solves a certain number of problems but which is difficult to be practically represented, find here a relatively simple and completely different image, which gives a supplementary ex-

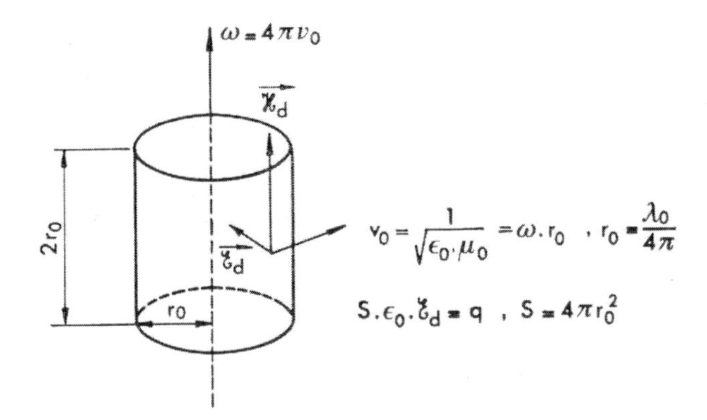

figure 2-12 The electron in Synergetics.

ample of the coherence and the richness of ideas that can bring the fact of taking into account the existence of ether.

The formation mode of the electron imagined by Vallée leads to give it the shape of a small cylinder, where energy is trapped and circulates at the light speed, and we find there, again, something familiar which brings us back to the gyrostats of Le Bon or the vortices of Maxwell. Besides, the possible shapes are in a very limited number, they are in fact those that we can observe when a phenomenon of cavitation happens in a fluid, that is when a sufficient energy creates at a given place a break in the continuity of the fluid in question, whatever the form this break takes: bubble, torus, cylinders and their associations, what else? Vallée, who didn't prejudge what are the right forms, has chosen, among these theoretical mechanical objects, the cylinder in rotation around its axis, model which allows him relatively simple mathematical developments, which lead to find again classical and fully agreed fundamental values, like the Bore magnetron, and calculate the value of the limit electric field:

$$\varepsilon_d = 38{,}67.10^{15}\, V/m$$

2-8 : Critical analysis of the theory

After having the best possible resumed what constitutes a theory which is as considerable as Relativity -let's notice, by the way, that we say “the” Relativity, when there are two of them, the restricted and the generalized, with incompatible hypothesis-, it's now time to look at it, not with the heart and the passion of a synergetician, but with the sharp eye of a critic.

To better understand the Vallée's methods and his way of reasoning, it's not useless to recall his education course and particularly his passage by Supelec. Since this High-School has become a ENSI, in 1958, the best electricity school of France recruits now its crew among the candidates to the general group of the High Schools, that is those who have followed, since the class called Special Mathematics, a mathematical undifferentiated formation. Before this date, the access to Supelec was prepared in a dedicated class-room, with students who had early chosen this channel, their channel. In other words, and this is the point where we wanted to go, they had got since a long time the vocation for this branch of the physics, which is not obligatory to-day: for example, you can integrate Supelec only because it's the only examination you have passed. We may say that after 1958, Supelec has preserved its rank, but lost a part of its soul. It remains, of course, the leader in its channels, but something is nowadays missing. Vallée's year was an ancient one, with enormous courses of electricity, even before entering the School, but nevertheless with a mathematical formation able to challenge any other engineering French High-School. For what more specially concerns R-L Vallée, it's established that he had, like his mentors Poincaré, Duhem, Langevin, Brillouin and others, an absolute confidence in mathematics, but only as a working tool. His big difference is that, for him, experiment was more important than mathematics, not the opposite. This being agreed, we must consider him, without any hesitation, as a practiser of theoretical physics, defined through the principles of Duhem, instead of a rational physicist, these two fundamentally different concepts of physics being explained at the beginning of chapter 3.

Let's now pass to the Synergetic Theory. Like his predecessor Tommasina, in order to beforehand define the space he's going to study, Vallée imagines the ether as a moving interlaced design of EM waves of which the whole constitutes a sort of matter, similar to an odd jelly, and which would so become, through a sort of miraculous transformation, the propagation medium of these same waves. That's very difficult to ingest, and may be that Vallée should have kept mute instead of claiming that a medium could be constituted by the waves which cross it. It's a definition which cannot be accepted in rational physics, or even by someone having a minimum of elementary logics, this last commanding to do a distinction: on one hand we have a real propagation medium, on the other hand vibrations which travel inside, full stop. A wave is an immaterial physical phenomenon which cannot be neither assimilated to its propagation medium, nor confused with any matter, so strange and magic should it be. In terms of calculation, this ambiguity has fortunately few influence on the validity of the results, first because it's only a theory, and second because that the true nature of the ether is of a minimum importance. In return, its status of stationary inertia medium, which is only defined by interpreting Maxwell's equations and making the electromagnetic field as the prime component of the diffuse medium, is introduced by its author only while occulting a possible specific mass, about which he never tells us anything, either as if it was not of importance, or more probably as if he wanted to avoid the subject.

It's very possible that this could be, at the contrary, very important, and that the introduction of this missing parameter could have led, for instance, to a value of the light celerity variation, seen as a function of the gravity, far above the quantity found, and far better in accordance with some other measurement results, about whom little is spoken. Nevertheless this notice, whenever so legitimate it could be, doesn't call into question the most of the acquired knowledge. We simply may say that this last comes from exclusively taking into account the dynamic aspect of space, the only one considered by Vallée, without trying to know what can mean and bring, on another side, the presence of this well tempting and well mysterious specific mass. The question is precisely to know what can be the incidence of this oblivion, wanted or not, in the whole of the edifice constituted by the Synergetic Theory. We'll see later that another definition of the ether, completely mechanical, can lead, while keeping the acquired knowledge of Synergetics but making a selection inside, to a system of the World still more solid.

On the more general plane of the argumentation, we can regret a certain profusion of expressions like "*it seems legitimate to...*, or "*nothing prevent us to...*", or still "*it's very likely that...*", which can be interpreted as a lack of rigour and which, from that fact, are not to the taste of some demanding grumps. We may not forget that it is question of an exceptionally bold theory, moreover of still exceptional consequences, and that we may not demand that all should be demonstrated with an absolute certitude. Those who are awkward have just to read again Einstein, and particularly his first works: it's worst. But the scientific sphere is totally similar to this of the theatre: on one side are the authors, those who write and create, on the other side the critics, far more numerous and, in general, unable to do what the previous have done, but who have a destruction capacity that they use at will. If you really want to know if a theory like Vallée's one is founded, there is only one mean: be seated in a quiet room, at the desk, open a new copybook, take a pencil, begin to read, line after line, the corresponding book, calmly, with patience, writing it again like if you were the author, and let the new ideas slowly and cautiously penetrate in your poor mind, softened by the consumer society. Those who are able to do that have already did so for Relativity, and from that are now anti-relativists. But there is a big problem: the book we are thinking of, devoted to the stake, is now unfindable. What can be done to reissue it?

Let's now come back to the consequences which follow the hypothesis done by Vallée on the electromagnetic nature of the ether.

Say that the diffuse medium is made of the whole of its waves, is giving them a material existence. It's conferring them the rank of physical objects, despite of the fact that they are only mathematical ones: that's to be debated. But also, in the same time, it's also considering that the ether is an exclusively “dynamic” entity, of which the properties, established by the Synergetic Theory, only come from movement, without leaving a place for a possible static aspect which automatically would imply, concerning a fluid, its characteristic parameters, and first its specific mass and its compressibility. But all this doesn’t thwart reaching the notion of mass, even if we imagine the ether like Vallée does, and this will be made by the concept of radiation pressure MBL (Maxwell-Bartoli-Lebedev), which will actually be the hidden motor of the theory. Indeed a pressure, being a force, is by definition something able to give an acceleration to a mass, what allows Vallée's space to be coherent with the method he uses and the calculations he does. But we must agree that there is a certain embarrassment, a feeling of dissatisfaction and this of having perhaps missed still more important developments. Nevertheless, it remains that the photon model keeps available, and it wouldn't be not so bad if it was the only concept representative of the theory.

2-9 : Conclusion of chapter 2

The large scale demolition work endeavoured by the authorities against the Synergetic Theory -and his author- has partly succeeded, but not enough to annihilate it. Vallée having died before he had the possibility of harvesting the fruits of his efforts, the facilitator of this crazy story has disappeared, and nobody will be able to valuably replace him, so strong could be his personality.

It's not a reason to give up and let the oblivion recover such a work. Many of his supporters, those who were his students in the years 80, are still alive and in a good form, and all of them has only one idea in mind: find the genius experiment which will not be explained except by Synergetics, the invention which will bring us, for example, the definite, costless, non polluting solution to the energetic world problem. We still find to-day, on Internet, many sites where independent searchers describe extraordinarily simple equipments which give befuddling results. Among them, there are sure a certain number of charlatans or malicious jokers, who since the beginning use of the mystery surrounding Synergetics to sow discord, either members of anarchist tagging groups, or more simply driven by the taste for hoaxing, and we must not be influenced by these clowns: the Synergetic Theory is a very serious thing, which must be taken as such.

Synergetics was born in the CEA, and it seems that it should be the right place for a new birth. The capture of the energy by the production of radio-active β isobars has been the object of a patent -CEDIR process-, of which a copy has been sent to the Sciences Academy, but also, let's not forget it, has been realised in one of the Tokamaks of Fontenay-aux-Roses. Apart from these last, which have been put in pieces but which are very easy to reconstruct, all the necessary to re-initiate a systematic study of the Synergetic Theory and its main application exists. A good well-tailored program of only a few years could raise the doubt about all this affair, made nebulous artificially, and definitely give a ruling on the validity of Vallée's works. So, why is this not done? Why is there not in this country any willingness to seriously investigate in the tracks of the parallel research, in front of the sterility of the official one? Who benefits from the crime?

It's important to understand that, if we seriously want to initiate the production of electron-providers isobars, simply by pursuing the experiment where it was stopped, and if it is obviously shown that the process works, it could quickly lead to the industrialisation of a non-polluting new source of energy, quasi-permanent and cost-free. As a consequence, there could be a social disruption and a big risk that the economical machine seizes up, such that all its structures fly into pieces: no more useless atomic stations, fuel, gas, advent of the all-electric, and the end, not only of the profits, but of all the very activities of the energy lobbies. This also means the suppression of millions of jobs, too rapidly for any government to be able to face it. It's the reason for which no government, no state-controlled industry, neither is ready to confront this kind of situation, nor wishes to be obliged to deal with. There should happen an international political cataclysm so that a new drastic necessity, involving the survival of the nation, leads to realise what will sure be done some day, but probably only when all the fossil energy sources will be on the verge of being tapped out. In the meantime, the industry managers in charge of the branch of the energy will drive their activity on, while ensuring the preservation of the social peace. General de Gaulle, who had a particularly developed sense of the metaphor, used to say that Paris was a village of a few thousands inhabitants who govern France. This is the key of all the mysteries and a quite convenient answer to all the "Vallée cases", those at the end of which were buried an invention or a discovery capable, directly or non-directly, to be a danger for the large fortunes of the Nation.

It seems that, right now, we have largely diverted from science, by evoking coercive actions, furthermore not proved, from nebular authorities against poor misunderstood and persecuted scientists. What could be a paranoiac manifestation, an obsession of the plot theory, unfortunately is not a fiction. The whole story of sciences is marked by this sort of events, and what happened to Vallée also happened before to many others. In the case of Galilée, the oppressor was the catholic church, and the reason of state was the spiritual power instead of this of money, but apart from that, the two situations are very similar. By the way, the best known scientist among the astronomers was not to be pitied, his predecessor in front of the inquisitive court, Giordano Bruno, having been calcined for the same reasons. It took three centuries to the Vatican to officially recognize the turpitude of the judgement, stated by a college of murderous imbeciles. How long will it take so that Vallée was rehabilitated, he also, or so that somebody, somewhere, take the time to revisit the Synergetic Theory and realize that there is in that thesis a fabulous path of investigation? Or that a President have the good idea of naming to the Research Ministry someone who knows physics just a bit? Or that competent physicists -there are a lot, indeed- join up in a group and take the power to define by themselves their programs in fundamental research?

Synergetic Theory is not perfect and obviously contains errors, but no more and no less than Relativity at its beginning. Quite less, indeed. So is it not a reason not to know it, even less because it has succeeded in exciting the curiosity of hundreds, may be thousands, of young brains hungry of novelties. It's not possible to capture the attention of physics lovers for several following years, at a regular rate in evening courses, without something consequent to be discovered. It's not possible that all this people, after having done by themselves, with the help of monitors, tens of calculation pages without finding any fault, could all of them have been wrong or simply made blind by a sort of mysticism or snobbery of the new. A course of physics is not a sermon in a church, anyone keeps his intellectual free will and the possibility of all checking by himself. Counterfeiting is not possible. Perhaps there is inside Synergetics a crucial error not yet discovered, but if time passes without discovering it, may be we'd better agree that there is none, and consequently that this theory is the starting point, already well endeavoured, of a new physics capable of opening the eyes of sleepy scientists and make the drowsy fundamental Research start again.

Chapter 3 : the Ether (2)

3-1 : Rational physics

There are two kinds of physics : theoretical physics and rational physics. The first one, also called mathematical physics, is the official physics. It is taught in all classrooms, from elementary school to the third term in scientific Universities, in High Schools and everywhere else. In all these courses, we can see the dose of mathematics progressively increase, from the lowest level to the highest, until it occupies the majority of the text. Its best definition has been given by Pierre Duhem in "La Théorie Physique":

"*A physical theory is not an explanation. It's a system of mathematical propositions, deduced from a small number of principles, which have the aim of representing as simply, as completely and as exactly possible, a whole of experimental laws.*"

In front, the rational physics has not a legal existence. It however is practised almost everywhere, as soon as the necessities of the discussions and the desire of understanding lead to get rid of the iron collar of mathematics. Its in fact, principally, the physics of engineers, who do their job in a sphere where we can find a very miscellaneous scale of knowledge and languages, any way far more than in the University sphere, and in which the rigorous expression of mathematics is not enough to let the ideas be transmitted. The adjective "rational" doesn't mean that this physics pretends to be more rational than the other, the theoretical. That would be incorrect. It simply means that it principally trusts in reasoning, in the most general sense of this term, and permanently tries to establish causal relationships between facts, which is not the case of theoretical physics. Through its willingness of seeing beyond appearances, it is close to metaphysics, but remains different by its character of immediate usefulness. The engineer is permanently in the professional obligation of understanding, as far as this verb has a well defined signification, when the theoretical physicist, instead, lets mathematics do its job to get the essence of the hypothesis, while checking the results only when the calculation is ending.

One could wonder why making such a distinction between what simply could be two diverting, but necessarily close, aspects of physics. In the reality, the two concepts are very different in the ways they are approached, and their language also. To illustrate, let's take the example of the physical phenomenon called "wave". A theoretical physicist will say, for instance -it's the definition given in a University second degree course, excellent by the way, by the professor Lumbroso-:

"*A wave is a physical quantity -scalar or vectorial-, which depends on time and space coordinates, and which is the solution of a partial derivatives equation called wave equation.*"

A rational physicist will rather say:

"*A wave is a vibration which propagates in a medium.*"

This first example perfectly illustrates both the difference of language, and also the ways that one and the other use to tackle a given phenomenon. On one side, it's the immediate modelling and the refusal, non expressed but nevertheless evident, to get embarrassed with causality problems, considered superfluous. On the other side, it's the care, all the same immediate, to keep in permanent contact with reality, and go forwards with vigilance by reshaping at each step, if necessary, the causal chain in which the phenomenon under study has been placed.

We'll notice the richness of the implications contained in the second definition: first, that there cannot be a wave without a propagation medium. This evidence, that Einstein had not wished to take into account in 1905, was imposed to him ten years later, when the General Relativity replaced the Special one. Then, that a deformation always automatically propagates: as soon as it is produced, in the limits of the elastic deformations, as soon as it is enforced to a material body, either solid, liquid or gaseous, it is transmitted by degrees, with the sound velocity, in the material in question, and in several directions. The number of these directions depends of the body's geometry and on the application parameters of the deformation (place, force, magnitude, etc,...), and we call them propagation modes.

If we want to illustrate in another way what rational physics and theoretical physics are, referring to sciences history, we may do a parallel between the first one with the Britannic science, especially the 19th century Scottish one, and the other with the French physics of the same period, approximately. William Thomson, Faraday, Maxwell, Tyndall, were the representatives of a physics which absolutely needed a mechanical representation of invisible phenomena, heat, gravity, electromagnetism, to make them visible at least for the imagination. It's the first named in the list who is the leader of the group, and the reference man elected by the followers. Examples of mechanical models in electricity have already been given about Maxwell's works, let's not forget that it is this kind of exercise which is at the origin of notion of “displacement current”, and which allowed him to complete the well-known fundamental equations of electrostatics. All these people already used rational physics, without knowing it, so natural it was for them.

In front of them were the French scientific school, essentially represented by Henri Poincaré and Pierre Duhem, the most inflexible in their strong line. These two castigated the English school, of which the fantasy, for them, is opposing the rigour, not only of the French, but also German, Swiss, and roughly all the North Europe scientists. Duhem uses Pascal's classification to separate the wheat from the chaff: on one side, at the North of the Channel, the “weak but large” minds, on the other side us and our continental close neighbours, with “strong but narrow” brains. The first ones misuse mechanical comparisons when it is question of representing physical phenomena, on the contrary the second ones are stubborn people who use the rigour of mathematics to establish certitudes, even while not understanding them, but with the incomparable satisfaction of having put a law of nature in equation.

In a long chapter of “La Theorie Physique”, titled “abstruse theories and mechanical models”, Duhem details the subject at great length, to the point that he ends up rambling on. Probably only one paragraph would have been enough to express the principal theme, may be this one:

“*So those who, in France or in Germany, have founded the mathematical physics, the Laplace, the Fourier, the Cauchy, the Ampere, the Gauss, the Franz Neumann, were constructing with an extreme care the bridge made to join the starting point of the theory, the definition of the parameters it has to deal with, the justification of the hypothesis which will bear its deductions, to the way where will take place its algebraic development. From where these preambles, models of clearness and method, by which begin most of their reports.*

These preambles, consecrated to put in equation a physical theory, we would vainly look for them in the writings of English authors”.

And giving then, as an example, Maxwell and his electromagnetic theory! But Poincaré is not indebted to criticize in the same way: when someone reads his Treatise of Electricity and Optics, he may have the impression that he has written it with this of Maxwell just beside him, and that his first work was to note, line by line, all what seemed not to be in conformance with the French rigour. The opinion he expresses is indeed cited as an example in the book of Duhem, so happy to see his way of thinking cautioned by the most famous French mathematician. The most significant passage is in the introduction:

“*The first time a French reader opens the book of Maxwell, a feeling of discomfort, and even often of suspicion, is mixed with his admiration. It's only after a long duration commerce and at the price of many efforts that this feeling dissipates. Some eminent minds keep it still.*

Why do the ideas of the English scientist have so much difficulty to be acclimated by us? It's probably that the education received by most of the educated French prepare them to appreciate precision and logic before any other quality".

We notice here the extreme modesty of Poincaré, first towards himself -we suppose he thinks being among the eminent minds-, then to the others "educated French", only ones to be entitled for setting a judgement on Maxwell's works and find the weaknesses. Poincaré and Duhem, same fight! But after all, this is only a question of form. For the content, those who has read "la théorie des tourbillons"[1] must have gotten a more precise idea of the sympathetic beauty of the French rigour, and probably all the best appreciate at the right level the style of Maxwell, who never calls for elitism but who is of a sound humility when it is question of introducing some phenomenon to the reader. But Poincaré, who cannot do without expressing his disapproval, drives the nail a bit farther:

"*So, opening Maxwell, a Frenchman expects to find a theoretical whole as logic and as precise than the physical optics founded on the hypothesis of the ether; this way, he's preparing a deception that I'd want to save the reader in noticing to him of what he must look for in Maxwell and what he couldn't find.*
Maxwell doesn't give any mechanical explanation of electricity and magnetism; it's enough for him to demonstrate that this explanation is possible".

So what? Is this so bad? And first, what does allow Poincaré to judge what a French reader will think when he reads Maxwell? In what hermetic sphere does he live to have the pretension of supposing that everyone thinks like him? What is this claim to want to supervise physics, he who is only a mathematician? Because a large part of the present problem of physics is there: since Duhem and Poincaré, and after them many successors and adherents, have put their nose in this activity and have decided to appropriate it, physics indeed belongs to mathematicians. This annexation

1 Vortices theory

is introduced by them as, at last, the happy arrival of serious people in a domain polluted by eccentrics. As a result, the school books on physics are today books of mathematics which demoralize, often definitely, a big part of young people who nevertheless feel in themselves a core for being searcher.

This is why we need another physics, and calling it rational physics is not less founded than calling theoretical the mathematical physics. Rational physics wants to be a well-balanced approach of the study of the Nature phenomena, and makes a distinction between the process which leads to a law, expressed by formulas and equations, and the causal explanation of these laws, which constitutes the true goal of physics and represent, when it is mastered, the real level of our knowledge. From what precedes, this physics has much similarity with the English science, so despised by the French mathematicians, but it is also more complete, and probably capable of being defined with a better precision than the last, of which the existence has not ever been, neither really assumed, nor claimed by those practising. Let's also notice, and it this not a superfluous detail, that the scientists criticised by Duhem and Poincaré were also very good mathematicians, and their calculations themselves have not been contested, so far, by their censors.

The method which characterizes rational physics, and which makes the difference with the other, cannot be too summarized. Especially, its softness forbids a too stiff definition which could make of it the systematic opposite of the theoretical physics. In particular, the rational physics don't repel mathematics, which remain the privileged vocabulary of exact sciences and their writing language, but keep them at the level of a simple tool, necessary of course, but having no predictive power. Their acknowledged utility is, once a model is defined, to take the maximum of it, integrally. The main problem of the physicist is precisely to correctly design the model. And this last is never definite: it's not because the mathematical treatment of a model leads to experimentally checked consequences that this model is correct or complete. A wrong model can give a right result, theoretical physics is full of examples. Rational physics have a philosophy which, on the particular point of the modelling, which is, whether one likes it or not, the elaboration of a hypothetical mechanism which simulates the phenomenon under study, is diametrically opposed to the rules prescribed by Duhem. It will always consider any model as a provisional one, as long as the complete causal network on which it is built will not be perfectly established, and this can be eternally lasting. To perpetually call a model in question is a normal attitude, usual and compulsory in rational physics. It's a necessity which is not contested by honest physicists, and which doesn't accommodate, it's a completely random example, of the unsatisfying declarations of the relativists, who every day see in the scientific reviews a brilliant confirmation of their great theory.

Another important point: the “physical” model. When it is question of a model, in theoretical physics, it is agreed and understood that this model is conceived on the basis of mathematical objects, and that its architecture is built by using these objects, all of them being fictional and sometimes incoherent: lines and planes infinitely thin, punctual masses and consequently with an infinite density, mechanical works evaluated between a point and the infinity -how to reach infinity?-, mysterious forces postulated to justify a noted but not understood equilibrium, etc... Such models will be thus called “mathematical” models. Rational physics, on the contrary and if possible, will try to build “physical” models, where points, lines and surfaces will have three dimensions, and thus well identified classical characteristics which could be taken into account if necessary: specific mass, elasticity, compressibility, etc...

Lastly, following the process of the theoretical physics described by Duhem, it is prescribed that the model, as soon as possible since it is put in equations, is forgotten to let mathematics work to the end, that is when there is nothing more to get from. In rational physics, on the contrary, the calculation progression is constantly supervised in order to detect the least incompatibility with the starting hypothesis, the least anomaly, the least suspect result, and there is no hesitation to go back to possibly correct the initial model. It's longer, but you risk less mistaking.

It results from this comparison and these notices that rational physics couldn't be other than etherist: one cannot imagine a propagation without a medium, and in this case one cannot do as if it wouldn't exist. Moreover, its investigating character will lead it to revisit the too simple models of the theoretical physics, like this of the kinetic theory of gases, to take out other conclusions from facts observation and correct some ill-advised orientations which have led to mistakes or oblivions.

1906 17 December, Henri Poincaré, who was paying homage to Pierre Curie, who had recently died, declared this at the opening of the annual public session of the Academy of Sciences:

"*True physicists like Curie neither look inside themselves, nor to the surface of things, they know how to see under things.*

Mathematics are sometimes a trouble, or even a danger when, by the precision of their language, they lead us to affirm more than we really know".

What a change in attitude, a few years before his death, by the one who so openly disapproved the so-called "English" method! Because if there was in France someone who practised it, it was indeed the Curie couple. And what an avowal, that we can guess sincere, on the omnipotence of mathematics, from the best of its emblematic representatives! But also, what a release to see that an intelligent mind can, one day, do a zest of self-criticism and open his mind to reflections, belated but honest, on the too clear-cut strong lines in the domain of physics.

3-2 : The kinetic theory of gases revisited

There are, in our habits of language, some mostly popular expressions which are a challenge to the common sense. Among them this one, that anybody has used one day or another: "*that's the exception which confirms the rule*". An exception has never confirmed a rule, at the contrary it weakens it, sometimes definitely. When, in physics, we fall on an exception, we must at once hurry up on it and consider it as a present, a jewel to be preciously kept. We must then quickly study it thoroughly, and try to see what is hidden behind, because it is possible that it is the sign that we have mistaken somewhere, at the precise place it has appeared, where we had fabricated certitudes. But it also can happen that we create this exception, or this lacuna, by awkwardness, or by a lack of care in the elaboration of a model. When this happens in physics, the consequences can be incalculable. The kinetic theory of gases is a striking example of a slapdash project, but so anchored in the habits that it is generally considered as a successful modelling, simply because it leads to previsions which are confirmed by the observation. Well, we already have noticed that it is not because a theory is in accordance with facts that it is a proof of its well-founded, from where the comparing, from the point of view of the reasoning logics, with the famous exceptions which never confirm rules.

The beginning of the kinetic theory of gases consists in showing them to us as clouds of particles which knock together and bounce on the walls of the containers, “like ping-pong balls” they say. How many times haven’t we eared that in the fifth form physics courses? That's indeed a conventional image which, at first, is only there to set a quick picture of the phenomenon that the professor is going to treat, except that there is a little problem: these molecules which endlessly bounce on the walls and so create the pressure exerting onto, this is by definition the perpetual movement. But the perpetual movement, in physics, is something reputed impossible, and may be the same professor, another day, will teach that to the same students, hoping that none of them will notice a contradiction with another page of the course and will not ask any embarrassing question. One may say that this is looking for the detail, being over-critical, and look as narrow-minded as the theoretical physicists previously pointed out for their obstinate attitude. After all, what is so terrible in choosing a model as the simplest possible? Well, let's go a bit further to get rid with this critic, and let's try to examine the phenomenon with more investigating eyes. For a tiny molecule of gas, at its scale, the wall of the container, even perfectly polished, must probably resemble more the Briton coast than the surface of a frost lake seen from far. So is it completely irrational to say and to believe that it could run back, after having met the wall, like a ping-pong ball, or all the same like a billiard ball ideally bouncing on a perfect cushion without loosing kinetic energy. We rather imagine that the poor margin molecules find, in the gigantic cracks of the container walls, tortuous traps from where few of them will escape, and even in bad condition. Consequently, it is very likely that there is, at the level of the walls, an energetic loss, a decrease of kinetic energy, quite similar to sound waves absorbed by the pyramidal picks of an anechoic chamber.

In parallel with this new look over gas molecules striking a container wall, the professor we talked about higher will tell his students, yet another day but in the same course, that a gas is a viscous fluid, that its flow in a tube is laminar, and that the internal friction between layers show a partly transformation of kinetic energy into heat, which is in total contradiction with the hypothesis of perfectly elastic shocks. So we have, about of the gas structure, two incompatible pictures, following we consider either the mac-

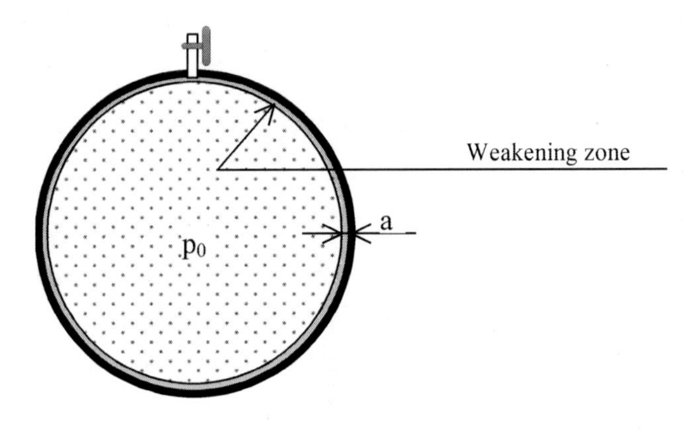

Figure 3-1

roscopic aspect or the microscopic one.

So we can see, in this precise example, that rational physics is not a school which splits hairs, but which breaks down with care and attention to detail the phenomenon it has decided to study, so as to identify, if possible and in the limits of the reasonable, all the parameters chosen to build the physical model, especially when the theoretical one appears too simple. In the present case, this model is not only too simple, but wrong: we must take into account the limit conditions, that means at the contact with the walls, and see what consequences are to follow. But among these, there is one which is perfectly obvious: if there is, either absorption of molecules into the walls, or loss of kinetic energy, or even both, the pressure in the container **should decrease**. And it doesn't, it remains constant. This "non-event", deliberately not seen or, in the best case, swept off with the back of the hand in the classical theory, is at the contrary of prime and determining importance. Indeed, if we investigate a bit more deeply in the behaviour of a gas, and if we try to go beyond the appearance, we find ourselves in front of an alternative the two branches of which are contradictory: on one hand, we have an ultra-simple schema which leads to absurdities, that are cheerfully passed under silence, on the other hand we are in front of the obligation to find an explanation to the absence of a phenomenon the prevision of which seems to be logical, but which refuses to exist. And anyone who wants to be a physicist has not the right to let such a question without answer, contrarily to the habits.

How do we explain that a phenomenon which might happen doesn't happen? There are only two possibilities: either we have done a reasoning mistake, and we must look for the fault, or we have found the weak point of the theory, and we have to elucidate. First of all let's try, starting from the former notices, to quantify this pressure decrease, which would be the consequence of an energy decreasing in the peripheral layer of a gas, supposed contained in a sphere of a radius R , and set to a pressure p_0 . We'll also suppose, as a starting point, that the external layer, where acts the energetic erosion, has a thickness a ,equal to the free average course of the molecules (figure 3-1). Here the loss may theoretically vary from 1 to 99% -0% means perpetual movement, 100% is a total absorption, equivalent to a phenomenon of evaporation-. We are only looking for an order of magnitude, just to clarify expectations. The volume of the considered layer is $dv = 4\pi R^2 a$, and the impact surface of the peripheral molecules is $4\pi R^2$. The matter in hand is now to determine the rate of the pressure decrease, and for that a hypothesis must be done on its causal process. We'll choose to say that the molecules which strike the walls, either are trapped inside, or bounce with a lower speed, and that the result is a decrease of their kinetic energy in the peripheral layer, in a proportion that we'll arbitrarily set to 50% to begin with. It results from this hypothesis an instant decrease inside this layer, where all happens as if a certain number of molecules were disappearing -while précising that they don't really do-, at least not all of them.

In normal conditions, the pressure is proportionate to the number and the kinetic energy of the molecules in the container. Besides, we borrow from Rocard (Thermodynamique, Masson, p299) the expression of the number of molecules Δn which come striking the wall surface unit per second: $\frac{1}{4}\nu\bar{c}$ in CGS system, be $2,5.10^8\nu\bar{c}$ in IS (International System). ν is the number of molecules per volume unit.

So we may write: $dn = 2,5.10^8\nu\bar{c}.dt$.

From elsewhere we have:

$$\frac{dp}{p} = -\frac{dn}{2n} = -\frac{0,5.4\pi R^2 a}{\frac{4}{3}\pi R^3}dt = -1,5\frac{a}{R}dt$$

We may then define a law for the pressure decrease, a decrement such that: $\frac{dp}{p} = -1,5\frac{a}{R}dt$

which, integrating, will give $Log\,p = -1,5\frac{a}{R}t + Cte$, or still $p = e^{-1,5\frac{a}{R}t} + C$. The integration constant is determined, as usual, through the initial conditions: $t = 0 \rightarrow p = p_0$, from where finally:

$p = p_0 e^{-1,5\frac{a}{R}t}$ or $Log\frac{p_0}{p} = 1,5\frac{a}{R}t$

Numerical application: let's take a rubberised sphere having a 25 cm radius, that is a volume of 33.5l and with a constitution approximately similar to this of a car tire of an average size. It is inflated with air, assimilated for the circumstance to pure nitrogen, at a pressure of 5kg/cm². The average free course is $a = 5.10^{-8}m$. Let's calculate the necessary time, with the given hypothesis, for the pressure to pass from 4 to 2kg/cm^2, which leads to the relation:

$$Log\frac{p_0}{p} = Log2 = 0,69 \, .$$

The result is: $t = \dfrac{0,69 \times 0,25}{1,5 \times 5.10^{-8}} = 2,3.10^6 s$, in other terms about 27 days.

Let's sum up and translate: supposing that the molecules of air contained in the tire of a car leave 50% of their energy when they bounce on the envelop, this tire inflated at a pressure of 4 kg might be found at 3 kg about 11 days later and at 2 kg the space of a lunar month. Nobody has noted that, evidently. If we add to this example another one, to pass to a more general vision of the phenomenon, we can notice that, taking several identical containers made of polished metal, but some of them plated inside with particularly viscous substances, like jam or fresh tar or any other in this category, we cannot see the least pressure loss in any of them. And someone would want us to ingest that molecules bounce as easily on fresh glue or salted butter than on polished steel? It is obvious that the usual hypothesis on the kinetic theory of gases is completely unrealistic and proceeds only on a puerile desire of all simplifying to the extreme, trusting in the sole mathematics to find the solution. This is forgetting that if these last can effectively extract the whole quintessence of a model, they cannot be predictive, which means that if the model is more or less unfinished, they will lead at a time to a wrong result, despite the fact that this result comes from a perfect demonstration. It's there a new occasion, that shouldn't be missed, to appreciate the difference of reasoning between theoretical physics and rational physics, and also the possibilities offered by one and the other, in the precise example of the kinetic theory of gases.

The first one, the theoretical physics, will say: being given that nothing happens, we may suppose that molecules actually bounce without any energy loss against any partition. The concordance of the theoretical consequences with the facts will prove the well-founded of the hypothesis. This is a currently taught bullshit: concordance with facts never proves the well-founded of a theory, but simply that there is no contradiction so far. This is very different. Indeed the kinetic theory of gases, such that it is taught since nearly two centuries, has given a sufficient number of results not to be contested. However it's laying on a basic schema which challenges logics, but theoreticians pretend not to see the enormity of the contradiction which goes with: molecules which strike one another perpetually, without their speed decreases, so without friction, and which on another side become collectively viscous as soon as the gas is moving! All this is not serious.

The second, the rational physics, will wonder this: nothing happens, but something **should** happen, and this something, which has been quantified in the case of a car tire, is not without importance. On the contrary it's a considerable phenomenon, which indeed seems to be systematically ignored, but the non-existence of which must be caused by another yet hidden phenomenon. On the plane of appearances, it seems not contestable that molecules indefinitely keep their kinetic energy, since a gas well shut up in an airtight container also keeps its pressure.

This means that there is something else which gives this energy to the molecules, in other words this means that there are other forces in play. Now, according to the extreme simplicity of the model chosen in the kinetic theory, if we reject the idea of a self-maintaining of their perpetual movement, we have only one solution to save us: there exists external forces which create this movement and which are the actual causes of the molecular agitation. These invisible forces are not bound to the container where is the gas we are looking at, but to the whole space, in which we may imagine permanent waves which strike the molecules from every side, and which maintain the random movements that we usually observe. But we perfectly know this space, so well described by Tommasina and Vallée and defined by them as the ether, peopled with its double infinity of EM waves, and which comes back again to bring its physical lighting to a phenomenon left, since now, in the hands of mathematicians.

So we find ourselves again, with an evident happiness, in an acknowledged country, but what precedes has the enormous advantage of starting from a well classical chapter of ordinary physics to make the link, after a critical analysis of the conventional exposition, with rational physics and offer to the etherists a more credible and more deductive introduction to the theme of ether. This justified introduction to the propagation medium of EM waves completes this, more intuitive, of the reference authors, and allow a reinforcement of its character of necessity. It creates a footbridge between traditional physics, the one that we call theoretical, and etherist physics, that we call rational. The fact of attributing the first causal role to the ether, in the case of the kinetic theory of gases, clears up all the usually camouflaged contradictions and allows a more satisfying view of the phenomenon. Of course, apart from the notices made higher to justify its re-visitation, the behaviour of a gas has a mysterious character on which teachers don't insist enough in the physics tuition. For instance, the fact that there is always the same number of molecules in a given volume, whatever the considered gaseous body, is obviously a puzzling subject which should lead to numerous and enormous questions: let it be hydrogen or mercury vapours, with respective specific masses of 0.09 kg/ m^3 and 13.6 kg/m^3, that is to say a ratio of 150, the pressure is the same for a given volume at a constant temperature. That's something extraordinary which poses a problem! That could mean, it's an idea like another, that the accelerating forces which are at the origin of the molecular agitation are so enormous that they don't take into account the difference between molecules masses, these ones only acting through their very inertia and consequently their speed, which any way is agreed in the classical theory.

So we are at a crosspoint and placed in front of two interpretations, completely different, of a phenomenon of which the apparent simplicity has, for a long time, procrastinated the true interrogations on its real nature. The unrest of gases is not a phenomenon in itself, but the consequence and the revealed evidence, under the condition of being correctly interpreted, of the existence of a substrate as active as invisible and intangible, which is its real cause. It is also the real cause of other important phenomena, separated the ones from the other since now, like Brownian movement, temperature or capillarity, as well as many others in the atomic and nuclear domains.

3-3 : The first motor

Young kids used to ask extending questions, after which each answer of an adult leads to another "why", unceasingly, until we surrender through an "*it's so*" or a "*you'll know when you are grown up*", or still "*it's too difficult to explain*", which marks the limit both of our knowledge and our patience. Physicists almost do so, except that, as responsible adults, they use more sophisticated terms, to well show that they have turned childhood, and that now they know a number of things. Nevertheless, if we scratch a bit, we quickly see that the heart of their own questions is not really more mature than those they were asking when they were younger, and that the basic questions are still without answers. But all these interrogations, formulated or not, have still a precise goal, which is to find the original cause, the so simple, so basic and so general phenomenon that all the others must be its consequences, directly or with more or less successive intermediaries. All the philosophers, and all the physicists who have philosophised, have sooner or later sacrificed to this inevitable exercise which consists in going back, as far as possible, in the causal chains, in the hope of at last finding this "first motor" which is at the origin of all the others, through a process of induction. And all of them found themselves surveying the infinitely small, wether it be the vortices of Descartes or these of Maxwell, the monads of Leibniz, the "ponctules" of Nodon or the "cirons" of Pascal, with the same misfortune, but apparently none of them thought of a phenomenon instead of an object.

All what precedes, in this book, is intended to prepare the reader, the most progressively possible, to eventually admit a truth which goes against, not only what he has been taught, but also the evidences suggested by his senses. It's the non-mastering of these last, and the too strong confidence we have in them, which are at the basis of our "good sense" mistakes, worsened by an education which, despite itself, begins its work by teaching us this of our predecessors. This aspect of things will be developed further, so fundamental it is in the comprehension of our behaviour and our evolution, if evolution there is. Because it seems well, when we read for example the physics by Aristotle, and then the Special Relativity, that intelligence and analysing capacity had not changed since two millenniums: the first of the two had not at his disposal all the help which the second could rest on, and his feelings on the structure of the Universe, in this context, are not of a lower logical level. To come back to what is most interesting for us, that is explaining the "rational physics" version for the notion of first motor, it's time now, after having paved the way, to set first the definite postulations from which the new thesis will be developed, and which will become their "principles". These last will not be surprising, if the former chapters have been well assimilated, but they are going to pass from a suggesting form to a clear and non ambiguous statement:

1- The ether exists. It's a perfect fluid, which propagates EM waves without weakening them, and which has a very high specific mass. The way of representing the movement of a material body in such a medium will be explained further.

2- It is permanently crossed by a double infinity -in directions and frequencies- of EM waves which create a radiation pressure that we'll call MBL (Maxwell-Bartoli-Lebedev), which acts in any point of space and ensure, among others, the cohesion of the matter.

3- The ether being present everywhere, including inside the matter, wether inert or living, there is a permanent interaction between them: the ether can carry away bodies, bodies can carry away the ether.

So are laid down, without a detailed development yet, the basis of a new physics which will grow up on the ruins of a shaky construction, this of relativist physics, and of which the quasi-totality will be destroyed with an open and fierce happiness, while keeping and explaining in another way its incontestable experimental acquired knowledge, or more exactly those which are relevant to it, this being agreed with a certain and consistent exaggeration. One easily recognises in the preceding definitions of the ether a miscellany of these of Tommasina and Vallée, to which has been added the notion of a high specific mass, idea which has only been touched on, may be even avoided, by these two mentors to whom, once more and nevertheless, we pay homage for their visionary intuition.

Something important still misses us before we begin to define a new system of the World, that's analysing our ordinary gripping of the notion of matter, and to pick out from its revised and corrected aspects what allows us to better understand its interaction with the ether. And before all, we must remind and insist on this point, as much as necessary, that matter, essentially and before all, is made of vacuum.

Let's imagine a car driver rambling on a country road and seeing, at the horizon, a wooded hill. These woods appear to him, from the place he is on, like an opaque and thick scarf, full and not penetrable, without any solution of continuity. As he gets closer, the details begin to be progressively visible. They turn the massive aspect into a whole of which he can now distinguish the elements which, at a distance of a hundred meters, reveals themselves as being the trunks of trees. If the man stops his car and continues his way on foot to the previously impenetrable body of the forest, near at hand this last will be transformed into a free space where walking straight forwards is possible, under the condition of avoiding the trunks from time to time. What seemed from far to be a fully occupied space is transformed, seen from its vicinity, in a quasi-total vacuum, where the effective volume of the trees is only a very small part of the very space.

This is the way we must imagine matter, only imagine because we cannot, like in the previous example, be in the same size ratio and draw nearer to the surface so close as to be able to see between atoms. In this order of ideas, a TV advertisement on behalf of an insurance company demonstrates what audiovisual can do for teaching. We see, like would see Superman rushing down from space, a blue planet progressively growing in the visual field. At the beginning we can see its cloudy atmosphere, that we cross at a high speed to approach the ground, where we now can see the inhabitants and other living beings. We then accelerate and dive onto a human, more precisely on his forearm. We are now at the contact with the skin, among hairs and pores, like if we were a little fly, and then we penetrate the flesh to begin a travel inside the living matter. At each shot, which replaces the previous one at the same running speed, we switch scale to adjust our size to the neighbouring, and soon we are in the middle of globules, like in “le voyage fantastique”[2] by Azimov. Then we pass to the molecular scale, then the atomic one, then the sub-atomic one, and we now discover the ultimate elements which constitute matter, small worms frenetically agitated in... the vacuum! When you see that, you are balancing between two feelings: admiration first, for what picture technicians, who have realised there a little masterpiece of work, are able to do, and also the regret not to find this kind of illustration in the first form physics courses, just when pupils are discovering physics and when this kind of presentation could the most easily go into the young memories. Instead of which we un-

figure 3-2 : the share of vacuum in matter.
Arrangement of carbon atoms, reduced to their atomic nucleus and strongly expanded to make them visible.

fortunately see that this technique is reserved to advertisement.

The fact remains that the message to be passed, if we come back to the physics, is still the same: matter, from the volumetric point of view, is essentially made of vacuum, a vacuum that the ether has filled up since the beginning of time, which is very far in the past. We may then begin, through all these examples and

2 Fantastic voyage.

these comparisons, to admit the possibility that a matter so constituted can move in a dense fluid, moreover perfect, that is without friction. Nevertheless let's try to precise the orders of magnitude, it's essential in physics.

We find in the book "chimie générale", by Gallais and Rumeau, precise indications on the inter-atomic distances of some substances, especially crystals with simple structures like carbon, for which the volume attributed to each atom is 5.7Å^3 (angström cube, 1 angström = 10^{-10} m). Let's say that, for solids in general, the distance between atoms or molecules is of the order of a few Å. But it is inside the very atom that there is the biggest proportion of "vacuum" (figure 3-2). It is attributed to Rutherford a planetary model of the atom (1909) -despite of the fact that Jean Perrin proposed it eight years before, but no matter-. In this model, where the mass is concentrated in the nucleus, this of the electrons being negligible in comparison, Rutherford applied himself to determine at what distance from the nucleus an alpha particle should pass for it to be diverted. He could this way approximate the diameter of the nucleus to a value between 10^{-15} and 10^{-14} m, for an atomic sphere of somewhat 10^{-10} m, which corresponds to a volumetric ratio in the order of 10^{15} !

So we can see that the very matter, that is what we attribute a mass to, occupies a ridiculous volume compared to the actual one taken by any physical body, whatever the state in which we find it, solid, liquid or gaseous. Consequently we can understand that this kind of structure could move in a fluid that always all physicists have wrongly supposed as having no mass, or nearly none. The resistance produced by a material body, seen this way, when it is moving in a fluid, and especially when this fluid is perfect, must consequently be very weak. So there is really no obstacle, except that of our habits and that of what our eyes suggest us, to attribute to the ether a considerable specific mass, able to carry away suns and planets in other circumstances. It's a reality to which we now must adhere, and it is a fundamental basis of rational physics. But for now, the problem to solve is that of the first motor.

Any material body, from the moment it exists, is submitted to the MBL pressure, wherever it is. This pressure, caused by the etheric vibrations, called EM waves and made evident by the Brownian movement and other phenomena to be re-interpreted, must be considered as the only true cohesion force of the matter, and so eliminates the Van der Waals forces, fictive inter-molecular or inter-atomic attraction forces, to which this role was before attached. We once more notice, on this occasion, the simplistic method of the theoretical physics: some force is missing to explain matter cohesion? No problem: we right now create an attraction force which will do the job, and we add it to the panoply of the mathematician. What was done.

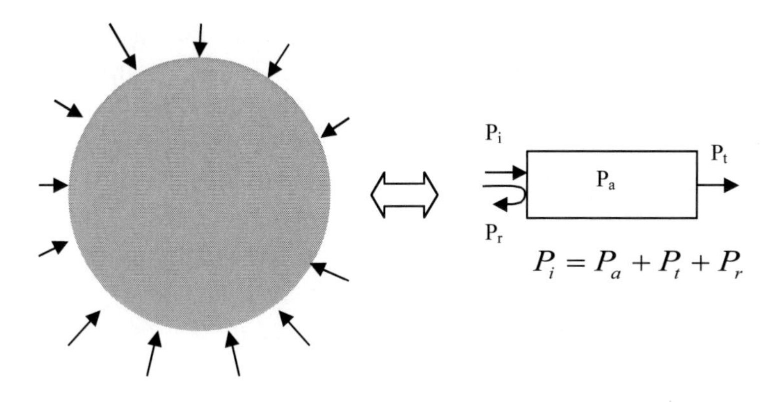

figure 3-3 : a body in space and its EM equivalent

Let's although agree that a force exerted by an external pressure is a more reasonable and more credible concept, no? Newton himself shouldn't have been against. Maybe it's time, now, to slip into the text a new iconoclast idea, which will be taken again later, but to which we'd better be prepared the sooner possible, and which will make electromagnetism explode, while lighting it with a new light, if we may do this audacious pun. That is to say that EM waves are not of a particular essence, but on the contrary totally mechanical waves, of which the sole originality is to propagate in a medium which integrally imbibes space but which cannot be seen by our eyes, and where the place reserved to what we usually call matter is given short schrift. It's important to well see, indeed, that since its distinct origins in electrostatics and magneto-statics, the electromagnetism, such we know it today, is the continuation of a science where all the causes are invisible and not directly perceptible to our senses, in spite of the acuteness of these. So it was immediately necessary, to put all that in equations, to imagine new representations in regard of those of classical mechanics, and give them a name. And this is how were born fields, charges, potentials, EM waves and all the vectorial arsenal which represents them so well, and which cements them in an irreproachable representation, unfortunately disconnected from any visible reality.

We must refuse to accept this appearance of fatality, which lead us to a too complicated physics and, finally, ends up misguiding us. Figure 3-3 shows a solid or liquid spherical body -but it can have any shape-, isolated in space, that is far from any mass, and submitted to the MBL pressure, represented by the arrows placed all around. On the right, we find the schematic electrical representation, that EM specialists know well, of a part of its surface and what there is behind. It's question of what the people concerned call a quadrupole, which is supposed modelling an electric circuit. The spherical body may represent anything, from a tennis ball to a star, it is constituted of a normal composed matter, that is a matter which, from an electric point of view, is neither a very good conductor nor a very bad one. The equivalent quadrupole receives a given incident power, noted P_i, of which a part P_r is reflected, another part P_a absorbed and transformed into heat, and a third part P_t transmitted elsewhere in space. Being given the principle of energy conservation we have:

$$P_i = P_r + P_a + P_t$$

This general formula is often used in electromagnetism and may be interpreted in several ways. In particular, it indicates that some part of the energy carried by the EM waves, which goes striking the body's surface, stays inside this body and turns into heat, through an agitation of the constituting molecules. These last have got a certain degree of freedom, which explains the properties of deformability and elasticity. This is available for all bodies. So we have there a basic phenomenon which takes part in the notion of first motor: a given body, through its very existence, by the fact that it is immersed in an energetic medium, acquires a temperature, this word pointing out the measurement of its internal agitation. We can read in physics books that molecular or atomic agitation is caused by the temperature: what really happens is exactly the contrary. Heat is the result of the absorption of a part of the ether EM which cross a body, and let's remind that these waves are mechanical vibrations, completely similar to sound waves, which are the cause of the existence of a temperature. This last, that we evaluate with the help of a thermometer, is the translation or the transposition of an invisible vibratory phenomenon into a continuous phenomenon made visible by this device.

At the moment, this new introduction of the matter doesn't overturn anything, although it is totally uncommon by its dependence to the existence of the ether. But it's worth being patient and well digest, step by step, all the non-conventional ideas which will lead, we promise it, to a fully unprecedented system of the World which will dissipate all the mysteries of Relativity. About this last, starting from now, we'll not give it any longer the adjectives of “special” or “generalised”, we'll consider it as an official, unique and confuse theory, result of a long following of avatars after which we don't exactly know what are the hypothesis on which it is based nowadays. We however can catch a glimpse of what the presence of the universal “medium”, by its energetic characteristic, is going to result in unforeseeable consequences for relativists. This question of temperature is indeed fundamental, because it introduces something almost living in a matter so far considered as inert. Moreover, this phenomenon of automatic capture of the wave energy in space has some prolongations that a short reflection will make evident.

First, we have to wonder if the phenomenon of acquiring an internal energy and the temperature coming with it, by an external cause, has limits. We hardly imagine, indeed, that the temperature of a body could indefinitely increase, although the external source being constant and inexhaustible, while next to that, the daily seasoning tells us the contrary. It is because another phenomenon originates at the same time, and is its direct consequence: as soon as a body has acquired a temperature, it begins to give back to space a part of what this last gave it: it radiates. The wavelengths of the two sorts of radiations, the incident radiations which causes the inter-atomic vibration and the compensating radiation which evacuates the heat produced by this radiations, have completely different orders of magnitude. The first ones are very short and correspond to the Compton frequencies, that is both the supposed frequencies of the electrons rotation around the nucleus and the upper frequencies, the second ones are known since the age of Maxwell and correspond to caloric or infra-red rays.

We can then wonder what is the compared importance of these two sorts of radiations, both EM, and if there could be a compensation and an equilibrium between them. Engineers who design transformers know the answer. A transformer is a device which, besides transforming voltages, heats inside because of ohmic -resistive- losses in the ferro-magnetic material, which is never perfect. If we refer to our body, alone in space, the origin of the internal heating is not at all the same, but this has no importance. What is important is that the transformer sees its temperature increase the same way, and that it radiates infra-red rays. So it has a tendency to thus balance its heating, but this is only possible if its dimensions are not too large. As a matter of fact, the internal heating depends on the volume, while the radiation depends on the surface, almost proportionally. It's the reason for which big transformers are obliged to be cooled with refrigerant fluids, drawing in their heart, when the small ones, those which are in our ordinary electronic equipments, remain warm without the help of a supplementary system: radiation, conduction and natural convection are enough to exactly compensate their internal heating, and they reach an equilibrium at a certain temperature. A slightly bigger transformer will also reach an equilibrium, but at a higher temperature. A too big transformer will burn or melt down.

For a body isolated in space, there is neither conduction nor natural convection, and radiation only can compensate the permanent input flow of the external energy. Apart from that difference, its case is very similar to that of transformers, and we may edict a universal law, of a surprising simplicity: the bigger a star of a given density is, the hotter it is, and of course the more it radiates, which is very important for its neighbours, if it has got some. So we can see that there is here a phenomenon that can't be ignored, and which also has a universal range and constitutes a first motor. It's a transformation process which exists because the ether exists, with the properties that have been exposed higher, and which concerns matter in its whole, without any possible exception. And if we think of a solar system, we may have the presentiment that there is, in all that precedes, all that is necessary to elaborate a completely new cosmological theory, of which the simplicity and the logic are going to reduce to nothing all the classical phantasms already expressed on this so important subject.

So that is, roughly, the precision that rational physics can bring to this notion of first motor, which has drilled the brains of so many philosophers and searchers, of whom the reflections on this subject have globally only been flights of fancy. They have mistaken in their approach of space by a wrong vision of its nature, or more simply by avoiding the question to stay, instead, in their traditional epistemological blind ways, where the influence of religions has made their work still complicated. Thus the first motor is no longer a thing, a being or a microscopic mechanism that we should look for in the infinitesimally small, except that it is not forbidden to think that several first motors can exist together, but it becomes obvious that it is in the infinitesimally large that lays the mechanism of the birth and the continuous renewing of the Universe, of the planetary movement and of the eternity.

3-4 : Cosmology

Now that the essential principle of the first motor, at the cosmic sense, has been put into evidence, with the hope that the proposed arguments are solid enough for convincing, and at least to initiate a debate on the validity of the reasoning, it only remains to take out the logical consequences.

If we suppose that the body, *a priori* ordinary, of the figure 3-3, is an astral body, it is submitted, like the others in the same environment, to a continuous bombing of mechanical EM waves, which give it at every moment a certain quantity of energy and, consequently, of heat. Its temperature continuously tends to increase, and this increase is only slowed by its infra-red radiation, which sends back to the ether a part of what it has received, and which mixes with the other radiations. From what was explained or suggested just before, we must think that below a certain dimension, let's say a certain diameter if we may consider that almost all the cosmic objects are spherical, there is at a time an equilibrium between the energy in and the energy out. But if we have to deal with something bigger, what happens? Moreover, does this question of size really represent the right parameter to differentiate suns and planets between them? We know that the planets of our solar system have not got the same volumetric mass, except for the so-called earthly planets, although there was not, so far, a possibility of direct measurement, for example by the weighing of samples, which could confirm the astronomers calculations. We would hardly think that a gaseous or liquid planet of the diameter of the Earth could behave like it for the problem of holding back heat. The criteria to be chosen, for the question that interests us right now, is probably more the mass, but not necessarily the same mass we are thinking about when we are weighing an object. Besides, maybe it could be good to make intervene the properties of heat propagation and specific heat, but finally that probably will not bring a big change. Indeed, the thermodynamics parameters only concern the speed and the propagation modes of heat in bodies, so we'll keep for the moment the notion of mass, weighing or inert, to associate it, for matter in general, to the faculty of accumulating energy.

But let's come back to this fundamental problem of thermal equilibrium: this last has probably a limit, and it is evident that, starting from a certain size, considering for example a sun of an average density, it is no more possible to evacuate as much energy than the one absorbed. What happens then? To what value can the temperature increase? What does become this prisoned energy which is unceasingly growing? The researchers of the controlled fusion branch often use the image of the solar furnace to describe the costly device which, according to them, might reproduce it in miniature and domesticate it. They argue, as the basic mechanism, a fusion of light elements, only possible in precise temperature and pressure conditions, these last being naturally present only inside our central star and its brothers in the Cosmos. This being said, in the very improbable case of success, it's to be agreed that this heat, produced by copying an extraordinary phenomenon, will be used, like in atomic plants, in a very ordinary and bestial way: water will be first heated and transformed into steam. This steam, overheated and put at a good pressure will set into motion a turbine, which by its rotation will drive an alternator, which at the end will give the wanted electric power. The fact remains any way that it would be a beautiful result, which theoretically would solve all the ecological problems of our age, but for the moment, the adventure that began in the 70's has not been successful, in spite of the billions invested.

Awaiting the advent of this always postponed technological promise, our sun imperturbably continues to receive its energetic food, which has already transformed it into a fantastic and enormous laboratory where the conditions of temperature and pressure are exactly these which are convenient for all the imaginable atomic and nuclear reactions, including those which make transmutations and creation of matter. In this context we are almost convinced, and we don't see well what else could happen, that the energetic gain is continuously transformed in matter gain, in other words in mass gain, what will give still more appetite to an already hungry entity. So we have there a phenomenon which cannot do anything but amplifying, and which is partly fed by itself as soon as the considered star exceeds a certain mass.

A body in space resembles a sponge in water, and this analogy is so strong that we'll use it in several ways, essentially to help us being conscious that we usually rarely see the things as they are, and that truth is generally opposite to the appearances. The mass of a sponge in water is first the mass of the contained water, and the same way it's going to become evident, with the support of all the notices and comparisons made so far, that the mass of a material body, or more exactly the mass that we attribute to it, is the mass of the ether contained in its volume, and not something which really belongs to it.

So we have now at our disposal a new and not conventional explanation of the origin of the solar fire. But there are not only suns in the Universe, there are also planets which turns around them. When an astral body has not reached what we may call the "critical mass", expression generally used by the specialists of the atom when they describe the conditions of a chain reaction, but that we confiscate with a light heart, so much it sticks well to the phenomena described here-above, we can conceive that the fact of agreeing on the existence of this EM energy, without thinking at the moment of the accumulation effect, means any way a theoretical contribution never taken into account in the other cosmological theories, and will have necessarily important consequences, to say the least. One of them directly concerns our planet, about which we don't know, *a priori*, if its mass is above or below this critical mass. But even if it is below, all the computations to quantify its supposed cooling must be revised downwards: Earth, from the EM energy it receives, cools far less that the usual present hypothesis have foreseen, through calculations which, besides, can be contested. It's well sure, if our hypothesis are right, that the astronomers who will honour us by studying them will also be, first in the obligation to check again their works, second in front of new outlooks which should open to them a gigantic field of investigation, with distressing consequences on their way of conceiving how the Cosmos works.

One of these consequences concerns the rotation of suns and planets. It is not contested that almost all the known planets are rotating on themselves, like the other stars out of our system. It's something which has become so obvious and natural that nobody seems having the idea of asking a question which, on the contrary, should be the first to come in mind: why do they turn? Where might the invisible forces come from, which keep this universal manège everywhere in state of gyration? In our local earthly system, we only know one exception: the Moon, which always show the same side to us. From what another question: why the Moon? What does it have special not to do like the others? We see well, once more, that there are a plenty of non solved mysteries, not in the infinite depths that our cosmic searchers try to survey, at the limit of their observation instruments, but right here, under our nose, just under the so thin blanket of our knowledge. And still here and now, it's worth realising that the fact of denying the existence of the ether has led us, from generation to generation, to reason in a way which has forgotten the benchmarks of the elementary logics, making instead a completely modelled environment which challenges the casual good sense. Other basic question: why are suns and planets spherical? What would prevent them, in an empty universe, to have another form?

In the active ether, which now is our reality, all these questions have each a quasi-immediate answer and propel us into a World where the vortex has a preponderant role. We'll show, indeed, that Kepler's laws are whirling laws, and that the fact of taking them such lead to a physical model of any solar system which is far more complete that all those that were proposed till now. Let's first notice that the MBL pressure acts, not only onto the external surface of the bodies, ***but also, by continuity, on the ether that is inside***. This simple notice results in several major consequences, which lead to a system of the World quite similar to this of Descartes, but added with both corrections and justifications which will change everything.

3-5 : The gravity

How can we imagine the ether, once we have agreed for its existence? How could we make a representation of this fluid that our senses can't detect, but which does everything, which is behind everything, which gives us existence, mass and energy? Is it a powder, a liquid, does it change following the circumstances, has it got a temperature? It's likely that our physical constitution and the specificity of our natural senses forever forbid us to be in the situation of answering these questions, so prisoners we are in our corporal envelop, which most often, by its material exigences, commands to our actions. But we must keep the hope that, one day, in a more scientific society, our descendants will have, through a better-ordered education, the intellectual weapons to go much further in the knowledge of how the big machine that we call Cosmos works.

Contrarily to what Maxwell was thinking, it's perhaps not necessary to attribute a proper movement to the ultimate element constituting the ether. The fact of being crossed by invisible mechanics waves, that one day have been called "electromagnetic" by the physicists, can indeed be enough to explain the origin of the "vacuum energy", of which the existence is not contested any longer.

In this case, the simplest conception we could have of the ether would be to see it as a sort of powder, made of rigid and perfectly spherical balls, of an incredibly small dimension, in front of which an atom would look like a giant galaxy. Nodon, in his book “éléments d'astronomie”[3], has already proposed this concept and named “ponctules” these ultimate divisions of the universal fluid:

“*The ponctule of ether is an extraordinary small particle having an extraordinary high quantum of energy.*”

We'll use again this term, which seems to be well born and which is not given any precision about its geometry, but we'll suppress its intrinsic energetic character to transfer the origin of this property, undeniable, to the permanent presence of EM waves in space. And as we are paying homage to Albert Nodon, an advised etherist more, let's seize the opportunity for relaying two citations that he does himself about Newton, in the aim of comforting his personal thesis on gravitation:

“*I understand, through the word attraction, the effect that do bodies to approach one another, either it be the result of the action of elements which shake one another by emanation, or by the action of the ether, the air or any other corporal or not corporal medium which pushes one towards the other, in a way or another way, all the bodies which are swimming inside! I have explained since now the sky phenomena and those of the sea by the gravitation force; but I don't identify anywhere the cause of the gravitation.*”

This passage is taken out from “*Principes mathématiques de la philosophie naturelle*”. It shows, contrarily to what a majority of scientific students think, that Newton not only never believed in the existence of attraction forces, but admitted well the possibility of this of the ether. It is in a letter to reverend Bentley, taken out from the “Scientific Papers” of Maxwell (vol. 2, p315), that the message is the clearest:

---

3 See the biblio.

"It is inconceivable that inanimate brute matter should, without the mediation of something else, which is not material, operate upon and affect other matter without mutual contact, as it must do if gravitation, in the sense of Epicurus, be essential and inherent in it.......That gravity should be innate, inherent, and essential to matter, so that one body can act upon another at a distance, through a vacuum, without the mediation of anything else, by and through which their action and force may be conveyed from one to another, is to me so great an absurdity, that I believe no man who has in philosophical matters a competent faculty of thinking can ever fall into it."

Gravity is the best- known form of the gravitation forces, because it is local and obviously acts on ourselves and our immediate environment. But it is also unique, because the force which makes two bodies going one towards the other is of a universal nature. Now that all the necessary callings in mind are done, it's time to try proposing an analytic description of what really happens when we see that we are lying flat to the ground, and when this ground seems to irresistibly attract any object that our hand releases. Let's remind the givens of the problem:

1- The ether is a powder made of spherical ponctules, not deformable and with a very high volumetric mass. We also may say that it is a gas, if we suppress to this word the idea of lightness. It still can be a liquid, like said Descartes. Any way, it's a fluid.

2- It is unceasingly crossed by a double infinity -in direction and in frequency- of mechanical waves that have been called EM (electromagnetic) and which create a MBL[4] pressure onto the surface of any matter.

3- Attraction forces don't exist. They are the symbolic representation of non elucidated phenomena, which have the property of making closer any two material bodies, and which have been gathered under the vocable of gravitation.

These principles having been stated, it becomes obvious that the hypothesis done before on the structure of the fluid "ether" is not much different of this by Descartes or Huygens, with the little balls and their agitation. But the main difference, which com-

4 Maxwell Bartoli Lebedev

pletely changes the things, is the presence of the EM waves. It's these waves which create the agitation of the elementary particles, which is well real and which is no longer a disorganized movement, but a local deformation due to the crossing waves, these waves that are put into evidence by the Brownian movement and which create the MBL pressure. So we have gotten, as a basic tool, an old ether rejuvenated by the intuition of Tommasina and Vallée, who added to it what it was missing so that the 17th century scientists, who were not aware neither of the Brownian movement, nor of the electromagnetism, could go further: the existence of vibrations.

This being agreed, what precisely happens when an individual who tries to jump sees that there is an omnipresent force which inexorably pulls him down to the ground? Saying that he is in a field of forces doesn't solve anything, because nobody knows what is hidden behind these terms: a field of forces is a zone of space where there are forces, full point. It doesn't matter making subdivisions to better tackle the problem, separate the electric field from the magnetic one, the gravity field from the gravitation field, nothing works, mystery stays along. At the end, only one "natural" explanation comes to the mind of a candid who simply has his capacity of reasoning. Curiously, it's exactly the same that was cautiously advanced as a proposal by particularly intuitive physicists like Huygens or Roberval, that if an object that we release starts moving, it is because something carries it away. If we don't agree on the existence of the ether, which is a *sine qua non* condition, the attempt stops there. If we do, things become of an extraordinary simplicity. Swept off fields, potentials, vectors and all the arsenal of theoretical physics, a field of forces becomes a zone of space where there is a **flow of ether**, which has the property of carrying away, with it, any matter it finds on its way, the same way that the water of a river carries away what we let fall into it.

This simple image of reality generally makes bursting out laughing those of the physicists who only believe in the power of the mathematics, and feel easy only in their both rigorous and virtual world: that's too simple, if it was true it would be known since a long time, etc... That's forgetting that the history of sciences is paved of a number of thesis, abandoned at some moment as too simple, like for example the heliocentric theory, and which were born a second time, centuries after they were set aside, to allow doing a new step forwards. We always can oppose to the thesis of "it's too simple" the thesis of "it's too complicated", Nature showing us that its fundamental laws, that we discover with so much lowness and hesitation, are of an extreme simplicity, which doesn't prevent them from giving place to wrong interpretations and conclusions, or at least too intricate. Newton's law is a very good example. The fact remains that there are only two possibilities to tackle the problem of gravity: either we admit that it is an etheric flow, or it is only justified by the equations of

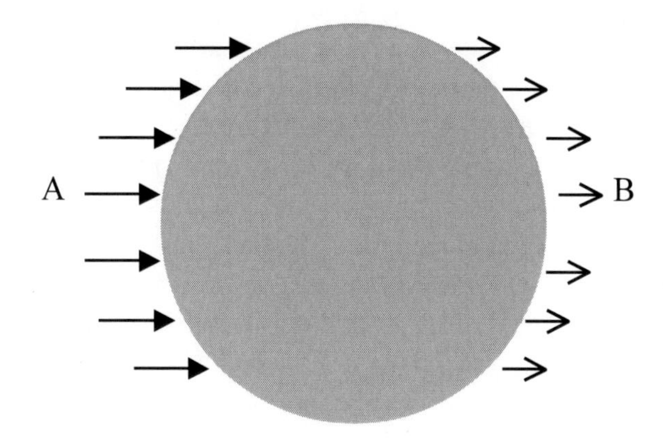

Figure 3-4

the theoretical physics, and in this case goodbye comprehension, killed by a complete addiction to mathematics.

So there is, we make of that a definite hypothesis, an etheric flow which flattens any object onto the ground, on our Earth and on all the other planets in the Universe, with a force which directly depends on their mass. It cannot be the MBL pressure itself, since this one is applied onto all the sides of a body, as well upwards as downwards, but the pressing force exerted on the part of ether which is inside the astral body, at the surface of which we are looking at, and which carries some proportion of the internal fluid towards the centre. By continuity, the part of the fluid which moves inside carries with it a part of the fluid which is outside, and which would carries away any other body if it was not stopped by the solid surface (figure 3-4). The problem becomes less simple when we take into account what happens then at the point B, diametrically opposed to the point A where we are, and where we must agree that there is also a flow of the same nature and the same efficaciousness, in the opposite sense, and which should annihilate the effects of the first. But let's not forget that crossing the star cannot been made without an absorption of energy, and this causes that the ponctules which go out, probably have a lower speed than those which go in the other direction. So we must see the total movement as the superposition of two particles movements, in two opposite senses, the addition of which give a final resulting inwards component, so that we are carried away towards the centre of the Earth, a particular point where the effect becomes null. This idea of two contrary currents can interrogate the non-physicists, who might conclude, *a priori*, that if the MBL pressure is the same at each point of the sphere surface, there would be a balance and no movement at all. It's then enough to imagine an Underground entrance, with people who gets out, crossing and knocking over those who come in, to have a good image of what must be with the ponctules of the ether. This furiously resembles the complex movements of the molecules of two different gases that we introduce in the same container, and that we find, passed a certain time, perfectly mixed. On figure 3-4 have been represented the in-flow and the out-flow by distinct arrows, to well show that the first is more important that the second, but it's worth imagining a superposition on this image of another, symmetrical, and where would be shown the flow going in at B and going out at A.

We find again, here, one of the master ideas of the Synergetic Theory, but we went there by a completely different way of which, for the moment, we have successfully excluded any attempt to using the mathematical temptation. The image we propose of how gravity works has completely abandoned the electromagnetic basis of the former theories, to take with resolution the way of the fluid dynamics. Let's not forget, however, and this is very important, that the equations established by Maxwell take their origin, precisely, in the mathematical basis of that other branch of physics, at first sight independent but in fact very close. So we must not be surprised that two apparently different thought processes reveal themselves totally coherent and compatible. But the best is to come.

The universal attraction law, called Newton's law, is the expres-

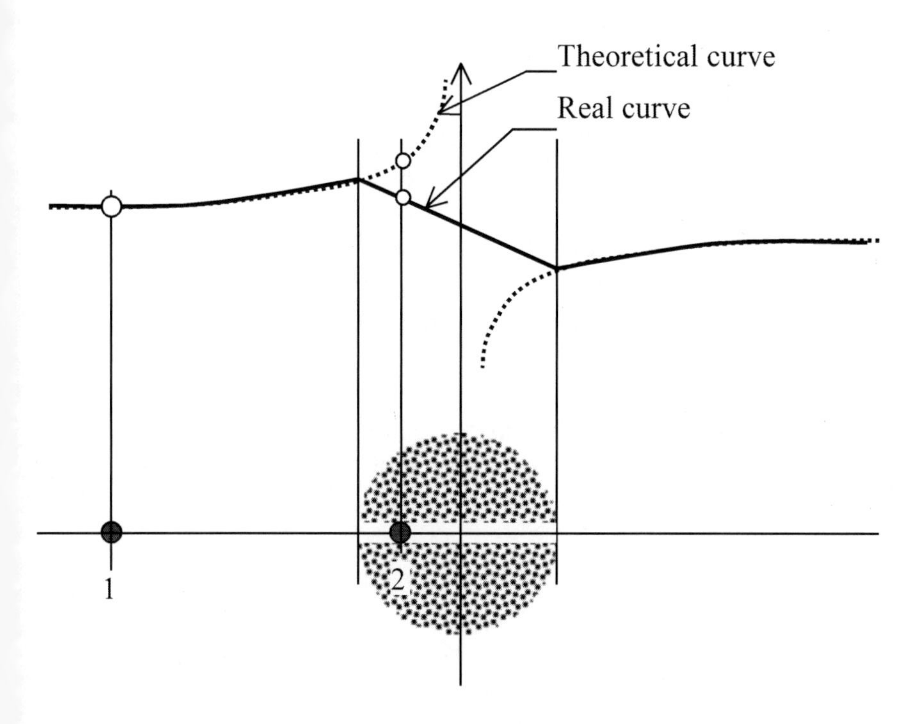

figure 3-5 : Newton's law

sion of the force supposed reciprocally exerting between two bodies by the formula:

$$F = G\frac{m.m'}{d^2}$$, where G is the gravitation constant.

In this formula, the masses m and m' of the two bodies are supposed punctual and concentrated to their gravity centres, with as a consequence a representative curve which is wrong for the values of d inside the bodies. In the figure 3-5, where we see at the centre a "big" spherical body and, on the middle horizontal line, a "small" one which comes from infinity on the left and can go until infinity on the right, crossing the centre by a narrow tunnel, small enough not to perturb the demonstration, we can see the real curve, in continuous line, and the theoretical one in broken line. We note that the two curves are not distinct until the surface -position 1-, but as soon as the small body has penetrated in the big one -position 2-, things are not according to previsions: the portion of the big body which is on the left of the small attracts it, and so partly compensates the effect of what is on its right. The total force doesn't increase any more, on the contrary it decreases, moreover linearly, to completely be annihilated at the very centre, where by symmetry there is a perfect compensation between all the elementary forces. We see well on the figure 3-5 that, for the position 2 of the small body, that there is now a noticeable difference between the theoretical value and the real value, and the more we approach the centre, the bigger the difference is, to finally reach a complete opposition. Newton's law, as it is expressed in the usual formula, is consequently no more available for vicinity phenomena, it is reserved to astronomical problems where we effectively may consider that planets, if it is question of planets, may be assimilated to punctual masses separated by big distances in regard of their own dimensions.

This being assumed, let's transport us in the domain of hydrodynamics. The previous description of the phenomenon of gravity having been made in a fluid, and this fluid being mobile relatively to the surface of the sphere called Earth, it indeed seems natural and evident to move towards a branch of tuition where interactions between the two forms of matter, liquid and solid, are known since a long time. In particular, the flow around a filled sphere, studied by Stokes in 1851, has many similarities with the case that interests us. The essential difference will be that the fluid in contact with Earth is a perfect fluid, although the dynamics of the fluids has got a very complete classification of real fluids, particularly as a function of their viscosity, the pressure and the ratio between viscosity and volumetric mass. As a matter of fact, we are not so interested in the whole of this too copious arsenal, which is made to cover a number of situations, like turbulent flows, which are out of the subject for the comparison with gravitation phenomena, qualified of "calm", if we may say. Their respective scales are for much in that. So we'll only remind, as an object to study, the case of a large laminar flow, in the middle of which is located a fixed and filled sphere. In hydrodynamics, if we want a flow to be laminar, that is without vortices, the number of Reynolds which characterises it must be under a given value, a

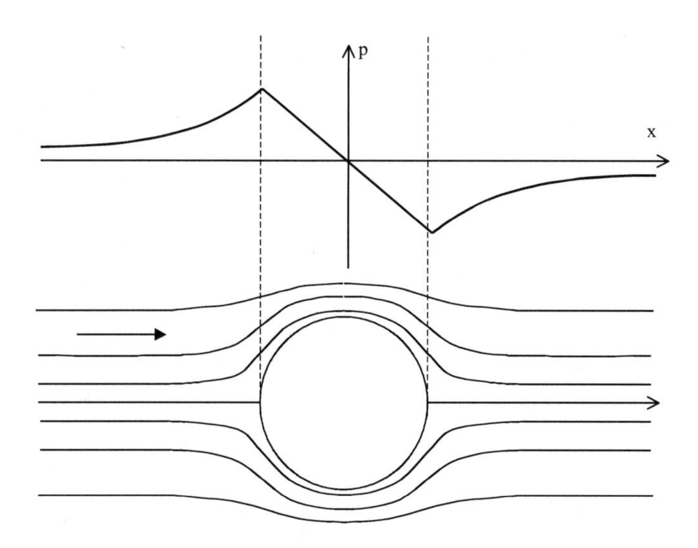

figure 3-6 : laminar flow around a sphere

condition that we'll supposed assumed and which, in the true experience and without getting into too much detail, corresponds to a certain viscosity and a not too high speed. Despite of these restricting conditions, it's a quite casual case, which is close enough to our etheric model of gravity to allow the comparison, since in both conditions we are in presence of a fixed sphere placed in the continuous flow of a laminar fluid.

The figure 3-6, reproduced from a book on mechanics (Mécanique de l'ingénieur de Bamberger, tome 4, p75), is so striking that it's almost not worth adding a comment. But in fact it's not useless to make this comment, because we are here in front of a determining bend in the comprehension of gravitation forces. The upper curve shows the intensity, as a function of the distance to the sphere centre, of the resultant, on the x axis, of the forces exerted on this sphere by the fluid. Saying that there is an analogy with the real curve of gravity, in the figure 3-5, would be too few, and it would be really dishonest not to agree. Whatever it is, when a scientist is confronted to so similar representative curves, despite concerning phenomena reputed without any relation between them, his intellectual duty is to enforce himself to an exhaustive and serious complementary investigation, in order to see either if it is there only coincidence, or at the contrary if something important is hidden behind.

A grumpy man will at once notice, and we cannot really blame him, that in the case of the sphere placed in the moving fluid (figure 3-6), this last licks the sphere considered as not penetrable, instead of what, in the etheric interpretation of gravity (figure 3-5), ether crosses the sphere. The two phenomena would then be no more comparable, and not to be assimilated. But if they are not, despite the appearance, the similarity degree of the curves is such that it would be difficult to believe in an effect of random. On one hand we don't know if, in the etheric model of gravity, there is not a skirting, maybe partial, of the terrestrial globe, what would suit us well and which is possible. On the other hand, we don't know, either, being given that no theoretical study has been done on this model, of course, if by chance this study could not lead to the same representative curve, justifying it quasi-definitely. We are impatiently waiting for a big mathematician brain who would accept to tackle the problem and dissipate the possible remaining doubts we could still have. In that hope, anyone can make his opinion.

We are now in possession of a fantastic tool to revisit all the cosmologies proposed to us till now, and then to build another one, both simpler and more complete.

3-6 : The solar system

Why are suns and planets spherical, and why are all of them in rotation? It seems that these questions are never asked, and that the evidence of facts, with which we have so long been living, is printed in our brains in a way which is both natural and definite, without any more interrogation. Why does it turn? it's so. There is however food for thought. For example, a sun is considered by astrophysicists as a gaseous ball, for Earth its volcanic activity shows us that, inside, it is mostly a liquid in fusion. How then is it possible, taking into account that they are in rotation, the centrifuge force doesn't transform them into a flatter and flatter disk, into a sort of magmatic pancake of an unceasingly increased diameter, to finish in an endlessly enlarging ring? It's well a bloody question, this one! But nevertheless, it doesn't seem to be preventing many people from sleeping well. Specialists have many more important things to do, in the vein of hunting new far galaxies or tracking down the hidden black mass, than losing a precious time in trying to solve problems reserved to 10 year old children -Groucho Marx would say: “go quickly look for a 10 years old kid”!-. Liquid spheres remain spherical? It's because there are cohesion forces which maintain them so! It's no more complicate in theoretical physics: if we only have to invent a fictive force to get rid of an embarrassing question, no problem, we know how to do that, it's a part of the method.

Let's be serious. This way of getting rid of the true problems begins to be more and more irritating. If, at least, those who use theoretical physics were modest and had a minimum of reserve, they could be half-forgiven for their pretentious behaviour, half only because the explanations they give of the great mysteries of the World are often within a hair's-breadth of stupidity. But no: there is today an always increasing megalomania from a growing number of research managers, generally officiating in big state organisations, and presenting to the public, when they successfully infiltrate the media windows, bombastic images of their activities, from the infinitely small to the infinitely large. Their performances and their messages don't generally leave traces in the large public, because what they tell us, pretending to be super-brains who condescend, from time to time, to surreptitiously leave their sanctuaries and make the good people profit of their advancing, is either incomprehensible or without any interest. Who can have a passion for astrophysics today? Who really understands the popularisation articles on this subject? Do the very authors of these articles even understand what they write? Let's be forgiven for this new fit of anger against official authorities, but we must become well conscious that, when science loses its tuition values, which are of prime importance in our kind of civilisation, because of an elitist diverting that cuts it off to the average man, they are

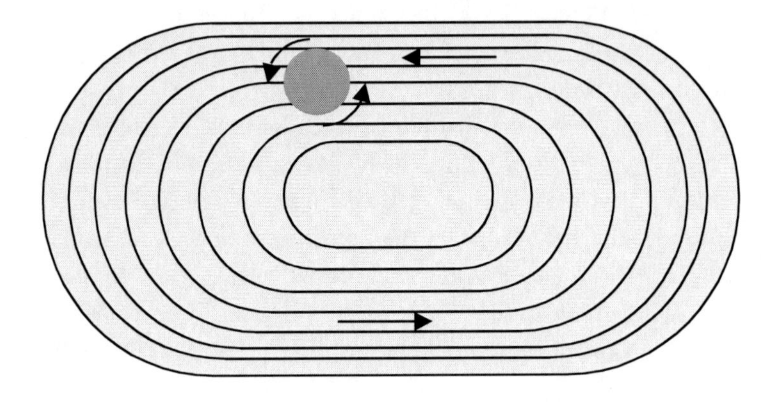

figure 3-7

always very soon replaced, in the collective unconscious, by the obscurantist temptations where charlatanism, religions and all the manifestations of the gregarious mysticism rush with celerity and delectation. More than the education, science is the affair, and also the ownership, of every one of us. We must fight.

But let's come back to our familiar stars and their obstinacy to remain spherical in spite of the centrifugate force. From the previously stated hypothesis on the constitution of the vacuum, it's well understandable that an astral body, internally fluid, naturally takes a spherical shape: the MBL pressure acts in a homogeneous manner onto its surface, and the final equilibrium of the forces, even if its chronological history has passed, before, by a different shape, cannot be conceived on another geometry, which corresponds to the smallest surface for a given volume. Besides, it seems that all is turning in the cosmos, despite nothing, in the stars constitution, precisely explains why they are in rotation. This general phenomenon is a fact that astronomers are obliged to agree with, but they leave this impression that none of them is interested in looking for its causes, except if they consider that there is not here any problem, why not? In any case, that's a subject we never see in astronomy magazines. However, there must be an explanation? Let's have a look.

Supposing that an astral body, originally fixed and submitted to the MBL pressure, the least mass dis-symmetry around its gravity centre, the least lack of homogeneity, will cause the creation of a

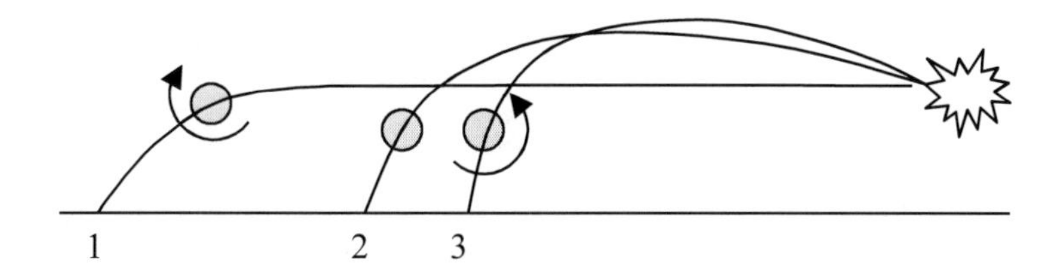

figure 3-8

torque. Even if we had imagined a perfect sphere, constituted of a unique and well defined matter, the least eddy of the ether all around would be enough, one day or another, or more exactly one millennium or another, to initialize by a little shaking or a light touching, at the cosmic scale, a beginning of rotation. After that, if the body is in relative movement referring to the ether, this movement can only amplify itself. Current life is full of similar schemas or facts which, seen at our scale, can give us a more familiar and understandable image of the phenomenon: let's imagine, for example, a spherical sponge of such a density that it remains at constant depth beneath the surface, in a bath. If we put the water in movement along the inner side, the different layers will progressively get this movement, from one to the following, from the periphery towards the centre (figure 3-7). The sponge will be carried away by the created current, but as the water speed goes down nearer to the centre, its volume will be submitted to unequal forces of which the unbalance will create a torque, so that, in the same time that it will be moving along, it also will be rotating on itself. This is an example of a fact which can be transposed in the space and the ether to make understand how a star can be put in rotation. But there are others, and this is not the main one.

The just above experiment describes one of the two particular situations of the interaction between a fluid and a porous sphere, this where the sphere is carried away by the fluid which transmits it the double movement of translation and rotation. If the fluid is now no more laminar, but is moving at constant flow, the sphere and itself have the same speed, the relative displacement is null

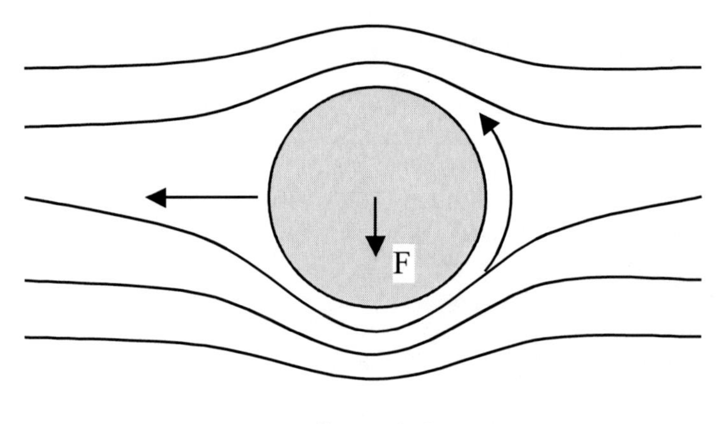

figure 3-9

and nothing happens. The other case is when the sphere is moving relatively to the fluid, with two possibilities: with or without rotation. If there is no rotation, the sphere can only have a linear movement, because it exists no force to divert it, insofar as the fluid can be considered as being locally at rest. If the sphere is in rotation, things are different. The behaviour of a sphere both in rotation and in movement with regard to a fixed fluid can be seen, to take a well-known example, like that of a tennis ball, to which a player can give a spin, following the way he strikes the ball. Everyone knows that, with the same striking power, following the spin you put on the ball, the trajectory will be different and more or less long: the player knows how to do it, but if he's not a physicist, which generally is the case, he doesn't know, in return, that he is mastering the interaction between a fixed fluid and a sphere in rotation, in the occurrence the air and the tennis ball. Figure 3-8 resumes and illustrates the three possible cases:

-1 sliced ball (downwards spin). The trajectory is made longer and can even be almost horizontal in the first part. The ball will rather go backwards when landing.

-2 direct striking: parabolic trajectory, only braked by friction on the air.

-3 lifted ball (upwards spin). The trajectory is shortened. When the ball bounces on the ground, it's escaping forwards.

The behaviour of a tennis ball is only an example, among others, of what we call in physics Magnus effect, which describes how a lateral force is appearing when a circular section object, sphere or cylinder, in rotation around its axis, has been given, with respect to the fluid in which it moves, a relative speed perpendicular to the axis (figure 3-9). There is however another approach to explain the behaviour of the tennis ball as a function of the rotation sense relatively to the motionless medium, very different in that it doesn't use the Magnus effect, but as much interesting: when there is no rotation at all, the air brakes the ball equally on its upside and downside. There is then a simple braking, symmetrical up and down. If there is rotation, the interaction is not the same upside and downside. Contrarily to what we could *a priori* think, this interaction is weaker when the relative speed is higher. It's a phenomenon well-known in mechanics, where it has been experimentally established that the sliding coefficient is, in general, lower than the friction coefficient: consequently, to ensure that a car stops the fastest possible, you must avoid sliding. To come back to the Magnus effect, the commander Cousteau had made built in the years 50 a special ship, the Alcyon, reproducing an attempt made some decennials before, and which instead of sails had been equipped with two vertical cylinders put in rotation by a small auxiliary engine. This system allowed the ship steering its course with a side wind with the same speed that would have had a sailing of a ten times bigger surface, and sailing before the wind.

The practical examples here-above described allow putting into evidence a way of interaction between a fluid and a sphere which is directly accessible to our experience and our comprehension. Their aim is to make easier understanding a generalisation of the principles we have put ahead, and their direct transposition from the ordinary and quotidian domain to this which better interests us: the ether and the stars. But something is still missing, which is indeed the most important, because this is the phenomenon which, once we have well assimilated and well modelled it, will lead us to the essential mechanism acting in any solar system, the vortex, to which we'll attach the adjective of "astronomical", to well distinguish it from the other kinds of whirlpools.

When you turn over a bottle, full of water, in order to empty it without any special precaution, water flows out gobbling, thwarted by the air which tries to do the inverted way and, as it can, makes its course upwards, where its lighter density calls it. If then we do it again, but if just before we give a rotative movement to the bottle, with a convenient gesture of the wrist, the flow

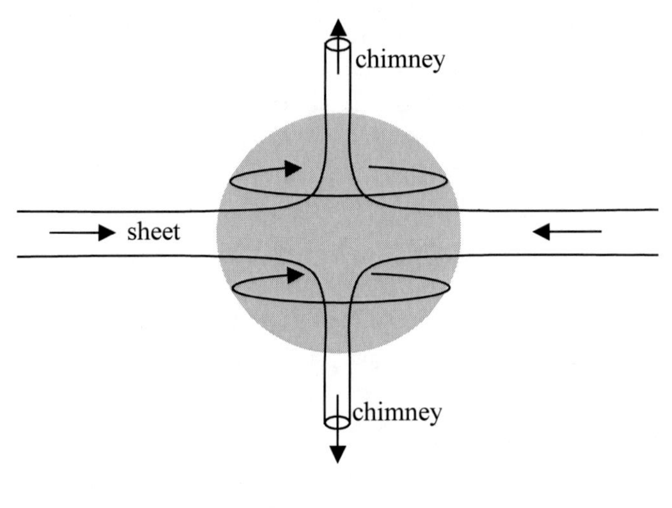

figure 3-10

immediately goes to a whirling state and a vortex is made. Water now flows in helix and escapes from the neck keeping the contact with it, and letting the air going upwards in the centre, taking a path where there is now no more trouble. If in the first case the bottle is emptied in five seconds, it will do it in hardly four in the second one. In the same order of ideas, a thick thread rolled around a radio antenna for car allows this last to be less noisy by suppressing turbulences. For Nature, which is a big lazy girl, there never will be hesitation: the fastest way is always the best and the smartest. The simple experiment related above reveals a general law on fluids behaviour, which is the following: when an aspiration point is created inside a resting fluid, it creates a movement which will be a whirling one, except if something is done to stop it.

Let's now come back to our sphere of the previous paragraph. This sphere will now be particularised and systematically identified to, either a sun, or one of its planets, and let's stay tuned as we start the analysis of what happens when it is in rotation on itself in the ether. The fact that this movement exists creates, compared to the static case, a modification in the way the ether takes to go in and out the sphere. When this last was motionless -very rare and even improbable case-, there was a balanced repartition of the etheric flow onto all its surface. When it turns around its axis, a possibility is offered to the ether to form a vortex, what it immediately will do: the entering flow will concentrate in the equatorial plane, perpendicularly to the rotation axis, and the exiting flow is ejected following the very axis, symmetrically in the two directions (figure 3-10). This disposition will spontaneously be realised because it is the one which allows the fastest transfer of the ether in the sphere, from that there is no reverse antagonist flow, like in the gobbling bottle, but compatible ponctules flows. It's quite similar to what was going on in the second bottle of the former paragraph, except that we are now in presence, instead of a simple vortex -the one well-known in terrestrial physics- of a double vortex, perfectly symmetric, that we have called “astronomical” to specially characterize it in regards to others, which generally are semi-vortices. This vortex will be identified by its “sheet”, quasi-flat layer along which the ether penetrates into the sphere by its equatorial plane, and two symmetrical “chimneys”, perpendicular to the sheet, by which it escapes. It is evident, for someone who knows about fluids behaviour, that this particular movement of the etheric mass inside the sphere will not prevent it to rotate, but at the contrary will boost it. So we have got soon a model which allows us to integrate the simple rotation of a sun or a planet into a far more important phenomenon, to a more complete mechanism of which the functioning can finally be unveiled and explained. It was enough, for that, to make visible, by the reasoning and the abstraction of a drawing, what is naturally invisible to us. But how many detours, how many notices, how much weavering to arrive to a so simple presentation of what is, it's probability being now so likely, the heart of the prime mechanism of cosmos, the first motor of the stars big show!

Let' resume: the ether feeds a sun in rotation by continuously injecting vibratory energy, which cannot be any more evacuated since a certain ratio mass/diameter, and in return the sun sucks up

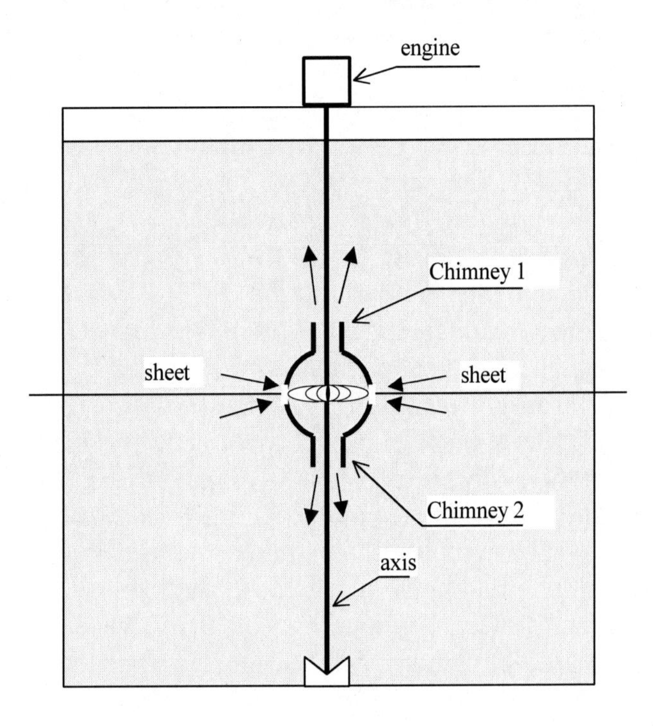

figure 3-11

a sheet of ether in its equatorial plane and rejects the same volume by the two opposite chimneys, each of the two spitting its flow through a pole.

To get a more conventional and more easily digestible idea of all that, let's imagine the following equipment (figure 3-11): in a sort of big cubic -it's easier to make it so- aquarium, full of water, a metallic rod, placed along the vertical axis and linked to an electric engine at its upper end, bears a small paddle turbine. This last is placed onto the rod at half-hight inside a small fixed sphere, with holes at the level of the turbine and ended up and down by two short tubes. The paddles are symmetrical and eject water equally upwards and downwards. When the system is put in rotation, the paddle wheel aspires water from outside the sphere, making a sheet, and spits it upwards and downwards through the tubes: that's the mechanical reproduction, visible and miniaturized, of a sun at work. Of course, this small dummy model doesn't tell us all, this would be too easy. In particular, it completely occults the real trajectory and the actual shape of the etheric flow in the vicinity of the rotation centre: in the mechanical model realised in the aquarium, the totality of the water flow is guided by a rigid construction, while a sun is a like big sponge which, despite constituting the cosmic turbine we imagine, or the engine if we prefer, drives only a fraction of the moving part of the ether, when another part, may be the most important, avoids it as it is carried away by the flow that really crosses the sun volume. All being weighed, however, it's very likely that exactly the same thing happens in the aquarium.

In either case, we are dealing with a whirling rating, symmetrical relatively to the central plane of the sheet. This last must be seen as a rolling up of successive joined layers, globally analogue to the furrow of a vinyl disc, and submitted to the same notice concerning both the language and the vision of the things: on a disc side, there is only one furrow, even if we say “furrows”, and the same, if we want to represent a vortex by laminar layers, we must see well that there is only one layer which is rolling up to the centre of the vortex. This double, or astronomical, or complete, or symmetrical vortex, following the name we may give it, can be seen as the juxtaposition of two elementary vortices, one of them symmetrical to the other relatively to the median plane of the sheet, and apparently very similar to this that we can see forming in a flat-bottom kitchen sink when it empties through the central plug. Rankine having been one of the first to try to put it in equation, it's often called Rankine's vortex, as a homage paid to an eclectic scientist -like a majority of true physicists-. In this kind of vortex, it is assumed that the layer which is rolling up on itself has a constant section, which wants to traduce a conservation of the carried mass all along this sort of ribbon, and the angular speed at any point of the whirlpool is inversely proportional to the distance to the centre. The variation law of the angular speed can be written in the differential form:

$$\frac{d\alpha}{dt} = \frac{-ds}{R.dt} = \frac{-v}{R} \qquad \text{with } v = \frac{T}{2\pi R},$$

which leads to the relation $T = KR$ or $T^2 = KR^2$, T being the revolution time of an element ds of the ribbon on a complete lap, this one considered as shut on itself in first approximation, so as

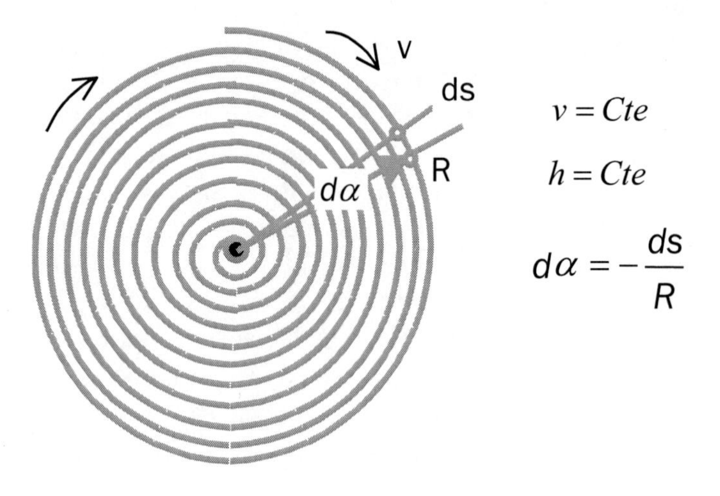

Figure 3.12

to be in accordance to the present theory, following which planets are moving along closed trajectories, which not only is wrong but also excludes considering any evolution of the system.

So we can see that this model of vortex, seen as a ribbon which rolls up on itself, constituting a sheet coming from the infinity and evacuating by two symmetrical chimneys -or “filaments”, as traditionally called-, perpendicular to the sheet and going the same to infinity, leads to a law which is not this of Kepler - $T^2 = KR^3$ -. This means that the physical model so far proposed doesn't correctly represent the movement of a planet in a solar system, and

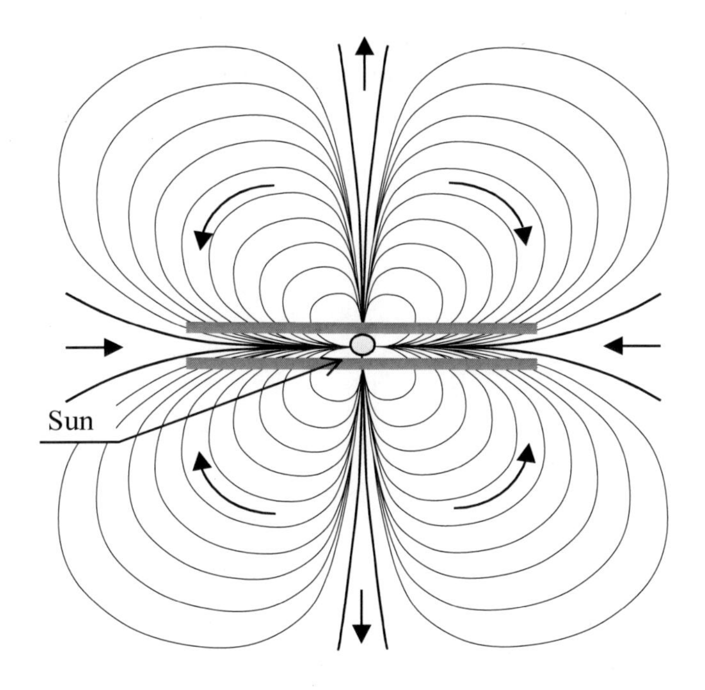

figure 3-13

we are immediately seeing why.

In an astronomical vortex, the heart, which is also the engine, behaves with the ether like a sort of vacuum-cleaner: there is a movement of the fluid towards the centre, it is aspired in a sheet which rolls up on itself like in a Rankine vortex, and it is symmetrically rejected through the poles that we have also called chimneys. The matter in hand being a fluid in motion, it thus comes that there are classical phenomena to which the ether, even as the universal fluid, cannot escape. In particular, there is a depression in the aspiration zone and an over-pressure is the exhaust zone, and as a result there is an apparition of forces which will tend to do so that the fluid went from the high pressure zone to the low pressure zone: that's a well-known phenomenon which is one of the basis of meteorology, to take the most familiar example. But it also exists in air or sub-marine acoustics: when the membrane of a loud-speaker, for instance, is moving forwards, it compresses the air, which has a tendency to pass behind, where there is, simultaneously, a depression caused by the same movement, and if the loud-speaker has not been equipped with a system able to compensate it, by artificially increasing the distance between the two sides, it happens what is called an acoustical short-circuit, which lowers the radiation and the efficiency of the loud-speaker, which can even be definitely deteriorated: beyond a given power, the vibrations amplitude reaches such a value that the membrane

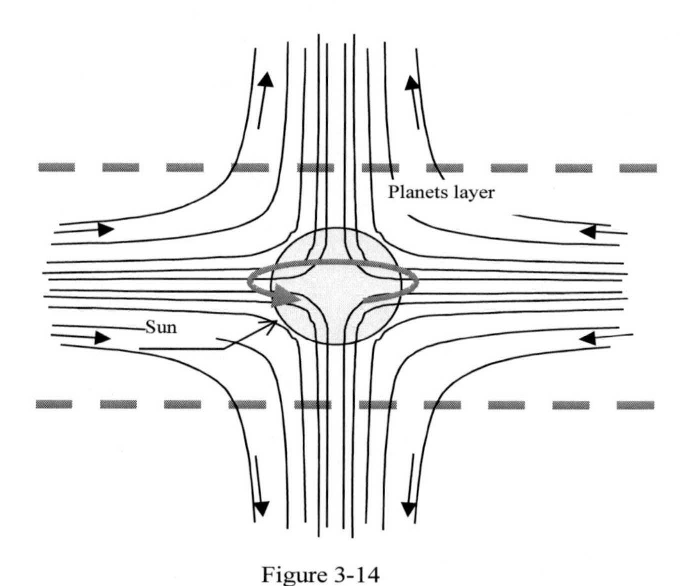

Figure 3-14

is torn out. As a function of all these notices, our solar system must be given another model than our first “pancake-ribbon”, this of figure 3-12, which doesn't concord with the complementary phenomenon of the entry-exit communication, and which so cannot work any longer. We see in the same time that the astronomical vortex is different from that of Rankine, and that it will be necessary to carefully characterize it, so as to make evident what is to change.

The figure 3-13 shows what happens to the solar system when we use the new hypothesis. It's first an overthrow of the habits of thinking. Not only it -the solar system- is not limited to the plane disc where we can observe our familiar planets, but the main part of the volume it occupies in its revolution is elsewhere. We have limited on the drawing, between the two thick dotted lines which are above and below the sun, still and more than never the centre of the system, the flat layer where the astronomers have so far confined the planets, among which we'll reinstate the unhappy Pluto, victim of an unjust and arbitrary segregation: even if it is smaller than the Moon, it is nowadays so much in the family that it would very sad to get rid of it, without allowing for the moral prejudice to those who have taken the time of learning by heart the names of our nine next-door neighbours. We could think that

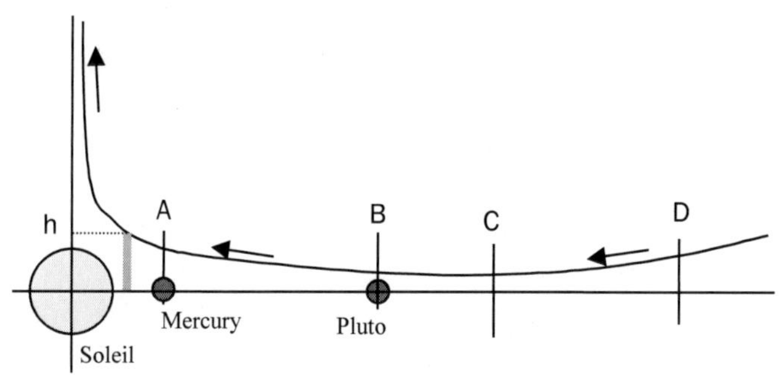

Figure 3-15

astronomers are today fed up with boredom, and attempting by all means to build the event, the novelty, that their fastidious observations don't give them any longer. But let's stop speaking ill and let's come back to our new-look solar system. In this new version, where we discover that the zone which contains planets is now only a small part of the rotating mass, the work is now to find the variation law of the speed as a function of the distance to the centre. For that, we have to apply to the precedent model an adjustment which will take account of the new disposition, while keeping the same principle of the ribbon which rolls up on itself.

What is the main difference between a double vortex of Rankine and the astronomical vortex shown in figure 3-13? The answer is evident in the drawing 3-14: in the planets zone, which interests us first because it is there that the Kepler's laws are applying, there is a regular decrease of the thickness, or the height if we prefer, of the whole of the current lines, as we get farther from the centre, in other words the Sun. Actually, by comparison with the experimental mounting of the figure 3-12, where the fluid current is guided, at the centre, by a material object, in the case of the Sun we are in the free space, and there is necessarily a driving of the neighbouring layers by the sheet, and these layers will bypass the sun to join again the polar jet a bit farer. In electromagnetism, in the way of Maxwell or Faraday, we would say that the flow density increases with the distance. We may still say that here, in the same way, reminding in this occasion that the mathematics of hydrodynamics belongs to the basis of the electromagnetic theories, and that it is consequently perfectly logical to take out analogies from them. If we keep the model of the Faraday's lines of force, which are for him the best mean for representing the ether, this last being necessarily the same than the one we are talking about at the moment and which carries away the planets, we must suppose that these lines are perpendicular to the representative ribbon -ribbons?-, which makes closed trajectories which are buckled on themselves after a complicated way in the vortex. The shape of the lines of force in the vicinity of the sun will allow us to give a justification for two phenomena, the first of them being known by everyone but never explained, that is Kepler's laws, and the other, never studied nor even mentioned, being the miraculous sphericity of suns or other astral rotating bodies, liquid or gaseous, that a mysterious force prevents from turning into pancakes.

So we take again our so practical ribbon, we still let it roll up on itself like before, but now we are giving it the shape which correspond to what we may deduce from the figure 3-15, that is with a variable height increasing as going towards the Sun. the figure 3-14, which has been drawn at an intermediary scale between this of the 3-12 and this of the 3-13, is the best matched to the definition and to the modelling of the most interesting zone, this of our "classical" solar system. Starting from the Sun and going towards

Pluto, we have first a regularly decreasing portion between A and B, which passes by a minimum at point C, to then find another part where, conversely, height increases with the distance -point D-. We have to well remember that the curve drawn represents the height h of the ribbon physical model, of which it is the radial section, that is more precisely the section by a perpendicular

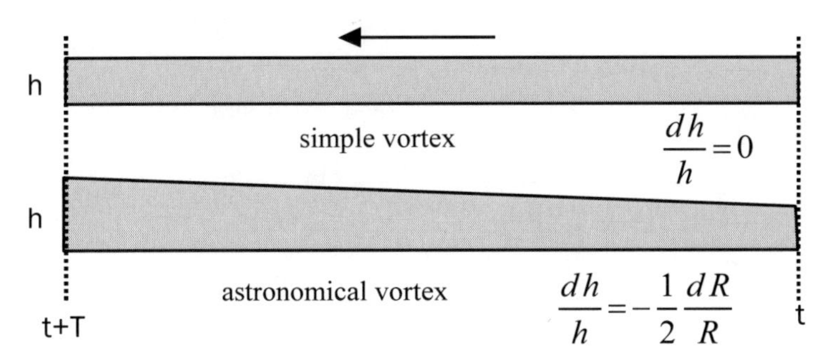

Figure 3-16 full lap unfold

plane passing by the Sun centre. The fact of situating the minimum further than Pluto, starting from the Sun, comes from an evident notice, when we consider the figure 3-14, to which the figure 3-15 is only the blowing-up of the upper right quarter, in its nearest part relatively to the Sun: this minimum must absolutely exist, given the global shape of the system. We can therefore see, if we accept this modelling, based on the simplest logics, that what we usually call "the" solar system is only a very small part of an incomparably bigger phenomenon, and that, accordingly, it's probable that the Kepler's laws, as we know them, are not available everywhere.

We have now all that is necessary to quantify, and see where this schema will drive us, knowing that the aim is to find again the Kepler's laws, through full deductive ways. The fact of keeping the ribbon physical model, now completed with a variable height which regularly increases when we are nearing the Sun, will allow us to keep a parallel with the classical Rankine type vortex, and then point out both similitude and differences. The figure 3-16 represents, in both cases, the ribbon portion which corresponds to a full revolution, that is when the angle α of the figure 3-12 has varied of 2π. The arrow indicates the rolling up sense when we are nearing Sun. Besides, given that there is no loss of mass during the whirling movement, we must take it into account and say that the flow remains constant all along the rib-

Rankine vortex	astronomical vortex
$h = Cte \qquad dh = 0$	$\frac{dh}{h} = -\frac{dR}{R}$
$v = \frac{ds}{dt} = Cte$	$vh = Cte$
$T = kR$	$T = kR\sqrt{R}$

Tableau 3-17

bon. As the height of this last increases going towards the centre of the vortex, this means that speed lowers. It then remains to discover following which law.

h is an inverse and continuous function of R: when R decreases, h increases, and we may write, supposing that the variation law is simple: $\frac{dh}{h} = -k\frac{dR}{R}$

To express that flow remains constant, the speed v must vary conversely to height. So we'll write:

$$\frac{dv}{v} = -\frac{dh}{h} = k\frac{dR}{R}, \text{ and } vh = Cte$$

We don't *a priori* know the value for k. So it could be, if we want to be rigorous, any function of R. But probably it's worth starting from the simplest, to go after, if necessary, to something more complicated in case of snag, and we'll not only suppose that it is a constant, but moreover that this constant, one or less, has a simple value (1, ½, 1/3, etc...), like the Nature usually proposes us in its most general laws. Then it only remains to make successive attempts to see if, by some happy chance, one among this values couldn't lead us to what we are looking for. The value 1 corresponding to the Rankine vortex and to the law $T^2 = KR^2$, we'll try for example $k = \frac{1}{2}$. Then it comes: $\frac{dv}{v} = \frac{1}{2}\frac{dR}{R}$, which gives, after integration: $v = k'\sqrt{R}$.

Let's now do a short parenthesis on the modelling of a vortex by a ribbon: when it is question of a ***permanent*** vortex, this ribbon is always buckled on itself, in one way or another. Let's take for example a bucket with a hole in the middle of the bottom, and in which the draining is made by a whirlpool, continuously fed at its upper part. The mass of liquid which goes away is, at any time, exactly equal to this which comes in, and this mass is not created, it already existed. Whatever the distance travelled through before, the water which goes away will join the water which comes in: at the shortest, there is somewhere a recuperation system, with a pump which directly makes the feedback, visibly. At the longest, the water which falls on the ground is mixed to this of rain and spilling before taking part in the natural cycle: evaporation, rain, capturing in the tank in which we find the water which will feed the bucket, and the system is also buckled on itself, of course by a less evident way but as continuously as in the shortest conditions, without neither adding nor loss. This notice justifies the conservation of the moving mass, in other words the constancy of the flow. This being noted, let's compare the relations which define the two types of vortices (tables 3-17): the relation $vh = Cte$ means, h increasing when we go towards the centre, that h decreases in the same proportion. But we know that proportion: it's $\sqrt{R}$. If the rotation speed, at a given distance from the centre, is lowered by $\sqrt{R}$ comparatively to the constant height vortex, this means that the revolution time will be increased of the same ratio. So we may write, T being this revolution time: $T = kR\sqrt{R}$, or still, using the regular form: $T^2 = kR^3$.

This is the third Kepler's law.

3-7 : Kepler's laws revisited

The grumpy adherents of the universal mathematics will probably say that there is not a true demonstration in what precedes. In rational physics, it is a demonstration and even a typical example of its processes. Let's never forget that mathematics are only one particular application of logics, general branch of learning in which it could be profitable to trust more, insofar as it is correctly practised. Let's briefly remind what process was followed to reach the result of the preceding paragraph:

First, rebuttal of the hypothesis of the kinetic theory of gases, refusal of a model which implicates a perpetual movement of molecules, without any damping, even against the walls of the container.

Then deduce from this correction the existence of an ether and, simultaneously, its energetic nature, unique reasonable hypothesis to explain the endless movement of the molecules, which only can be caused by an external action.

Then try to explain, in this new context deprived of the right of residence in the official physics, how a solar system works, through the analysis of the whirling movement created by its sun.

At last propose a physical model of the astronomical vortex and establish its mathematical laws, so as to check if it leads to results in accordance with observation, and if there are no contradictions with facts.

Now that all that is done, the time has come to make the balance of the operation. The new cosmogony model incontestably brings something new by comparison with all the previous ones, but above all it's not gratuitous. It's not the fruit of a hazardous hypothesis like those with which we have so far been fed, because all the previous, apart from the too sophisticated one by Descartes, have been developed on the basis of an empty space which didn't give them many possibilities. This didn't prevent some of them to be formalized, by lack of challengers and also because it was necessary to have one. It's this way that in 1810, Hassenfratz, "instituter of physics" like they said in that time, was teaching at the Polytechnique high-school a "celestial" physics course which was exactly the fax-simile of the "System of the World" by Laplace.

The cosmogonic theory which is expounded here, and which has got its roots in the Vallée's Synergetic Theory, has for itself to develop in a medium which, although obstinately denied by the official science, allows, by the very fact of its existence, justifications of a so evident and direct logics that it could and should be in the grasp of anyone. When it is approached by the right side, physics is of an incredible simplicity for what concerns the explanation of all phenomena, and it is evident, at least for the physicists who ask themselves existential questions, that the liberating existence of the ether cannot remain contested for a long time. For the question of the mathematical modelling, it probably will be another game, but the acquired knowledge of hydrodynamics is ready to be developed and fit with this problem.

The big, the enormous, the definite difference between this theory and the former ones, is that it sets in an incontestable manner that the third Kepler's law is a whirling law, put into evidence by a simple physical model. So is it foreseeable that the solar system vision will be upset, since it must now be considered as a living and evolving system, and not like a manège fixed for the eternity. According to the analytic process which has formerly been used to come to this conclusion, and which has stated on two types of vortices, the Rankine vortex and the astronomical vortex, we besides can see that the ribbon model can lead to other models of whirlpools, of still simple expressions, by modifying the variation law of the height as a function of the distance to the centre. But one only of these processes leads to the Kepler's law, such as we are taught. It's now a question to see what consequences it causes on the validity of the other Kepler's laws, and if the new theory is compatible with the observation. So let's remind these laws, at the number of three, of which we'll choose the definition, for its concision, in the volume consecrated to astronomy of the Bordas encyclopedia:

1ᵗ law: planets have around the Sun elliptic orbits of which the Sun occupies a focus.

2ᵈ law: areas swept by the vector radius are equal for equal times.

3ᵈ law: the squared revolution times are proportionate to the cube of the half-axis of the orbits.

We feel well, in these expressions, that they are the products of the mathematical physics school, which describes a phenomenon without explaining it, that is without introducing first its mechanism. In front, the rational physics proposes something very different, and may be that Kepler himself, who has deduced the laws which bear his name from the notes of Thyco-Brahe, wouldn't have been against. This is how it expresses these laws:

1t law: planets are driven in a whirling system of which the Sun is the central motor, following not complete orbits that we can see, in first approximation, as imperfect circles.

Comment: the solar system, in this theory, is no more a fixed phenomenon but takes a changing character, in accordance to the more general vision of a universe in perpetual transformation, such that astrophysicists usually present it to us, with its stars which turn, rotate, were born and will die. The orbit of a planet is an Archimedes spiral -that is with a constant thread-, more or less deformed.

2d law: for an angular variation of 2π of the vector radius, the tangential speed is constant in first approximation.

Comment: to be rigorous, if we remember the modelling of the solar vortex by a rolled up ribbon, of which the height varies with the radius, the speed decreases when nearing the centre, but for one revolution only this variation is of the second order and can be neglected.

3d law: the squared revolution periods are proportionate to the cube of the radius of the equivalent circles.

Comment: there is no change comparatively to the classical expression of this third law, except that it is now applied to a non-closed curve, that we can momentously see as an approximative circle, and except that it is demonstrated in a way which doesn't imply the classical laws of the universal attraction.

Call it what we want, for instance that there is few difference between this new expression of Kepler's laws and the classical one, this is not so surprising, as it is apparently impossible to put in evidence the simple fact that an orbit is not closed in on itself. Astronomers seem not to have observed this, as long as they have been surveying their sky, but have they done measurements for that? Did they only once have this idea? No way to know it. This impression that the average distance between Earth and Sun seems not to vary, knowing that it actually varies, let presage of a stupefying density for the ether, and for the complete solar system, in 3D as we speak nowadays, that the very solar vortex, as we know it and of a so minute volume in front of the volume of the whole, would have a big merit to make turn this gigantic manège by itself. But, despite of the apparent similarity of the laws, differently expressed in two totally divergent conceptions of cosmos, the differences between them are enormous.

The old theory registers and calculates. The new one explains and unveils. Maybe it's a bit pretentious, at first sight, to think being able to overturn ideas rooted centuries ago, especially if we take into account the fantastic inertia of dogmas and certainties of human specie. We only have to remember, if necessary, the story of Galilee and the blockade of the catholic church of that time to be well conscious of that, but there are many other examples, some of them having been more or less precisely related in the two first chapters of this book. We'll not go back to that once more, but simply it's good to be conscious and keep in mind that science, the true science, not to be confused with technology, progresses only toddling along.

Let's now come back to our reshaped solar systems, and let's try to make a balance sheet of what is brought by the presentation of cosmic phenomena in rational physics, where the ether has a so important and fundamental role. There are five points to be visited, five points which are never talked about in astronomy books, because they correspond to the famous questions without answers, and for that reason are never treated or asked. They are nevertheless capital:

- Question 1: why do planets turn around the Sun?
- Question 2: why are they in a same plane?
- Question 3: why are the rotating stars spherical?

- Question 4: why are the orbits of planets stable, when a planetary system, as it is presently defined, is eminently unstable?
- Question 5: why are planets so different?

The first question is so childish that nobody asks it, except may be ten years old kids -what a bloody nuisance, these!-. But we are so asleep by theoretical physics that it seems, some days, that we have lost any critical sense and any curiosity, and that the elementary logics abandon us. So let's go, since nobody wants to do it, for asking this first question: why do planets turn around the Sun? Does a theory exists that gives the answer? Is there a theory only wondering about it? Those who have assimilated the model proposed in this chapter should agree that they have now the answer. Only the fact of accepting the idea of the existence of an ether, this last being energetic, can lead to the concept of a whirling model of the World, which correctly explains the basic phenomena, like this one. We refer to the preceding paragraphs to justify this affirmation.

Second question: why are planets -approximately- in the same plane, which moreover is the equatorial plane of the Sun? Nothing, in the Newtonian theory on gravitation, calls for that well particular configuration, and nothing obligatorily leads to it. When Rutherford proposed the planetary model of the atom, there was not that condition: the atom was considered, by the creator of the model, as a miniaturized solar system, copied on the true one, but with orbits in any planes, like illustrated by the old ORTF[5], which had chosen this symbolic image as its logo. All the calculations made on a planetary system, starting from the universal gravitation law, apart from perturbations, could be done with orbits contained in any planes, with the only condition for these planes to contain the Sun. With the whirling model, this geometrical disposal becomes clear, evident and obligatory. We may even add that it is the only possible explanation.

Third question: why are astral bodies, in rotation or not, spherical? Which particularity can explain that the centrifugal force doesn't progressively flatten a liquid or gaseous sphere, to eventu-

5 French TV in the sixties.

ally make it turn into a ring which will vanish in the infinity? In theoretical physics, this question is not founded: the liquid structure resists the centrifugal force? That means that there is another force, centripetal, which exactly compensates it. We then only have to give a name to that force, for example cohesion force, why not, and the trick is done: we have added an item to the already well fed panoply of the imaginary forces, without trying to physically define it. It will be done later, if we have time

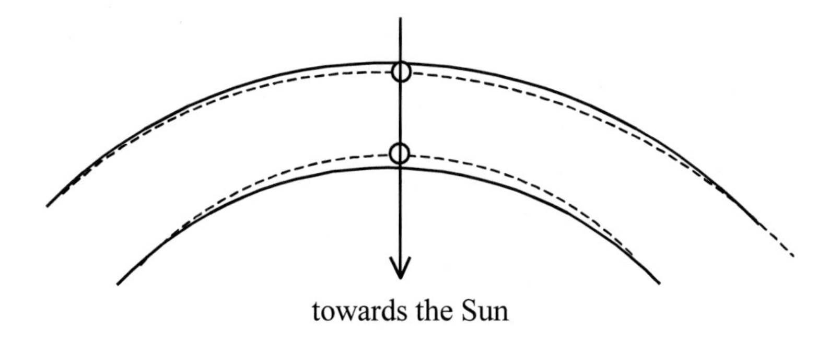

Figure 3-18

enough. But this new force will give the possibility to the mathematicians of the physics to exercise their favourite lonesome pleasure, and see flowering new equations, without the direct comprehension had progressed a centimetre. Figure 3-14 shows that it is not a mysterious force, but the incident flow of the equatorial sheet which acts on the liquid mass, repelling it inside and towards the poles, like an air-blow lifting a veil, compensating this way the centrifuge effect. Why is this compensation quasi-perfect, why is it necessary to make precise measurements to find that Earth and Sun are not exactly ideal spheres, it's there a matter for investigation and a new study path, full of promises. The theoretical study of whirlpools is only beginning, and taking in hands the whirling model of solar systems promises mathematicians future great moments of happiness.

Forth question, stability of the orbits. This question comes out as soon as we have an eye onto the perturbations problem. When we say "perturbations", in astronomy, it's the action of the presence of a planet on the orbit of another. It's while studying these perturbations that Le Verrier and Adams would have deduced the existence of an eighth planet -Neptune-, and Pluto would have been discovered by the same method. In fact the first calculations of Le Verrier and Adams were found wrong after verification, as for Tombaugh, the official discoverer of Pluto, he owed it to a student who was in charge of checking the photographic plates of the Lowell observatory, in Arizona, during the holidays of his boss, and who had noticed a change in the position of the future planet relatively to other fixed stars. Even if these discoveries are not the fact of pure random, they are not, either, the divine proofs of the omnipotence of the theoretical calculations. With a minimum of reflection, we can even guess that it is at the precise moment of the perturbation of a planet that it is enough looking further in this direction to find the possible disturber.

So there are in these discoveries only patience and observation, the two mammals of astronomy, and that's just fine. But let's go back to perturbations. When two planets, having neighbouring orbits, are in conjunction, which means that the line which joins them also passes through the sun, they exert on one another an "attraction" which tends to make them closer, and which thus slightly modifies their orbits: the most remote gets a bit closer to the Sun, the other momentously goes slightly further -figure 3-18-. Then, and here is the miracle, the two planets reach back their usual orbits, well calmly, and imperturbably takes again their usual courses. But there's a problem: a principle exists in mechanics, which wants that to annihilate the action of a force, it is necessary to create another force with the same intensity, with the inverted sign and which acts for the same time. In the case of our two planets, their mutual attraction increases when they get closer, passes by a maximum when their distance is the shortest, and then progressively decreases, but never with an inversion which would create a repulsion. So there is nothing, when we suppose this event coming in the absolute vacuum, which could justify a peaceful come back to an immutable orbit like it seems to be. On the contrary, this orbit should be constantly modified each time the two planets are close again, and this up to the collision: a solar system in vacuum is, if we don't believe in the attraction forces, eminently unstable, and the same notice can be made about the planetary model of atom, inspired by the previous one. If we now replace the phenomenon in an etheric context, all becomes clear and logical again: the ether, which possesses a certain elasticity, is compressed when the two planets are the closest possible and in conjunction, and then takes back its normal volume after the event, when the pressure decreases. The planets in question, on their part, don't see a change in their relative position in regards to the volume of ether which surrounds them and drives them. It's not, strictly speaking, the orbit of each planet which has been changed, it's the local structure of the swirling fluid which has been submitted to a compression between two mobile masses.

Fifth question: why are the planets of our system so different? That's again a never posed question, while there are a plenty of formation theories which make of them a united family having a common destiny. Why, in this case, are they not similar, like the members of a normally-built family? May we seriously believe that Venus and Jupiter have the same origin? What do Mercury and Saturn have in common? The diversity of masses, compositions, densities and aspects of the different planets make of them a true rotating museum, and it's not even possible to see in that diversity eventual steps of a common evolution, which could follow their position relatively to the Sun. So, from where do they come? A new glance to the figure 3-14 gives the answer, or at least a possible answer, but so plausible that it's worth taking it into consideration. The equatorial sheet of the solar vortex is a sort of vacuum-cleaner, as it is the entering flow of the system. The same way that a floating body in the vicinity of a whirlpool will eventually be attracted inside, an erratic planet being not too far from the solar system, if it has not a too important relative speed, will be captured like an insect by the funnel of an ant-lion. If this hypothesis is the right one, we already understand how the solar system, as we observe it today, has been formed and continues being formed, but with a so huge time scale, relatively to our miserable existence, that it seems to be, for us, a permanent and stable phenomenon, and that its changing character is beyond our mind. The diversity of our planets becomes then something evident and natural, and gives to the solar system a new aspect, this of a machine in perpetual evolution, driven by its central motor, which itself uses for its benefit the vibratory energy which surrounds it and gives it its power, both creative and destructive. When was the planet before Mercury swallowed? What will happen on Earth when Mercury will disappear? When will a new planet be visible to astronomers? All these questions will probably never have answers for us, so different is our time reference in regards to this of the Cosmos, and so short is the time at our disposal to solve the great problem of the continuation of the life, ours in particular, when the conditions on Earth will not allow it any longer. This is all that we have to mind as soon as now, we nomads who believe in the timelessness of things, when all is continuously changing in the Universe.

3-8 : The mass

We'll note that, in the whirling theory previously exposed, not once it was question of mass. The supporters of the Newtonian theory, however, use to establish the third Kepler's law from the universal gravitation law. As a matter of fact, by extraordinary, this last has not been used here. How was it possible to find again the Kepler's laws without using this principle, reputed fundamental and necessary? What must we conclude? And first, is there a conclusion to make? We have the right of posing a lot of questions about a notion which has always been one of the big philosophical problems of physics: what exactly is a mass, what then is the“mass”?

The only direct quantifying physical approach that we have to measure a mass, is its weight. The weight of an object is the force exerted on it in the terrestrial gravitational field, also called gravity. The same object imagined somewhere in space, far from any other body, has no more weight. However, it is agreed to say and to think that it has kept its mass, what implies two things: first, that this mass belongs to it, is fastened to it, that it is one of its characteristics, and then that it is something definitely acquired, except if we take off a part of its volume. To separate what is measurable from what is hypothetical, physicists have imagined two terms: weighing mass and inert mass. And however, in classical physics, to both is attached an equal mystery, that rational physics only will be able to clear up.

It's question of weighing mass in statics: in a laboratory, you weighs an object, you find a certain weight which is the force corresponding to the action of the Earth gravity onto its mass. We'll call this mass weighing mass. It's question of inert mass in dynamics, when we want to communicate an acceleration to an object by the mean of a force, and when we see a certain number of phenomena, of which the whole is called inertia, which become apparent in various ways. In particular, it's the opposition force which appears either as soon as we want to move a resting object, or when we want to change the speed of an object already moving. So we can see that there is, each time we tackle this question of mass, a fundamental ambiguity on its definition, of which the fuzzy character shows well the embarrassment of those of the physicists who took the risk of proposing a precise one. Some of them, like Henri Poincaré or Jean Perrin, even thought of the possibility to make the mass, failing better, being the theoretical coefficient which links force and acceleration, this last being in return directly measurable and consequently without mystery. The obstinate denying of the existence of the ether has made that, today, we are no further forward today than four centuries ago on the true nature of the mass, when things become extraordinary simple if we accept, may be a while only, to momently agree on that hypothesis.

Let's consider a metallic sphere laying on a horizontal plane, and thus motionless. It has got a weight, that a pair of scales can immediately give us. Now, let's vigorously push it: it takes from this thrust a movement and a kinetic energy, both of which being a function of the propelling force. What about its weight, the time it is rolling? Stupid question, isn't it? Maybe yes, maybe not: to weigh the ball, this last must be motionless, and if someone thinks of the contrary, he just has to imagine a system able to, let's be clear, allow a **direct** measurement of the weight of a moving object. So we may, at least in a first time, wonder what exactly are the inert mass and the weighing mass: does the first take the place of the second? Is it added to? Is it cut off from? Is there a more complicated process depending on speed? Or is all that a dummy problem because weighing mass and inert mass are the same thing? All this claptrap is not trivial, it helps being conscious that mass is a theoretical notion, which cannot be measured but only deduced. From that the apparently exaggerated, but in fact perfectly well-founded, cautiousness of Poincaré, Perrin and many others, who prefer keeping a mathematical definition not to be embarked in endless discussions.

In our etheric world, precisely, all these discussions become aimless, because answers come from themselves if we take pain to make a drawing, really or in our brain, of what we are wonder-

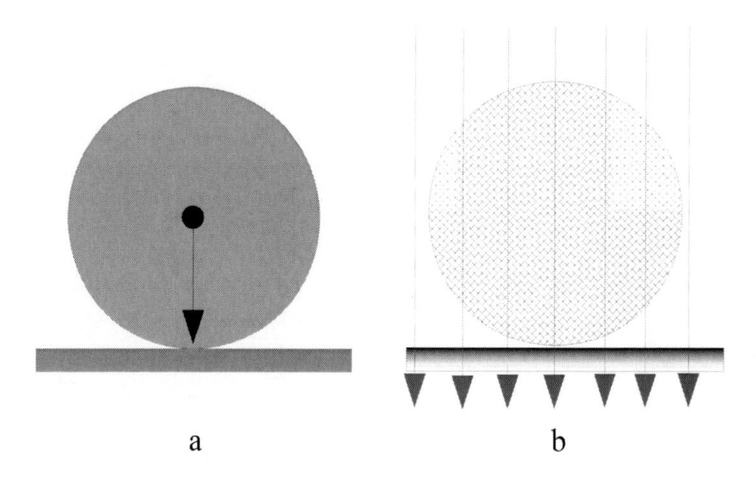

Figure 3-19

ing about. In actuality, when we keep in mind that matter is composed with small weightless domains, pressed the ones against the others and separate by considerable distances in regard to their own dimensions, and that this porous matter is bathing in a dense fluid which does have a specific mass, all the more enormous, it's very possible to make a visual interpretation of the ether/matter interaction phenomena, by analogy with the knowledge we have from the observation of quotidian facts, even if we are not conscious of that, in the domain of fluids. The hydraulic analogy is one of the most powerful and demonstrative in physics, unfortunately it is not exploited as it should be, means as a real tool, and not as a simple style exercise. In these optics, the figure 3-19 illustrates the two visions we can have of a weighing mass, following the type of physics we use: on the left (a) the traditional representation, this of a body having its own mass and which is attracted by the Earth. In the calculations, we consider that this mass is concentrated at the centre of gravity, to which is applied the gravity force, represented by a vertical downwards vector of which the length is proportional to the force. On the right (b), the same body seen in rational physics as it really is, that is a gathering of massless particles, maintained in a group by the fantastic MBL pressure. The flow of the ether which is the cause of the gravity crosses it, the same way that a water fall would cross a net, stretched across its trajectory, and try to carry it away. In this scenario, the weight of the body is the force applied to it by this flow which crosses it, and it is the same for any object, inert or living.

We find here again, more detailed, the materialisation of the idea, expressed by several scientists such as Huygens or others, that if an object spontaneously takes an acceleration when we abandon it at a certain place, that means that at this place pre-exists a sort of permanent flow, that we hesitate to call it by its true name, that name which has only brought troubles to those who have dared pronouncing or writing it. But however reality is well this one, despite of the fact that it has been hidden, since the beginning of electromagnetism, by the theoretical machinery built around the abstracted notions of fields and potentials, among others: any field of force is a place where is permanently acting a flow of the ether, of which the interaction with matter is the origin and the cause of all carrying force. When you drop an object on the ground, it's exactly the same than when a child leaves starting the paper-boat he had just placed onto the water, except that things are going on vertically instead of horizontally. Apart from this difference, it's the same physical phenomenon, that is carrying away of a light object by a heavy fluid.

If we somewhat think about it, the show proposed by the river Seine to someone on the bank is, on a lot of points and apart from the direction, very similar to this of Galileo, dropping objects of different weights from the hight of the Pisa Tower, or more probably on a slope of his fabrication, and showing that all of them arrived to the ground in the same time -approximately, to be honest!-. The Seine carries also with itself extremely various things, from dead rats to the very Parisian condoms, including sponges -very interesting, the sponges-, all that our beautiful capital leaves to its preferred river to carry it to the high sea, in a place where the vicinity of other presents coming from elsewhere will make them anonymous. But what is most important to see, is all that floating bazaar, as various as picturesque, sail together, with the same speed and keeping the same distance between them. Anyone physicist, let he be amateur or professional, who finds himself interrogating on these so simple, natural, instinctive and elementary observations, cannot do otherwise than let himself be oriented by powerful analogies, to the conclusion that gravity and universal attraction are, contrarily to what we are incited to believe, phenomena of a total clearness, since the moment they are considered as hydrodynamics facts.

This being said, in the same way that was introduced the notion of weighing mass, seen as a certain fluid flow crossing a fixed meshed structure, we have now to try to do the same with what the ancients called inert mass, the mass which is responsible of all the phenomena grouped in the global name of inertia. The most currently known experiment, which should be used as an introduction to this new version of the interaction matter/ether, is that of the shrimp-fisherman. This last holds as a tool a net, fixed onto a frame, the bottom of which scraping the sand, and which is added of a handle, long enough to avoid the fisherman not to be too tired when pushing the net. As soon as the net begins to move on, the water fills the net, which is then drawn and inflated, and takes the shape of a veil, as we know. When the speed is low, the resistance of the water, which passes inside the meshes, has a weak impact, but which rapidly increases with the thrust, to quickly reach such a force that even a very strong individual will not be able to heighten the speed. We may suppose that, at that moment, the water which goes inside the net cannot be evacuated any longer, and that this one -the net- makes its way by cutting into the liquid more than letting it cross its meshes. We must see the problem of the inert mass exactly from the same point of view, as represented in the figure 3-20: the spherical body, which has no mass but a quite rigid structure, moves in the ether like a net does in water, a part of the fluid crossing it and the rest passing around. The ratio between the two parts depends on the speed, and we suddenly discover there the justification of the relativist mathematical formula, painfully issued from a grinding of Lorentz's equations, which gives the mass variation as a function of the speed:

$$\frac{m}{m_0} = \frac{1}{\sqrt{1 - \frac{v^2}{c^2}}}$$

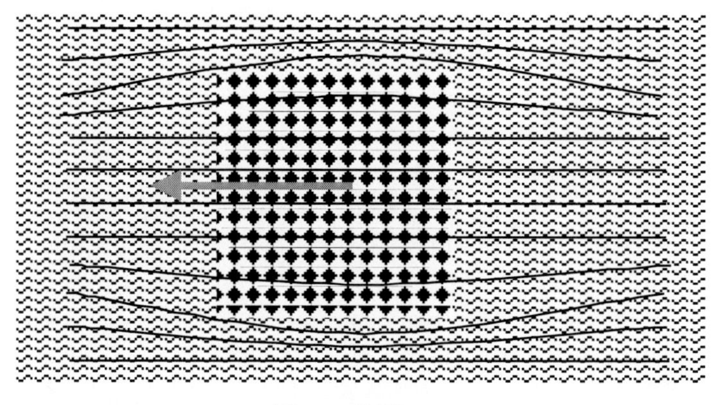

Figure 3-20

The faster the body moves, the stronger is the interaction with the ether, the harder it is for this last to cross the body's texture, and the more important is the quantity of ether stored.

When the speed is low, there is practically no bypass of around the mobile, it almost totally crosses the body. As the speed increases and gets closer and closer to this of light, skirting becomes more and more important and the moving body finally behaves, when we are close to c_0, as a cannonball, parting the fluid to cross it and keeping inside the quantity of the ether which corresponds to its volume. But the ether density is enormous, which explains what Relativity suggests when it says that the mass approaches infinity when the speed tends to that of light. In fact, the apparent mass of the body tends to a value corresponding to a full occupation of the ether in its volume, and this mass is not infinite.

Theoreticians will probably say that all this is too simplistic, because they hate simplicity: it's good to justify studies, and all these superhuman efforts to acquire the mathematical level which will distinguish them from ordinary people, and give them the status of an elite. Having done all that for nothing? Nonsense! And however, one thing leading to another, we must agree that admitting the existence of the ether leads to a schema of the Universe which is so sturdy and so simple, that the facility of understanding the physical phenomena, which goes with, shouldn't be ignored any longer. Let's say, calmly, that it would probably be profitable to take that into consideration, even to criticize, which would already be a big satisfaction.

3-9 : conclusion of chapter 3

It's not possible to progress, whatever any given field, without breaking something, without replacing, one day or another, a thing by another thing, a law by another law, a theory by another theory, a principle by another principle. The big fault of theoretical physics is to think that it cannot make mistakes, that each year going along cements, still stronger, what is considered as acquired, that knowledge branches are constantly adding, and support one another to make a more and more solid edifice. All that is bullshit. Science, like history, is stammering, losing its way, being polluted by errors and falsifications. It's a human activity, and man is not God, this god he has created to state an ideal and who is exempt from all our faults, supposedly. Man never stops mistaking, hesitating, venturing in blind ways that he creates himself, but his greatness, if he has got one, is not to become discouraged, to always have a piece of work in hand, to endeavour even without hope, and Marie Curie, the pragmatic obstinate, better incarnates physics than Henri Poincaré, the theoretician.

One cannot judge the iconoclast cosmology expounded in this chapter other than comparing it to the others, and making the balance sheet of which novelty it brings. It's the only swirling theory worthy of the name, that is using the vortex properties agreed by hydrodynamics, and not to visionary but confused images of Descartes's sky, where the vortices are only imprecise fluxes of which the sole quality, but it was already enormous for that time, was to fill space and give it life and structure. It will be kind to remind that, at the origin of the present approach, we find a critic of the kinetic theory of gases, in a branch of physics from which it was impossible to think that it could lead to cosmology. In fact this is very encouraging and revealing, because the chain of reasoning which has been proposed shows that there is, that there must absolutely be, a total coherence between all the branches of physics: to start from the microscopic analysis of gases behaviour and arrive to the solar system, it's food for thought, no?

The main revelation, in all the foregoing, is that the third Kepler's law is a swirling one, and consequently that the two others are obligatorily also swirling laws: they can't be else. Starting from this evidence, the cosmos takes a completely different expression than this presented to us so far, and we'll keep going, we'll see that later, to irremediably condemn the Big Bang theory in favour of a globally stable Universe, although nowhere immobile. The first consequence of the swirling theory is to predict an end to the existence of any planet, of any solar system, though both of them were considered since now as somewhat permanent: how the whole works, as previously analysed in this chapter, is such that all the planets will be successively captured in the central furnace, and will take part, as its food, in its future prosperity. It's impossible to know how many planets have known this end before Mercury, which inevitably will be the next, but it's sure that some have been concerned before, and may be that, why not, the last engulfing, by the cataclysm it has caused on Earth, could be a new explanation to the disappearing of dinosaurs, who knows? This new hypothesis doesn't seem, at first sight, more stupid than those that currently prevail, it only will be one more, at worst! But gone the carousel, either stupidly suspended in the vacuum for the eternity, or with an existence conditioned by this of the Sun, of which the programmed death would cause those of all the planets. Stop to the buckled trajectories, still determined by wrong-founded Kepler's laws, which are now seen as instant approximations of the general swirling law. Over, the certitudes on the universal attraction and the sanctification of Newton's law: planets are now no more than milestones, mobile marks which show us, where they are, like cork shavings on turning water, the indicators of the vortex which carries them into its heart.

It's worth well seeing, indeed, in this new approach of the cosmogony, that the mass only exists in the motive fluid, and that the mass attributed to each planet is only a mathematical parameter, of which the only virtue is to be in conformity, through the Newtonian principle of the universal attraction, with the Kepler's laws, in their classical formulation. If, by magic, we could instantaneously permute Venus and Jupiter, there would be, as the only change in the solar system, the ephemeral perturbations that these two planets exert on their neighbours. Each of the two would buckle its lap conformally to what imposes the vortex, governed by the swirling third Kepler's law. That means that Jupiter, put at the place of Venus, would have the revolution time of Venus, and that, conversely, Venus put at the place of Jupiter, would have the present revolution period of Jupiter.

So, this thorny and philosophical question of the mass and its nature finds here an answer of a surprising simplicity. Although suspected by a certain number of physicists, tormented between their own doubts and the certainties of the Establishment, this conception of the mass is, let's acknowledge it, hard to digest insofar as it is in total contradiction with what our five senses suggest us for our perception of the external world. In reality, it's only a question of education. If these notions were early inculcated, since the first physics courses, we would largely have the necessary time to assimilate them before reaching the status of adults. But they first might find the agreement of the ruling institutions and, taking into account what we know about it, it's not going to happen tomorrow.

It's not a reason for capitulating: the history of sciences shows us that, in spite of the fantastic inertia of the responsible institutions, in spite of the hierarchic blocking which prevails in all the social compartments, scientific or not, a theory which has more advantages than the present and official one will eventually win, one time or another. But there is a *sine qua non* condition for that: the new theory must bring something new, or lead to something new, with such an evidence that its exclusive role in this discovery might be stated beyond all question. It may be, either an invention materialized by a material object, a machine, a tool, which brings significant progress, or a simple answer to an embarrassing question, like these wondered by the searchers of the infinitely large or the infinitesimal small. Will the above theory fill these conditions? We'll see.

Waiting for that, we must now tackle another part of the job, the one which consists, after having proposed a new idea, in destroying those with which it is in conflict. That's not the less interesting, and also particularly enjoyable: it's worth a fourth chapter!

Chapter 4
The myths of the modern science

4-1 : Geniuses

Oh, geniuses! What could we do, we common people, without you? Behind this prostituted word, used anyhow and about anyone, is hidden a human necessity as old as the very species, this of believing in something above us, entity or being, we conscious, in the deepest of ourselves, that our mind is able to evolute toward a superior state. This superior state is not in our grasp for the moment, but we feel it, we guess it, through the discovery and the consciousness of our faults and imperfections. Realising that we have limits is sure a promise of future progress, of a possible access to an intellectual level that we can only vaguely imagine, failing better, and of which we make an ideal, or at least a step in the progress way. Then remains to know if the human species, such as we know it, has got the real possibility of reaching it, or if it must be replaced by another, and that "homo sapiens" leaves some day its place to "homo rationnalis", the same way that the man of Cro-Magnon has replaced the man of Neanderthal. That's the question.

Originally, the genius was someone gifted of superhuman powers, but he curiously could also be a slave to man, under the condition, for this last, of having got the know-how. May be he was, in this point of view, the ancestor of the robot, that is someone having a strength above ours, but ready to obey us since the moment we know the way to command him. Nowadays, in the current language, he simply is an over-gifted man, or at least so considered, to whom the public rumour, the media, an exceptional success, have attributed capacities over the average in a particular domain. This places him in a category which is situated still over the elite, this one finally being only constituted with good students, who have learnt their courses by heart through the price of a fierce education and, supplied with a weakness-less memory, are able to recite them at any moment.

We must remember that man is a gregarious animal which, it's a common law to all these types of species, needs a leader, in fact one per speciality, per group, per any association of individuals, will he be wanted or not, permanent or transient. We already are here in front of one of the human equations among the most recurrent, this of the hierarchy, based on a scale of values which varies with the intellectual level of the society: at the beginning, the criteria is the physical force, like in the other animal species, afterwards we evolve, with the time, to other types of criteria, more theoretical, more developed, less evident besides, as logics and morale are progressing. Some of us understand very early this principle, specially if they are in a familial and social environment favourable and concerned, and their existence will consist in climbing the hierarchic grades as quickly as possible, going the fastest possible to the summits. They are the winners, the leaders, the bosses, like they use to call themselves. Besides them, there are the others, we normal, modest, simple attendants, who will realise too late this natural law and will pass our life in the herd, without being able to get out, may be also without willing it.

But for geniuses, more precisely those who are to-day named this way, it's still something else. Physiologically speaking, they are men or women, of course, but who have characteristics that the others, be they winners or losers, consider as exceptional and all the most incontestable. Because it's exactly there, in the minds of the others, in their imagination, in their phantasms, and also probably in their interest, that live and were born geniuses. Manner of God, who represents the sum of all the powers, the genius is a human invention which answers a precise wish of the so-named society: to have at our disposal a reference, a model, someone exceptional that we'll try to imitate, with in mind the hope of reaching a level of knowledge and capacities that we suppose he has, and that we have not. We hope then arriving one day at this status and acquire by this way power and consideration, one going with the other and reciprocally, as could have said the great philosopher Pierre Dac[1].

If we ask a passing-by to cite the name of a genius, we can bet ten to one that the first name which will come to his mind will be Einstein. He doesn't know Einstein, he knows nothing about his works, he wouldn't understand anything reading one of his books, but he's confident in what he has read or eared about him, in the respect shown both by the scientific sphere and the whole community, and it is out of question to suspect the opinion, necessarily well-founded, of a so big number of individuals: it's only there one of the aspects of what Marcel Boll[2] called "gregarious mysticism" in his book "Education du Jugement", key-book in the matter of logics and the reading of which should be obligatory in final year.

In a test proposed on Internet to evaluate our intellectual capacities, we can read this: "Einstein had a IQ of 160. What is yours?". This is the kind of ineptitude which quite well states the global level of the Internet pages, but which is not, however, without interest, if by chance we want to use it for a decrypting. First of all Einstein, who hardly spoke when he was 5 years old, has never passed a IQ test, which existed at this time only under an experimental form: when the Binet-Simon test was launched in France, in 1905, on request of the French government to try to evaluate the intelligence of pupils, Einstein was already 26 and had other fishes to fry than being submitted to this kind of test, especially because he was not living in France, and was moreover publishing his first work on the Special Relativity. Moreover, if he had passed it, he probably would have got only the pass mark, so much mediocre and anonymous he had been for his studies, but never 160 any way, even having won his Nobel prize a long time after. In spite of that, Einstein is now at the centre of such a legend that nobody, to-day, poses to himself the question of what he has really brought to humanity and what consisted his genius

1 Famous French humorist
2 French logician

in. The wording of his prize, attributed in 1921, is approximately the following: "...for the discovery of the photo-electric effect -in fact, it's just its explanation- and the whole of his work". The Nobel jury is a very pragmatic college, of a total cautiousness and rigour when it is question of exact sciences, and we can appreciate the skill and the subtlety with which he has, preserving its reputation, paid homage to the great scientist. This because it would not be fair to deny he was a great theoretician, all his life having been devoted to his passion: physics. When awarding him, first for a precise discovery, which by the way may be contested since the paternity of the photo-electric effect is also attributed to Hertz in 1886, while paying to him a fuzzy homage for a life fully dedicated to physics, the judges of the most prestigious international prize have avoided to commit themselves in clearly supporting and adhering the Relativity theory, the only work that hardly anyone could deny it's he's, but that he probably is the only one to understand, and even we could have doubts. Any way, it would have been very difficult to them to ignore a scientist honoured by such a majority in his community, from what this unusual wording of a skilled diplomacy. Whatever it is, for anyone, Einstein is a genius.

This being noted, what are the characteristics of a human genius? What is his difference with the others? Has he really got a superior intelligence, is he built differently of us, has he got something more...or less? Probably some people thinks so, because a few of them, among the scientific sphere, have emitted the idea that, after exhumation and reconditioning, thanks to not yet known techniques, we could open the brain of certain marking personalities such as Einstein, of course, persuaded that there should be something visible to see. They probably think of some supplementary convolution, or the abnormal size of a dedicated zone, something which could explain, anatomically and physiologically, an unjust superiority and would open the way to some genuine genetic manipulation. We simmer when we think that an illuminated politic could have the desire of promoting such an idea, terrible remembrance of a time that we try to forget, when lost doctors were leading guilty experiments in the insane hope of improving the human race. As for Einstein's brain, it escaped the incineration and is kept somewhere for a possible future analysis, when some new proceedings will make possible to find the revealing detail which, for the moment, is still lacking to the corpse breeders. Right now, only a slightly more developed hemisphere has been noticed, which is not so rare but has nevertheless given a path for incredible computations.

Let's remain serious. Men, all of them, have got the same brain, with the same possibilities, the same capacities and the same limits. "*Give me a homeless and fifteen years,* used to say this University professor, *and I make him a doctor of science*". Human brains are made, indeed, following a sort of production line, starting from an immutable number of chromosomes, except when there is a serious malformation. And like all the objects made in mass production, we may say that they all are the same, and all different, just like it is said for cats. All identical, because designed for the same use and starting from the same plans, to speak like in industry. All different, because it is impossible, even by cloning, that two perfectly identical individuals exist. But we absolutely must precise, speaking about the differences, that these last concern only minor characteristics, of which the essential is, for what is inside the skull, what the electronics specialists call "response time". In other words, the human brains are more or less fast, from an individual to another one, their memories work more or less quickly, but they all have the same potential and probably the same memory capacity. The intelligence, i.e. the faculty of understanding, is the same for all, it's a question of a general brain conformation, it's a species characteristic. Doubting on this affirmation is the cause of most of our evils, socially speaking, but it however is a status that we cannot fight against, because it makes the profit of a number, and we unfortunately must admit that our society, even unjust, has found in this biologically indefensible inequity, since a long time, a sort of balance which is what it is but, from the point of view of the social order, is better than nothing: hierarchy is the foundation of the order, whatever it is the criteria on which it is based.

Let's go back to geniuses: there are none, they don't exist, but however we know the process to build them, and many of us believe there are. Here is all the contradiction of a species which names itself rational, and which conversely makes a theatre without which, apparently, it couldn't live.

Einstein was an ordinary man, with an ordinary brain, nevertheless in good state. What made his difference with the others is his love for physics, this vocation which falls on some of us, without warning, and does so that they cannot look at things without trying to understand them. Those who are of this type are not so rare, but few of them will be in the right conditions to realize their gift: the birth, the familial sphere, lucky meetings, all the life parameters which do so that there are few winners.

For Einstein, the people he met, principally at the Polytechnicum of Zurich, was very influential on him, as they allowed him to elaborate his theory, particularly at the contact of his professor and mentor Minkowsky, father of the "space-time", but also this of several other students.

Meanwhile, in France, Henri Poincaré was also elaborating a "new mechanics", about which some science historians think it was ready before Einstein's publication, and treating the same subject. The fame could have chosen the first, more rigorous and excellent mathematician. It pointed the second, less clear but more skilled, and also specially supported by a PanGermanist club which had decided to block Poincaré's way by all the possible means. The first is almost forgotten, the second remains perched on his pedestal, although unstable, but that nobody dares shaking. All this because one day, what he has written about a new interpretation of Lorenz's equations had made enthusiastic Max Born, of whom the influence in the scientific sphere was already well established, as well as Ernst Mach, and that those two had audience enough to set or destroy the reputation of anyone. In return, say that you are the father of a new idea, for scientists who have a solid reputation, is always risked, and it is very probable that Einstein had been the right man at the right place for all the people concerned: on one hand for himself, with a support which was going to make of him the leader of the new physics, and on the other hand for his mentors, assured to launch in their environment a perfectly remote-controlled missile.

Germany and the northern countries were in that time at the centre of the revolutionary thought, in physics as well as in philosophy or in sociology, and the anti-French feelings, as much as the fact that Poincaré, the only advanced relativist theoretician in France, was a mathematician, and for that slightly set aside from their family by suspicious physicists, made that the new physics took form beyond the Rhin[3] and that Einstein, by the opinion of the others, was miraculously propelled to the head of a movement that all the famous physicists joined up soon, instead of staying on the platform.

So we can see, if we analyse without neither complaisance, nor *a-priori*, the case of the most uncontested "genius" of the 20th century, that there is nothing, in his history, neither able to make someone enthusiastic, nor some marking event related by the media of that period. The testimonies of surprise, then of admiration, came very progressively, such as a rumour which, starting from nothing, spread in an irrational manner and took importance only through the phantasms of the suckers. Only remains, right now, his discovery, which is not really a discovery but more a theory, and that we'll dissect further.

Concerning geniuses we have been, so far, keeping focused on the man who wears the best this name in the people imaginary. In the scientific domain, however, there are a lot of others of whom the intelligence was comparable to this of Einstein, for example all those who are cited in the first chapter, and many others more. Certain passages of Malebranche are treasures of logics, and we have the feeling, when we read the physics of Aristotle, that the brain of this last had not less possibilities than this of our reference. But other disciplines than science have also their geniuses: Arts and Letters are not in rest to suddenly glorify such one, male or female, who one morning offers to a befuddled public a painting, a play, a sculpture, a romance, some odd futility which will arouse the enthusiasm of the crews and elevate the author at the rank of genius. The realisation that causes this sudden veneration and will flatter the eye, the ear or the mind of the individual, at the point of amazing him beyond all moderation, will of course be qualified of "genius", which goes without saying.

In fact, each speciality of the human activity has got its genius or geniuses: genius of the finances, genius of the crime, genius of

3 Border river between France and Germany

informatics, etc...But the scientific domain has, from this point of view, a special character which formally distinguishes it from the others: it is question of the total opacity, not only on the reason for which a scientist is called genius, except if what he has laid is so strong that nobody understands all of it, but also of the circumstances which have brought him to this supreme distinction. When he is an artist, a painter for example, his work is accessible to the appreciation, competent or not, of anyone: it's enough to have a look, and we immediately decide if it is either beautiful or ugly, if it is incredible or terrible. We may find genius or horrific a painting by Picasso, following the taste and the culture of the ones and the others, but apart from the snobbery of the unconditional admirers, anyone, young, old, farmer, intellectual, is able to emit an opinion following what his eyes tell him.

In physics or in mathematics, it's different: when a scientist is suspected of being a genius, those who, as amateurs, are interested in sciences and wish knowing more on the well-founded of his work, are generally blocked in front of the impassable barrier of the mathematical and theoretical knowledge necessary for the comprehension of the subject. In the case of Einstein and his Relativity, we may think that only one person among ten thousands has got this level, in a country like ours. We must not forget that Einstein himself, momently stopped during his theoretical researches, could restart only with the help of new mathematical tools built by his friends. When a physicist feels obliged to appeal to maths to be able to progress, it's the sign that he has no more directive ideas, and that he trusts in the absolute rigour of this branch of logics to make emerge, from the equations, the providential parameter which will allow him to restart. When he is obliged, fully running out of ideas, to ask a mathematician to make something completely new, in order to extract himself from the inextricable slough where he is lost, it's then still more serious: it means that the supposed original starting idea has escaped him and led him to a wall. May be it's this last point which affords us to definitely refuse, not only to Einstein, but anyone else, Nobel prize or not, the status of genius, that must be always taken at the second degree, i.e. as a metaphor of homage and consideration from the public to someone respectful. To be respected because having proved his competence in his domain is not so bad, and has the merit of remaining human. So we'll say that the adjective "genius", too easy to use, can only be applied to some well particular idea or realisation, and that it simply means "exceptionally smart", expression too long to be pronounced.

When a renowned physician has passed fifty years of his life in theoretical researches, it's perfectly normal that his work, evidently personal in its main part, not to say lonely after a certain time, leads him to definitely leave contact with the others. This is not really wanted, but specially occurs because he has got such an advance in the speciality he has built that it is impossible for the others, at a given time, to follow him and, *a fortiori*, to recapture him. Having a founded judgement on his work is then practically impossible, if this work doesn't lead to some practical application that can be irrefutably related to it. If not, even if he is a genius, reason and cautiousness command to look at his theory with a sound mistrust, which is simply the sign of a good mental health. In all circumstances, we must keep our feet on the ground.

Those who have attended, at the "Collège de France", the lectures of scientists such as the awarded Pierre-Gilles de Gennes or Alain Connes, the first one with the Nobel prize and the second with the Fields medal, have kept the impression to be in presence of very accessible characters, simple, sympathetic, at disposal, answering stupid questions with an infinite patience, and completely at the opposite of what we could first imagine about so seasoned and so honoured people. Themselves don't think they are geniuses, despite of the fact that only a few people can follow their talks, even if they do their best to make an important effort of popularisation, which is the rule at the Collège, which has originally been created for the grand-public to be aware, free of charge, of the last scientific results and advances. In fact, their audience is usually composed with two well distinct parts: on one side the general public, more or less advised, but always curious, and on the other side the close colleagues or some students who come here to attend a sort of revising session, or ask questions that they might have kept during courses elsewhere, out of the Collège. The first part looses contact quickly, by lack of training, but stays any way to finally applaud. The second part surrounds the lecturer during he coffee-break, like ants-soldiers which mount guard around their queen to protect her from external aggressions. But whatever happens, how many of these various listeners have understood the message that the lecturer wants to transmit? Mystery. The two models of scientists who had just been cited above have probably intellectual possibilities of the same order than these of Einstein, may be still better, but both would have refused to be called geniuses, so overwhelmed they are through the satisfaction of good and well done work, and a healthy renown, made of consideration first.

As a conclusion of all that, it's recommended to suppress the word "genius" from the vocabulary, when it is used as a blind veneration testimony for a human being, that we dress with qualities he has never got. There are no geniuses, there are only genius ideas, now and then, coming from ordinary people, and that's not so bad, either. But we must refuse to be influenced, specially in the scientific domain and more particularly in advanced research, by legends partly built by the media, always starving of sensational and which, unfortunately, are the main source of information for us. To be interested in physics doesn't mean ingesting anything, instead of which it becomes a supplementary source of confusion and unanswered interrogations, and there are already such a lot of them that it is not necessary to add others. Physics is the soul, the heart, the strongest pillar of our civilisation. It must be shared by anyone, each of us has got the right to access it, whatever his level, and anyone should be able to have an opinion, a solidly well-founded advice, the possibility of debating on subjects that are infinitely more fascinating than the suppositories that the "twenty hours news" daily insert us. The way will be long.

4-2 : The theory of the Relativity

"*...the very principle of Relativity, when it is amalgamated with a physical law, moreover abusively generalised, the law of the absolute constancy of light velocity, gives birth to consequences that our reason cannot admit. It submits phenomena, as a matter of fact, to a reciprocity that leads to the statement of contradict-*

ory propositions, that is of logical impossibilities. A theory that has such consequences is necessarily wrong."

This citation of Raymond Leredu, Ecole Centrale engineer and author of "la piperie relativiste[4]", quite well resumes what all those who have tried to understand the Relativity, while leaving aside any *a-priori* in relation to the renown of his promoter, have concluded after having read it. Leredu was, with Bouasse and Gandillot, one of the most virulent opponent to Einstein between 1920 and 1930, but all of them, including the many others who are not cited here, have made an effort to bring a critical analysis, of which the very method, scrupulous and impartial as any constructive critic must be, merits attention and consideration. The sheeplike behaviour of scientists, all of them being ready to sail before the wind, has made that, to-day, the name of almost the whole of these spontaneous detractors has been kept unknown, and that nobody more tries to contest a theory of which the existence is enforced like a religious dogma, and which is good to cite it in each book of physics. It's indeed strongly recommended to add, in any serious publication, that the discovery you are relating and that you give to mankind confirms once more, in case of necessity, this marvellous theory of which the solidity is constantly reinforced over time.

Let's remain calm and quiet, as politics say, but we have the right to wonder what sort of event will make the scientific community awake, and at once decide to get rid of a theory which doesn't bring them anything consistent. Like Leredu said, "*Among all the equations which create an apparent bond between his kinematics and his dynamics, only the equations (...) allow Einstein to incorporate with his theory some concordances with astronomical and experimental facts. He is submitted, without knowing it, to the appealing of these concordances which come, it seems, to corroborate his ideas, and he accepts as a demonstration a proposition which is, in reality, empty of any mathematical meaning. And it's this way that Relativity, which was nothing, was going to find itself dressed with this deceitful brightness which, too long, will ensure its existence.*"

4 The relativist loading

No engineer will positively state that the relativist theory has helped him for anything in the exercise of his job. No individual is able to point out, in his home or in his immediate environment, some object, some device or some tool that he owes to the conception or the existence to the Relativity. No professor can simply and clearly answer the question: "*Excuse me Sir, what exactly is the Relativity?*". So, why some people continues bending in front of this magic name, which however doesn't represent anything concrete for anybody? May be a part of the answer is there: Relativity, that's not concrete. Despite all the efforts made to introduce it to us as one of the indispensable backgrounds of the modern science, it's only theoretical, an exercise of style, a baroque and wrong-balanced edifice, born from the group cogitation of a revolutionary team in Zurich, team for which physics was not, at the beginning, the main preoccupation. Of course they were physicists, Einstein, Adler, Minkowsky, Solovine, but also theoreticians for all of them, and not laboratory physicists, not practitioners who, like the Curie, were constantly keeping their hands onto the matter. Sheer brains they were, capable of discussing about anything, highly cultured, but missing something very important, what makes the seasoning of application engineers and laboratory searchers: the permanent contact with the matter. This contact is fundamental in the physicists formation, it allows them to permanently keep their feet on the ground, and is a sort of parapet against a too developed imagination. Of course this last is indispensable in any job where manual skill is not exclusive, where we must have a rigorous practice and a permanent reflection on the go at once, but we must keep it under control and never lose one's way in free flights of fancy. And, according to what has been said before, it's very tempting to consider the Theory of Relativity as the typical example of what is not to do.

Criticizing is easy, as we currently say. It's true if it's only systematic opposition, formal contradiction, but it is a consequent and useful work since the moment it's done in the strict rules of analysis. For that, it is necessary to proceed with method, not to divert, and attack the target by one side only, a well chosen and well identified side, so as to conduct with logics. It's the price for efficaciousness and credibility. In what concerns Einstein and Relativity, it seems that his own presentation of his theory be an excellent goal, that's why we are trying to carefully dissect the book where he explains himself the new principles of physics: "*La Theorie de la Relativité Restreinte et Générale, Gauthier-Villars/Bordas, 1976.*". This book, many times reissued, but which was originally written in 1916, applies, following the foreword of the author, to the average reader, having only a minimum of knowledge in mathematics, let's say this of a student who have just passed the baccalaureate. That's perfect, because in these preliminary conditions, he obliges himself to consent vulgarisation efforts which don't exist in higher level books, where mathematics take the ascendancy over the true explanations, and lead, as a consequence, to a drastic selection of the readers. So these last may hope, when they have chosen this book among others, to have the occasion of grasping the quintessence of the ideas born in the tormented brain of the genius physicist.

So let's begin by this foreword, where we immediately discover a honey and paternalist tone which will after impregnate the whole of the book, and where Sir Einstein begins speaking to the reader like a college professor in front of primary school students. The last lines are particularly characteristic: "*Let this small book be a stimulant for many readers and make them pass a happy moment.*"! Instead of happy hours, it will be for many of them a discouraging experiment which will end into the box-room. But let's go forwards and begin the first chapter. This chapter is built so as to introduce the doubt in the supposed certitudes of the reader, those he was taught in the matter of geometry, and also in basic notions he considered since now as indisputable. Einstein precisely intends to contest them, and that is a good departure when you want to propose new concepts: nothing to say against this principle. What is more embarrassing is to replace well clear certitudes by well fuzzy ideas, to destroy a universe of knowledge may be not so solid, that's true, but in order to replace it by a constantly increasing sum, as we go on in the book, of unanswered questions ("*...What do you mean by the affirmation that these propositions are true?,* does he write), only made to shake the fragile certainty that is the award for well learnt lessons. He prepares the ground, like a gardener who begins turning up his parcel before seeding. His aim is evident, once we have understood his tactics: bring doubt, get the reader to wonder if, effectively, it would not be right to reconsider the indemonstrable axioms of geometry, and ask himself if that intuition, which suggests these fundamentals in a so strong manner to us, is really so well-founded. We must never forget, concerning Einstein, that him and his friends of the Olympia Academy, that they had founded in the town of Berne, were revolutionaries who wanted to change the World, like any other revolutionaries, but including physics in their great projects for the improvement of the Society. That shows the place that takes dialectics in his general thought process, which on this point of view is very close to that of a politic, who is constantly trying to make his speech pass in a digestible way.

It's this way that he invites us to mind on the deep meaning of the word "true", which is an adjective that, nevertheless, even children understand without the least ambiguity. It's well evident that attacking such a simple and universal notion opens the way, after that, to any possibility, and it is in full knowledge of the case that Einstein endeavours this semantic analysis, with a quite rare level for a physics book. But it answers a precise goal, which is to ruin all the convictions, all the conventions, and to make a clean sweep, grouping them under the label Euclidean geometry, of all the classical notions incompatible with his own conception of space and time, which will be the background of the Relativity Theory:

"*Geometry starts from certain fundamental notions such as the point, the straight line, the plane, to which we are able to associate more or less clear notions, and certain simple propositions (axioms), that we agree seeing them, in virtue of these representations, as "true"...But we know since a long time that, not only we cannot answer this question* (are the axioms "true"?) *by the mean of the geometry methods, but that it has no sense in itself*".

We are already no longer in physics, but in a philosophic-metaphysical speech which gives the tone of the rest of the book, and which will oblige the reader asking questions of the same order, which normally would never come to his mind, and which make him lose all the certainties that his professors had tried to provide him for his scholar course. In addition, it's not forbidden to lead these speculative digressions when practising physics, but under the condition that it remains a short introduction, followed by a methodical study of a phenomenon. In the present case, it's very long, too long.

Though it's necessary to read all, so as to be able, at least, to credibly answer the question "*But, did you read Einstein?*", it is allowed not to lose too much time on chapters 2, 3, and 4, where he persists in stating the obvious, describing with a Proustian[5] meticulousness the different manners of measuring and situating things. All that to make us put the finger on that any localisation is a geometrical comparison with a fixed and rigid object, and that a fixed and rigid object is never completely fixed and rigid. Of course the idea is interesting, but it orientates us more toward the "knowledge without certainty" of Popper, than to an unexpected advance in physics. But, for Einstein, this preliminary work is indispensable. Indeed, this allows him to show us, eventually, the absolute necessity, that besides we knew before he told us, to precise the coordinates system when we want to study a body's movement: an object which is falling from a mobile which has a constant speed will have a linear trajectory for an observer bond to the mobile, but a parabolic one for another observer bond to the ground. That's perfectly true, but already known since the 17th century. We don't yet know, at the page 13, where Einstein wants to lead us, but he begins taking out from his pocket the object which is going to become afterwards so important for him: the clock. Actually, if the two previous observers know how to define, each of them, an apparent trajectory, it's because they have got a clock, that's so evident. But the coordinates system is also as important, thus he is obliged to personally define the system which was in vigour before he, Einstein, intervenes. The system in question is the Galilean one: "*A coordinates system of which the movement is such that, relatively to it, the inertia law remains available, is called "Galilean coordinates system". It's only for Galilean coordinates systems that the Galilee-Newton laws are available.*". It's therefore apparent that the association of measuring a time and a localisation is at the beginning of any physical measurement. We knew that before, but when he particularises

5 From Proust, sophisticated French author

the classical coordinates system, and so distinguishes it from another possible one, just by giving it the qualifier of Galilean, Einstein open the door to the possibility of other systems, in unlimited quantity, among which the Galilean system would be only an example. This is a dialectic stratagem assigned to destroy the universality of the system in use until now, the one we used to get our bearings in comparison with a trihedral rectangular axes system attached to a place.

We now arrive to the statement of the relativity principle: "*If K' is relatively to K a coordinates system which is making a uniform rotation-less movement, Nature phenomena happen, relatively to K', in conformity with the same general laws than relatively to K. We call this statement "relativity principle" (in the restricted meaning)*". We stand gaping and arms dangling: what does this mean? What's new? What does "in the restricted meaning" mean? What happens when the meaning is not restricted? The following lines don't bring us much clearing, we simply conclude that the relativity principle is the expression of a classical mechanics evidence, which was already known by Descartes, Huygens or Newton. It's to be noted that we are now at the page 17 and that, so far, we have not found neither any revelation, nor a clear introduction where would be drawn a precise goal, nor even any substance: only emptiness and verbiage. But let's cheer up and go forward. We are suddenly surprised to see arriving, after a sum of notices on the addition of speeds, on the notion of simultaneity and on this of time, the promise, page 23, that is coming soon the statement of a "Theory of the Restricted Relativity", which surprisingly is not the same than the "principle" wearing the same name. The preliminary arguments, full of common-places and strange reasoning, are of such a confusion that we find ourselves questioning at every line, wondering where the author wants to go, and feeling confusion invade us step by step. It's precisely then that we fall, on page 25, onto this extraordinary warning, concerning the perception of the simultaneity: "*If you don't spare me that with conviction, dear reader, it's not useful to continue.*"! We are dreaming! Could we imagine a professor telling his students: "*If you don't understand, the best for you is to go back home!*"? But here it's not any professor, this one has all the rights, including this of taking us as mugs.

Since there, if we effectively accept to sign a blank cheque, and if we are not too much demoralized by decrypting the thoughts of the master, we can nevertheless progress in the reading without trying to assimilate every proposition, but simply in order to know the final destination, and moreover to understand how and why this metaphysical jumble had successfully become a respected theory.

We now tackle the fundamental question of the time and its evaluation, but in fact Einstein has decided to put us in front of the problem he has posed to himself, in an arbitrary way and which can be resumed as follows: how make a time or distance measurement in a mobile from a fixed position. To give substance to this problem and give it a less theoretical aspect, he has built for himself a little panoply that he will use in a systematic way, and which is composed with a well straight embankment, with a fixed observer on it, a train which goes parallel to the embankment with a constant speed, with on board a second observer, and to finish with two or several clocks in the hands of the observers. The clocks are supposed to be synchronized and kept so, what with a minimum of reflection we have the right to consider as impossible, and the train is very very long, possibly infinitely long, which suppress it a beginning and an end. These particularities avoids both the reader and the author wondering afterwards what happens to its dimension when the last will tell us about length contraction on board. Thus we see all the simplicity of the problem posed by Einstein: the fixed observer has to measure the length of, let's say a standard metre, which is aboard the train, knowing that it is the other passenger/observer who makes the measurement. This last, consequently, must transmit to the first his result by the mean of a signal, that Einstein supposes to be the light. Simultaneously, with the same illuminated sign, he must also send all the elements that allows a verification of the measurement results on the embankment. If Einstein was never born and if, knowing what we presently know, a physicist of the 21th century dared propose such ideas, presented under this synthetic form, it's probable that we would have doubts on his mental health. But at the beginning of the 19th century, introduced this way by Einstein, the Relativity Theory has any way shown attraction enough, in the sphere of the thinkers in physics, may be by the mystery of its statement only, to cause a wave of curiosity which finally has led to what we know, a sort of confirmed religion which we'll have an enormous difficulty to get rid of. But

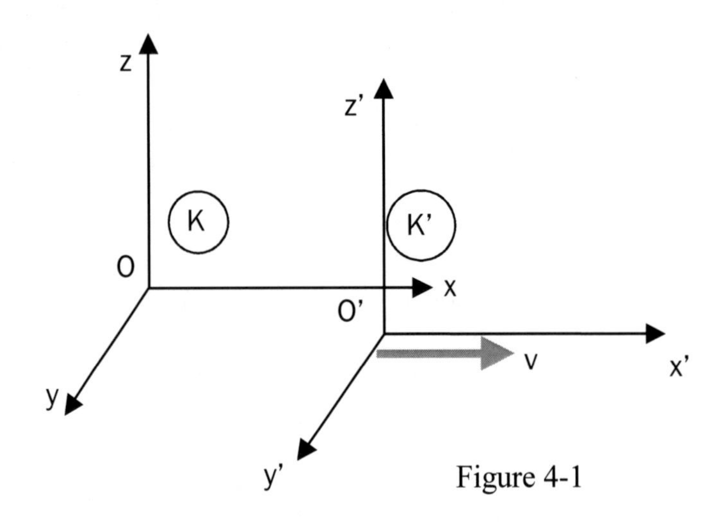

Figure 4-1

it's, or it was, the same for all the ideological fights of science: let's remind Newton and the corpuscular theory of the light.

Having reached this point in the analysis of the Relativity, as explained in the "general public" book by Einstein, let's repeat again that the question is not to demonstrate he's wrong, even if we'll do it if the opportunity comes, but first to try to understand him, only using what is supposed to be the best shared thing in the world and, besides, our only contesting weapon: the common sense. Doing that, we simply reply to his proposal of making us pass, like he says, a happy moment. For now, it's not precisely the case, and at the page 33, when we tackle the Lorenz's transform, we know well, after enquiry, that there have been many abandons. However, let's go on.

Einstein shows us two coordinates systems K and K', each of them represented by trihedral coordinate axis, the two trihedrons having their y and z axes parallel, and their x axis sliding along, one remaining parallel to the other. The problem is still the same: make a length measurement in K', while being in K. To do that, he will use two contrivances without which he couldn't develop his thesis. The first one consists in using, as a communication signal, an electromagnetic wave, which has as an indispensable property that its propagation speed is a universal constant. We'll come back on the justification he gives, but it's well evident, taking into account the notices brought in the previous chapters, that there are good grounds for contesting this affirmation, since the moment that the existence of the ether and its eminently variable nature have been stated. The second one, which is not exactly an artifice but something equivalent, is constituted by Lorenz's equations, reproduced hereunder:

$$x' = x$$

$$y' = y$$

$$z' = z$$

$$t' = \frac{t - \frac{v}{c^2}x}{\sqrt{1 - \frac{v^2}{c^2}}}$$

We particularly focus on the fourth equation, where we can notice the presence of a time t' different from t and which constitutes the big novelty in regards of the former habits, when t was considered as “the” unique time parameter of the mathematicians, this which was the translation in their language of the variability of a quantity. We also note that t' is dependant of x, even with a so weak coefficient that we'll quickly forget it. But however, it's embarrassing. In that text, the time t' appears suddenly, without any explanation, as if it was fully natural that in the K' system, it was enough to put an index onto each parameter to make it able to be related to K. This apparition of another time scale, as we cannot define it otherwise, -and by the way an infinity of others since we understand that every coordinates system has got its own-, opens the door to all the calculation fantasies we may wish. Indeed we are given, this way, a supplementary degree of freedom, which allows to completely hide the definition of the time by reducing it to a simple local parameter.

This is very typical of Einstein's methods: when there is something embarrassing, at the point of becoming a major obstacle in the proceeding of his statement, he flights over with a flashing velocity, suggesting that it's so evident that it doesn't need any explanation. In addition, and this is still more skilled, he only uses a system of equations already found by someone else, and on which, of course, he has no responsibility. He only uses Lorenz's work, his mistakes having been corrected by Poincaré, who had the elegance of letting the previous keeping the paternity. It is nevertheless not allowed to brush off this way, with the back of the hand, all that still constitutes one of the major problems for epistemology. We don't know what is the time, despite of the fact that we have a strong consciousness of it, but we only know how measuring time intervals, leaving besides the definition of the very word. And measuring a time interval, that is the duration which separates two dates, is always done the same way: we count a certain number of periods of a regular cyclic system, such as the revolution of a planet around the Sun or an atom vibration, and for that we use clocks. Thus the clocks we are told by Einstein, which have an absolutely fundamental role in his reasoning, are no more, no less, whatever the fabrication proceedings, than only counting devices which, for a given event to be measured, only register that a certain number of times a cyclic phenomenon has been repeated. Then we very well know that the second, the time unit in the IS system, which is given by a clock working following a certain principle, largely depends on local parameters, among which gravity, for example, and which are eminently variable with the location. It's evident for a simple pendulum, but it's also true, despite being less evident, for an electronic clock. We thus see, with a minimum of reflection, that clock synchronisation, which is a key-condition in the relativist theory, is something, in the theoretical meaning, impossible to realise, since two clocks placed in different places cannot, for all these reasons, give the same hour.

Another notice, which is never done because it leads to a blind way: in order to analyse the simultaneity of events in two coordinates systems moving one relative to the other, Einstein chooses the light, as the wave propagation phenomenon, to carry the information from a system to the other. But light is not the only mean to realise this indispensable communication: we may do the same general reasoning with a sound signal, adding the supposition that we work in an air of constant temperature and pressure, thus with a well defined propagation celerity, in any case as well defined as this of the light. The big advantage with sound waves, is that the ratio $\beta = \frac{v}{c}$ we find in the Lorenz's transforms under the form $\sqrt{1-\beta^2}$, can take far more important values and be directly measurable, contrarily to the light case, and that these values can be verified. We may so give to the Einstein's train a speed of 33 m/s, that is the tenth of the sound speed in air, which gives a β of 10^{-1}, instead of 10^{-7} with a light vibration, and make evident relativist effects which will so become easily measurable. But we already know, in this case, that the sound speed measured in the train will be the same than this measured on the embankment, since the air in the train is motionless relatively to the train itself, and the geometrical considerations made by Einstein when he uses light celerity, supposed being a universal constant, cannot be justified any longer. This is the reason for which he has chosen that last method, soft enough to warrant the impossibility of answering yes or no to his propositions, while carefully avoiding the alternative possibility, that he probably has also examined, and of which he has very quickly seen the danger of using it. In any case, there is no more a possibility of making appearing the notion of length contraction, if we use as the communication signal a sound transmission between two coordinates systems in relative movement. The aim of Einstein was so, not only to destroy the elementary notions of logic that professors have so much difficulty to insert in the brains of their pupils, but principally to introduce this wrong idea that light speed is a constant, and get rid, this way, of the obstacle constituted by the ether, of which the numerous attempts of description by his predecessors, with a number of contradictory properties, thwarted the clear expression of the major hypothesis that light would always propagate with the same celerity in all directions.

The scandal takes effect page 84, where we read with a sudden stupefaction, concerning the deviation of light rays by masses, that the speed of the light becomes variable! But the way of presenting this complete U-turn is absolutely spicy, because it shows the know-how of the famous scientist and his extraordinary poise to make us swallow anything:

"*...in conformity with the general relativity theory, the already often mentioned law of the constancy of light celerity in the vacuum, which is one of the fundamental suppositions of the restricted relativity theory, cannot pretend to an unlimited validity. Indeed, a curving of light rays cannot happen if light celerity doesn't varies with the place.*"!!

Could we imagine a student, preparing a doctoral thesis in France, developing his arguments at the last floor of the central tower of the Pierre et Marie Curie University, answering the jury who makes him notice that there is an obvious contradiction in his speech: "*Excuse me, Sir, but what I'm telling you cannot pre tend to an absolute validity. But it's only a detail which cannot put in question a three years work!*". No doubt that the candidate would have been kicked up in the backside, morally speaking of course, but when his name is Einstein, the ordinary laws are no more applicable. It's forbidden to touch geniuses. What is incoherence for the ordinary people becomes revelation for them.

We could go on this way till the end of the book, each page giving the occasion of debating, either because we find a sophism, or an illegible language, or even both at the same time. It's useless to continue an analysis which since the beginning, let's say the forty first pages, demonstrates a fantastic intellectual swindle. We find a typical example of this Machiavelli use of muddling page 32, when Einstein tells us about the measurement, made from an embankment, of a certain length on the floor of the moving train: "*It is not* a priori *proved that this last measurement will give the same result than the first one. The train length, measured from the embankment, can be different of this measured in the very train...If the traveller covers in the carriage the distance w in the time unit,* ***measured in the train****, that distance is not necessarily* *equal to w* ***when measured on the embankment***.". Try to explain that to a 10 years child, who has already difficulties in well possessing his basic logics! If you are not successful, if the kid looks at you with an odd eye, if he seems to wonder where you are coming from, that means he's not idiot and all the best hopes are allowed to him. If whenever he tells you that he understands, then beware: watch him over carefully, he can turn out badly and going to be relativist, which would be for him a big catastrophe.

We are stopping here, after about fifty pages, but for those who want to go forward, those who had not read Einstein before, or those who had read him with the eyes of Chimène[6], the rules of the game are now clear. It's in fact those of the 7 errors game: you open the book at any page, at random, or successively all pages if you are courageous enough, and you mark all that is suspect or strange. Then you look for the contradictions between the different pages, for example between page 22, where we are told that light celerity is constant, and page 84, where it's said that it couldn't be. If, at the price of a meritorious effort, you reach the end of the book, you'll be obliged to wonder why there has been so much leniency, from all level scientists, for this stack of contradictions, why a so large number of physicists gave it a plebiscite with so much conviction, and why we still emphasize this theory, despite of the fact that it leads nowhere.

But there is something still more severe. In the mind of the general public, the same than this pointed by Einstein as being able to read his relativist essay, all is mixed: Relativity, the Big Bang, the atomic bomb, the quantum theory, the black matter, the time travel, the Schrödinger's cat, etc... All these titles, casual in specialised reviews, are merging in a magma of confused ideas, that are considered as the children of the Relativity, as if Einstein was responsible of all that, as if he had changed physics. The writers of those articles never forget sacrificing to what has become a ritual, and which consists in adding, at the end of the text, that here is a supplementary confirmation of the Relativity Theory. We may notice, by the way, that we nowadays say "the" Relativity", without précising which of the two, restricted -or special- or general, forgetting that they are incompatible, specially by the dif-

6 Unconditional lover

ferent hypothesis made on the light celerity. Besides, when a given phenomenon "confirms" the Relativity Theory, it also confirms the Synergetic Theory, and may be other theories, more or less complete but not less available. To finish with, it is perfectly incorrect to use the verb "confirm" when there only is a "**compatibility**" with established facts, which is an obligatory condition for any theory.

It's for that reason, among others, that a theory cannot be "right" or "true". It can be founded, coherent, practical, smart, sophisticated, matched, all that you want except "true". Contrarily, it can be wrong if it has proven to be in contradiction with the facts. The supreme judge is thus, in the occurrence, the experimentation, and this is an occasion to remember that physics is, before all, an experimental science, of which the acquired knowledge has always obeyed the notion of proof: what is true is what has been proved by observation and experiment -it's another definition of the pragmatism-. This having been said, what new did Relativity bring in the 20th century physics? Of which verifiable revelation did it make us profit for our happiness? Is it possible that the diabolic invention of the Minkowsky's four dimensions space could have been the starting up of a collective furious madness, caused by the annexation of physics by the theoreticians? If we have a closer look at the scientific events of the two last centuries, in physics, there is only one discovery which really merits consideration and has changed our way of thinking: it's the variation of the mass with the speed. That the Theory of Relativity is in accordance with it is not contestable, but it is only a formal accordance: it doesn't explain anything, contrarily to the Synergetic Theory and, in this frame, to all the ideas that has been exposed in the previous chapter, where the etherist point of view for mass makes the phenomenon totally simple and evident. This property of the mass could only be put into evidence thanks to the technological progresses in atomic physics, and in particular in tracing and counting particles. It was noticed that an electron, moving with a speed close to this of light, had a larger, less curved trajectory, when it was deviated by an electric or magnetic field, than the one calculated from its static value. That's the only concrete fact which would lead, in physics, to the conclusion that only a variation of the inert mass could explain this behaviour. No need of Relativity for that, but the relativists have grabbed the idea to make of it, as they usually do, the exclusive confirmation of their damned theory. In the same time, it's fair to appreciate at its true worth this discovery, made by atomistic physicists, while we state that they never needed Relativity to do it: it's only the fruit, like for Marie curie when she discovered the radium, of a sound curiosity, combined with a long, frantic and sheer experimentation.

But why so much hate, will you say? Why systematically wanting oppose a movement of thought which seems to win a general adhesion since a so long time? Why even suspecting it, since it is taught everywhere? The principal reason is anger. Anger of seeing a majority of self-satisfied scientists burning incense before a worm-eaten theory, only by dogmatism. Anger of seeing the systematic denial of questioning oneself, from individuals who are in charge of promoting the knowledge and teach it to the young people. Anger of seeing the members of the social class which has the highest responsibility in the evolution of the ideas, this of the teaching establishment and of the industry managers, behave like yokels without imagination. Anger of seeing the systematic banishment of the contesters, from an authoritative Establishment, and the intellectual passivity of engineers and professors, who should, from time to time, rebel against the institutions, instead of resemble calves and propagate incredible bullshits about the Universe and its formation.

That's why.

But it seems, at the last news, that this position is somewhat moving. The simmering signs of a May 68[7] in physics are perhaps going on. A few Faculty professors, feeling ill at ease, begin to speak amongst themselves, four centuries after Descartes, about an ether which could be not as inconsistent as supposed, and of which the absence, or the lack of definition, begins to weigh. Because the ether is so indispensable that Einstein himself reintegrated it in the General Relativity, ten years after having made it disappear in the Special Relativity. Is all that well serious? The relativist constraints, which day after day weighs more and more in the wild imaginings of the cosmologists and the particle search-

7 Students riot in France

ers, at the two ends of the scale, is not for few in the physics slump. Why insist in believing in a theory which leads to cosmological speculations that we are not able to verify? Why remaining embarrassed with a tool that nobody uses? Because truth is right here: everyone speaks about Relativity, even out of the scientific sphere, but nobody uses it, for a very simple reason: it's strictly useless! It's a sacred text which, like in any religion, has been enriched by the hundreds of modifications that those who didn't understand it, on one hand, and those who didn't agree, on the other hand, felt obliged to bring into, with the result that it no longer resemble, to-day, what it was at the origin. There is a casual expression which so well describes this alteration phenomenon on a current of thought: everybody brings his stone to the edifice. But there is no edifice, there only is a stack of stones. We'll let Paul Collard ("les deux ethers", Chiron, 1925) pronounce the funeral oration:

"*Einstein didn't dare going back to the deep causes of phenomena. He only was satisfied giving of them a mathematical representation without looking for to determine their genuine nature. If he was close to set space into a concrete form, since he agrees it has gravitational and luminous properties, he only did that partly. His motionless space is neither abstracted nor concrete. It's only a word.*".

4-3 : The Big Bang

The Big Bang is the most insane invention in astronomy. Historically, it is related to astrophysics, and more particularly to spectrography, which makes possible the analysis of the light coming from a star, and then draw a sort of radiation map of this star. We can find on this map the characteristic rays of the constitutive elements of the matter, which makes possible to deduce the astral radiation source composition. It was noticed, at the beginning of the 20th century (Slipher, Friedmann), that the spectrum of far stars was regularly shifted toward larger wavelengths, that is toward red, in comparison with the same rays identified as coming from the same elements, but measured in a laboratory or coming from a close star. It's commonly agreed to afford the discovery paternity to the canon Georges Lemaître, member of the catholic University of Louvain (Belgium). Originally, The Big Bang is thus a Belgian story[8].

The shift of the spectrum toward red is one of the incontestable facts that astronomical spectroscopy has offered to observers. There are two possible explanations: gravitational force or Doppler effect. The first is a consequence of the light slowing by masses, the second this of a hypothetical radial movement of the stars. Then the question is to know what part must be attributed to each. And this choice is of a considerable importance, because it definitely orientates us toward a representation of the Universe which is somewhere between two limits: on one side a Universe in expansion, if we choose the Doppler effect only and neglect the gravitational effect, on the other side, with the contrary choice, a stable and immutable Universe, despite being nowhere resting. Curiously, knowing that the really scientific reasons were equivalent both sides, everybody has chosen the first scenario, which represented an overturning of the habits of thinking, and probably this reflex choice must have had, as probable explanation, the permanent desire of finding something new, at any price. This is a constant characteristic of mankind, which never stands immobility for a long time: things must move, jump, in the scientific sphere like in the others, and from this point of view a Universe in expansion offered much more promising and attractive outlooks than our good old ancient sky, faithful but boring with the time. Einstein and his flock -his flock principally, he not so much- immediately grasped the occasion to state that this phenomenon was foreseen by Relativity, of which it was -of course- a fantastic supplementary justification. And the astronomer Edwin Hubble was not late to join up the expansionists, and calculated the moving speed of far galaxies with the help of a constant that he determined, which now wears his name and which is regularly readjusted, year after year. All this in 1929. Physics was just enriched of a new constant: of course it was needed! But let's not poke fun at anyone, the case is dramatic. Indeed, once the preference has been given to the expansion hypothesis, you cannot dodge all a series of deductions which come so naturally that there is no reason to

8 French way for qualifying a poor joke

remain ignore them, and also no chance of passing besides: if the Universe is really in expansion, we have a furious desire of knowing since when and, to do that, way back in the time -easy with Einstein!-, to know how things were going on before, at the beginning. And there the human mind, as we know it so well, threaded by its warrior atavism and the films by George Lucas, couldn't imagine something better than, at the origin of the phenomenon, a fantastic explosion of which the galactic expansion would only be the remaining blow: the Big Bang was born! Alleluia!

The story doesn't stop here, it's just beginning. Thus the Big Bang is a philosophic-scientific fiction which states that, at one moment, which besides cannot be something else than the beginning of ages, a minuscule point, located nowhere in the infinity, suddenly turned into Matter in an incredible explosion. Perhaps incredible is not, by the way, the right term to use, because the physicists-astronomers-specialists talk about it without any problem, and even shower us with so many pointed details on what happened during the event, millisecond after millisecond, that we have the impression that they were on the site. The word matter has been wittingly written with a capital, because it is in fact "the" total mass of the Universe, knowing that there was nothing before. Good morning the principle of mass and energy conservation, but it's not so important, in a delirium phase all can be forgiven, the patient cannot be considered as responsible. But however let's do an effort and try to remain serious.

The phenomenon of the spectrum shift being the same in all directions, this statement makes us the centre of the explosion, as the galaxies go away the same way toward North and South, above and under, like would say any Shadok[9]. And thus the centre of the World. If that didn't shock the creationists, who saw there a supplementary proof of a divine presence and willingness, this genuine consequence of the Big Bang made the agnostic astronomers feeling awkward, as the fact of discovering that we are living at the very place where **all** would have been created was against a common and consensual intuition for a stable Universe, irrelevant of a particular event which would imply a time origin.

9 Stupid creature in a French cartoon

Because defining a time origin always calls for a blocking question, that automatically will be posed by this little devil of the ten years child, always here to bore us with his two-cent questions: but before? What was there before? If someone pretends to know an answer, be cautious and take your distances, he's ready for the asylum. Time infinity is perhaps the most rooted notion in the depth of our acquired knowledge, and having a doubt about that is a pathological sign. Apart from these epidermic considerations, the sudden creation of the totality of matter, starting from nothing, goes against the fundamental laws of both the physics and the simple common sense, about which we begin to wonder, when we are reading the most recent articles on cosmogony, if it's really the thing which is told the best shared in the world.

Only the two last arguments, primary but weighing, developed before, should have made condemned the very principle of the Big Bang, but the superior brains of the same-named research have nothing to do with ordinary people, with their too small intellectual means and their total and crass incapacity to abstraction. Nevertheless, this anthropocentric position was hard to digest, even in their closed sphere of free thinkers, and it was thus absolutely necessary to get free of those of their affirmations which made them too obviously and too evidently open to criticism. Thus they began, basing on soi-disant new more pointed calculations on the near cosmic environment, putting the centre of the explosion at a distance from Earth, by pushing it away of some ten thousands years of light, -but still in our galaxy-, not knowing what to do better. We must see well that, at the cosmic scale, this correction is perfectly negligible. To get an idea of that in an astronomical context, we only have to imagine the same correction applied to a far galaxy, seen from Earth and, as the case may be, if we consider either the edge or the centre: that's peanuts. So it was not a good solution to rub the embarrassing impression of being at the centre of the World. For years, indeed tens of years, the cosmologists of the Big Bang have cudgelled their brains to get out from an evidence which plaid against them. Especially since unlikelihood were adding to improbabilities.

The most enormous among these last comes from the Hubble's constant. This constant, introduced by this astronomer in 1929, is the ratio between the recession velocity of any astral body and its distance to the supposed centre of the initial explosion. At the risk of repeating, we'll incidentally notice that, being given that measuring the shift toward red is done from a terrestrial observatory, which register equivalent values whatever the targeting direction, this laboratory is obviously located at this centre, by reason of measurement symmetry, this measurement being independent of the direction chosen. The supporters of the Big Bang will never be able to dodge this evidence: if this origin were elsewhere, it would be quite easy to determine at least its direction. But that's not the essential. Hubble had given to his constant a first value of 500 km/s by Megaparsec -the parsec is about 3.26 years of light-. From rectification to rectification, it has been lowered to a tenth of its initial value, for at last being coming again to 75 km/s by Megaparsec. Meanwhile, the improvement of observation devices, and specially the launching of the spatial telescope, have made that we should now locate the most distant galaxies at such distances that their speed, computed from the constant, even revised, reach the light celerity! This last misadventure should have rung the knell of the Big Bang. Its creators and supporters, all of them relativists -it's important to remember that-, should have agreed that too much is too much, and that there are limits to unlikelihood, but no: in science as in politics, when you are defending an idea, you do it to the end, to the death if necessary, even if the rest of the mankind try to make you understand that you are wrong to the neck. It was absolutely necessary to react.

So the “expansionists” have thus found something else, in a way their last cartridges: it's not the galaxies which move away, it's the very space which expands, and this expansion is the cause of the apparent increase of inter-galactic distances. In other words, in order to try to understand this last avatar of the Universe expansion, if at a moment t we are located in a volume v, at the moment $t+dt$ the former volume has become $v+dv$, so that two objects previously distant of a length e are now distant of $e+de$, but as in the same time all of us have grown in the same proportion, there's no change in the close environment: that which was 10 cm long is still 10 cm long, as the centimetre has increased without we could notice it. We feel here, again, a perfume of General Relativity, and the presence of the cracked adepts of the Minkowsky's space-time, who come back to the scene of their crimes. The most curious, in this new fable, is that, if we go to the end of this new idea, and if we sail upstream along time, we may

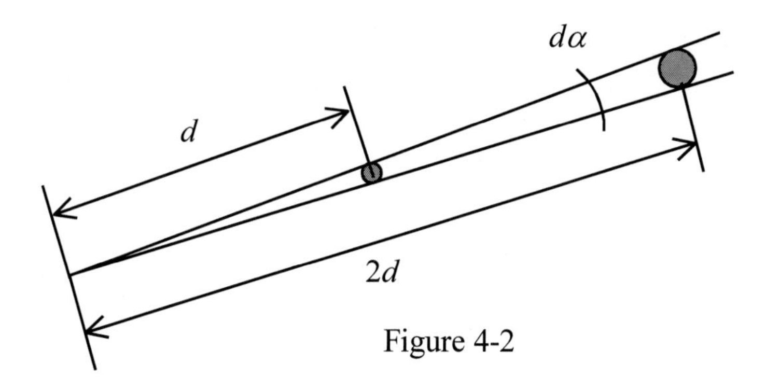

Figure 4-2

go as far as we can imagine in the past, we are still remaining in the same environmental conditions. These conditions cannot change, since we take part ourselves in the expansion, and as a consequence we cannot find any time origin: we can look around, no more Big Bang! And we find again, with a certain release, this time infinity, in the past as well as in the future, synonym of a familiar Universe, without any space or time limit, which refreshes our primary logic and takes back its legitimacy in front of the theoreticians madness.

This being assumed, we now have to come back to the problem of the shift toward red of the far galaxies spectrum. If we rely on the calculations based on the Hubble's constant, the most remote of them escape at the light speed, and thus have an infinite mass if we take into account the acquired lessons of the relativists. Because this problem really exists. But it was called for mind, at the beginning of this paragraph, that the Big Bang and the Universe expansion theories were born from a choice which has privileged the Doppler effect at the detriment of the slowing up of the light rays near masses. But this last is any way effective, and is not contested, neither by the Relativity, nor by the Synergetic Theory, no more by the rational physics. All the theories qualified of “the field" admits it as a theoretical consequence, confirmed by the observation. Thus the other branch of the alternative consists in conferring to the gravitational effect a preponderant action, which stands to reason if we become conscious of the enormity of the expansionist theory. This last should normally lead, at the price of a minimal critical mind, to condemn in final judgement the Doppler thesis, given all the incoherences it leads to. But it's probably more efficacious to explain that through a very simple principle, animated by figure 4-2.

All the astronomical observation instruments have an optical part, which determines their precision. This last is defined by the minimal solid angle under which you cannot separate an object from another, and that is generally called "resolution limit". A spectrograph, for example, which is placed at the eyepiece of a simple or astronomic telescope, will register the emitting spectrum of what will exactly be in the field of the strict optical part. At the limit of the resolution, this last will distinguish one sun only, and we'll be able to state that the registered spectrum corresponds to this sun, and this sun only. We then understand well that the distance to which we'll estimate the sun to be, in this observation condition of the limit angle, will be proportional to its real diameter. Said differently, as shows figure 4-2, if a spherical body of a diameter D is in this limit angle at the distance d, another spherical body, seen in the same conditions with the same apparent diameter, and being at the distance $2d$, will have a diameter of $2D$. Mass being proportional to the cube of the diameter, supposing that the most remote sun has the same composition than the smaller one, and thus the same density, that means that its mass is 8 times more important. Consequently, it will slower up light more than the small, and thus will cause a more important shift toward red. This is the other explanation, much simpler, of the spectral shift as a function of the distance.

We are not yet capable to calculate it, because nobody has tried to do it, but it is measurable. When the astronomers will have brought their attention onto this other causality, when they will accept to study it seriously, they obligatorily will find the missing relations between mass, distance and spectral shift. We'll close this paragraph by making notice that not all the galaxies increase their distance from us, and that Hubble's law is only applicable to these which are at the limit of the observation possibilities. Andromeda, that Belot[10] thought in 1924 being at 600 000 years of light and that we rather situate to-day at about 2 or 3 million, moves closer to us with a speed of 300 km/s instead of leaving us with a speed of 60 km/s, if we believe in Hubble's law. Other close galaxies have a spectrum which shifts not toward red, but at the contrary toward blue. So, even if it was founded, the expansion theory couldn't be generalized, and contrarily to what is commonly said, an exception never confirms a rule: it always weakens it, sometimes definitely.

4-4 : The attraction forces

We are now arriving to one of the most deep-rooted myths of physics, not only because we still believe in them to-day, even in the scientific sphere, but also because the common language carries it from century to century and comforts it in all the classes of the population, all over the world. We don't question ourselves any more, if one day we did it, when we say that a magnet “attracts” a piece of iron, or that the Moon “attracts” oceans water to cause tides. Of course, when we have a magnet in hand, the first sensation we have let us think that it has something in itself, in the heart of its matter, a particular power that most of the other objects, made of another matter, have not, and which enforces a piece of iron to come nearer. The magnet acts like an invisible prolonging of the hand, like if we tried to directly take the piece of iron, but without touching it, to finally go to the same result. The essential difference is that our hand can take any matter, provided it is prehensile, that is not too hot, not too cold, and not chemically aggressive, and not finally a stack of restrictions that don't necessarily affect the magnet, when this one only acts on very particular elements called ferromagnetic. Thus the force called “attraction force” of the magnet doesn't make a double-purpose tool with the human hand, it's rather complementary and will allow, for example, to quickly find a needle in a hay bundle, which generally is considered as very difficult, not to say im-

10 French astronomer

possible. The hand takes iron, the magnet attracts it, this is all that the current language remembers as a first description of the phenomenon. It's the appearance, for a child, that the magnet is magical, that it possesses a mysterious force hidden inside it.

The theoretical physics, as well, has given up since a long time knowing the genuine origin of that force, but has succeeded in creating the abstracted models necessary for its quantified study. That's its credo, its philosophy, its coherence. All the same we still remain hungry, even after magnificent equations, when you are given a formula which brings you a result of an incontestable utility, but where the true explanation of the phenomenon is missing, this that you are looking for, this which would at least give you the impression that you have understood. We there come back, again and always, to that fratricidal opposition between he two conceptions of physics, theoretical or rational, and, as usually, it's the rational physics which will lead us to the explanation of the fact. All this has already been said in the previous chapters, but we are here in a domain where we can see, more than elsewhere, the role of the language in the transmission of the basis of physics in what we often call, may be a bit too simply, the "collective unconscious", which in this particular case can be identified to innate, that is the inborn acquired knowledge. From Aristotle to De Gennes, numerous were the humanist scientists denouncing the disasters of language abuses, but perhaps is it Clemence Royer who has the best expressed it, at the page 41 of her "Constitution du Monde":

"*Science is made by scientists, it must not be made for them; it must be accessible to all the average minds. Mathematics, specially, are a powerful investigation mean. Only them can allow us formulating phenomena laws. But these laws must light phenomena, and not make them darker and more complicated. Mathematics are, before all, things of good sense and logics; and the greatest scientists may only pretend looking for shorter ways and simplified processes for the solution of problems or for clearer posing them. It happens that, when it is question of the most elementary, the most simple and the most general facts, the clever formulas of the algebraists are often wizard's books for only express La Palisse's truths*[11].

Accustomed to reduce all to points, lines and letters, of which the arbitrary value is agreed between themselves, they don't know how translating this volapuk[12] *at their use in the common language, without arbitrarily change the meaning of its words. Lost in the abstractions of arithmetic, geometry, algebra, they lose, with the clear notion of the concrete things, the only truly genuine, the faculty of describing them in an intelligible way for any mind.*"

And you can moreover notice that, when she was writing these lines in 1900, she didn't know the not yet born Relativity! With respect to what has a link with the phenomenon of bodies attraction, whether the general action of gravity forces or the particular ones of the electric and magnetic fields, the ordinary language carries some ways of thinking, some ways of seeing, which are printed in us since our early childhood, and which consequently are very difficult to modify. All the confusions come from the use we do of the verbs pull, attract, and the corresponding substantives. When you pull a chariot, that means that you are before it, in the direction of the movement. When you push it, you are behind. In fact, the physicist will tell you that, in both cases, you push it. When the individual who pushes is behind the pushed object, there is an indemonstrable evidence which makes that anybody agrees. When he is before, we don't say any more that he pushes, but that he pulls. In fact, if we analyse the action with the eye of the scientist, we notice that the part of us with which we pull is obligatorily, the reference sense being this of the movement, behind a determined part, rigidly linked to the rest, of the "pushed-pulled" object. If we focus on the exact application point of the force that makes the object going forward, we very well see that, in any possible configuration, this force pushes on the mobile to make it move. So, after mature consideration, we are obliged to agree on this: when we pull, we push. This come-back to a healthy and logical description of the reality is relatively easy to do, because we can establish it on a quasi-immediate perception

11 evidences

12 old complicated language

of our senses, essentially, here, seeing and touching, but without ever forgetting that it's the brain which manages.

Let's now pass to the magnet and the piece of iron. In this case, the senses are completely caught out, because seeing and touching are no more in accordance: we don't know any longer how to define an application point of that force which takes support on nothing, but which weighs on the fist and of which the existence is manifesting with a total evidence, though the continuity between the cause and the effect seems to be interrupted, may be non existing.

It's thus understandable that the average brain, ours indeed, incompletely formed to the notion of space, and deprived of sufficient knowledge for that concerns the ether, since this last has no legal existence, would have as a first reflex to see there an action proper to the magnet. Same sensation when we are holding a gyroscope in rotation, and that our hand tries to change its orientation: on what does it take support to resist this way to the request of our fist? What indeed is this apparent "solidity" of the vacuum, as astonishing and strong than its invisibility? Here is the crossroads, and here occurs the choice between theoretical physics and rational physics. The first one, as it systematically does, will edit the necessary quantity of hypothetical forces, and the potentials that come with, to be able to put the phenomenon into equations. The second one will try to make a schema for the movement of the ether responsible of the real force. Maxwell's position was between the two methods, using the tools and the mathematical machinery of fluid dynamics, but without trying to imagine, before, the macroscopic behaviour of the ether. His hypothesis on the fine structure of this last led him to such a complication that, finally, he was obliged, using a hypothesis directly coming from the notion of the ether, to classically edit the equations which made his glory. His idea was nevertheless greeted as a genius intuition, which has brought electromagnetism where it is nowadays, at the summit of its development but unfortunately confined in a blind though undeniable efficaciousness. We must agree that seeing in every invisible phenomenon a manifestation of the ether doesn't automatically lead to an exploitable schema, even if it opens the way to an extraordinary number of analogies, built on what we presently know, through our personal knowledge of the simple behaviour of water or atmosphere: as the solutions are so numerous to come, we suddenly discover a hidden world of which the strength and innumerable branches are such that we are first thoughtless. It's a new Universe which is suddenly revealed to brains asleep by secular myths, comforted from generation to generation by a too rigid and too self-assured tuition. It's however towards this direction that we must go on, and rebuild a physics that will not deny for that its mathematical acquired level, but that will become again presentable and attractive, which is absolutely necessary if we want it to remain the basis of our culture.

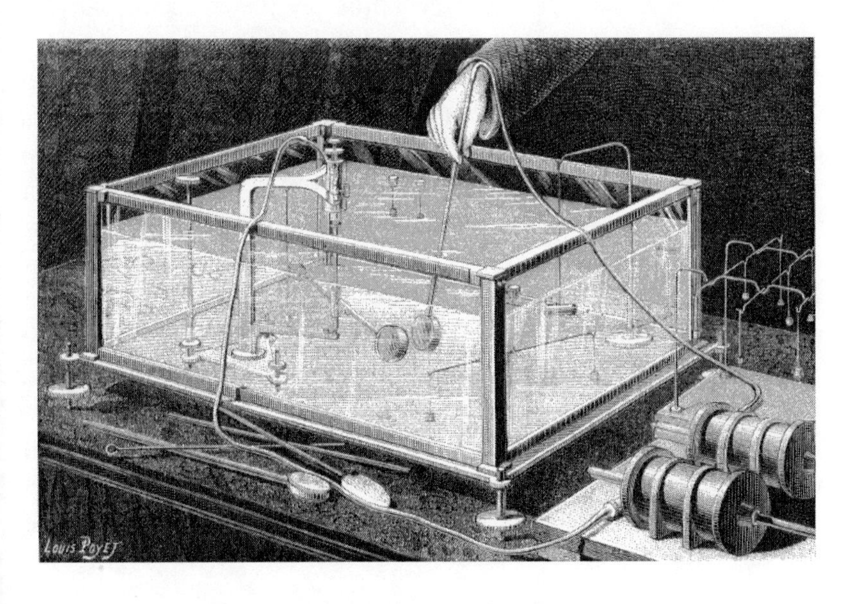

figure 4-3 : La Nature, Nov 12, 1881

Explaining all the natural phenomena by the interaction matter/ether, the ether being assumed as the universal fluid, needs before all four action processes:

- whirlpools or vortices
- continuous flows
- vibrations
- cavitation

This short classification of the basic phenomena assumes since the beginning, of course, that the Universe is mechanic, totally mechanic. Electromagnetism, to take this example only, is not something else than a theoretical presentation of phenomena which, despite of the fact that they are not detectable by our senses, must be classified as known manifestations of the attractive and repulsive action of forces, the attraction forces not existing as such, as really attractive, let's remember. Some scientists, contemporary of Maxwell at the end of the 19th century, attempted to establish close links between the observed electromagnetic phenomena and the classical mechanics, by the mean of particularly clever experiments, never mentioned in the courses of physics, but that, thanks to the CNAM[13], we can now easily find on Internet, in the archives of La Nature (cnum.cnam.fr). The first experiments, chronologically speaking, go back to 1881 and are the works of doctor Carl Anton Bjerknes, professor at the Norvegian University of Kristiana and a specialist of fluid dynamics. One of these experiments is illustrated in figure 4-3. We can see on this figure a mechanism, the heart of which is constituted by two disks immersed in a sort of aquarium. Those disks, which in reality are acoustic capsules, are put facing themselves by the experimenter, each of them being connected to a vibration pneumatic generator (the two cylinders down on the right). We are here in front of an experiment which is sensed to show how the vibratory energy present in a medium can give birth to forces which, following the relative phases of these vibrations, will be identified as either attractive or repulsive. Indeed, when phases are in concordance, the disks appeal one another. When there is phase opposition, they repel each other. The initial and final aim of the experiment was to provide a mechanical explanation of a possible origin of the magnets force. We may consider it as taking part, like the following, in a pedagogic equipment created to give to students or other viewers an analogical explanation, among several others, of the origin of magneto-static forces, and then, by generalizing, of the other electromagnetic forces. The message was the following: when two close objects, and only them in the vicinity, have a

13 Conservatoire National des Arts et Metiers, French institution

tendency either to get nearer or more distant without a visible cause, we have to look for an explanation in an energy which acts in the medium which is around them. We may notice, by this occasion, that no famous scientist, including Newton, has ever con-

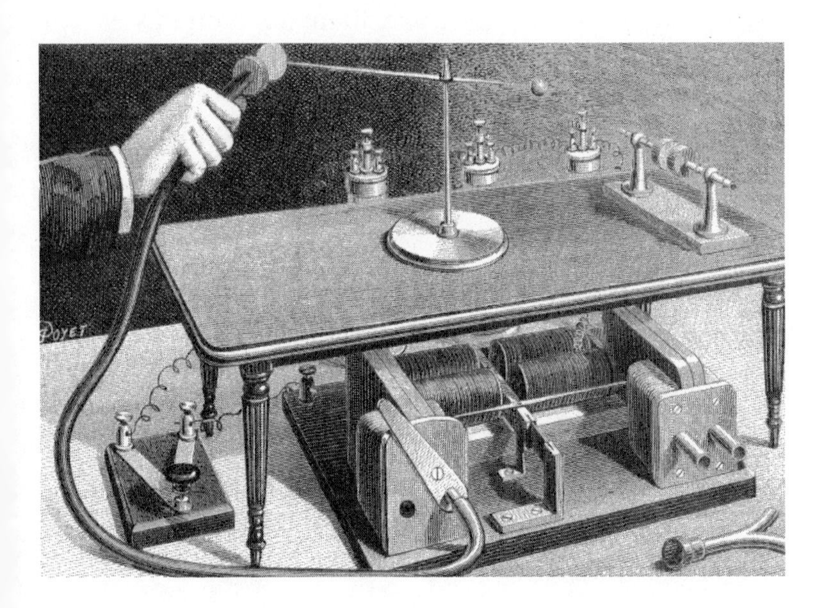

figure 4-4 : La Nature, July 8, 1882

tested the principle of continuity, which consists in considering that an action at a distance cannot be made else than progressively and continuously, along an uninterrupted material chain.

Almost simultaneously another physicist, Stroh, inspired by the work of Bjerknes, realised a bit later the same kind of experiment, but that time in air (figure 4-4), and obtained concordant results. Stroh had probably a more practical mind than Bjerknes, since the fact of working in an aerial medium brought a better experimental comfort and possibilities that couldn't provide the experiments made in a liquid medium. As a consequence that led to a mechanism both richer and easier to use, which led itself to a fundamental discovery on some interaction laws caused by fluid movements. The two mountings separately put into evidence that, following the phases of the two vibrating disks, these last were "attracting" when there was concordant phases and repelling when phases were in opposition. There was no other movement than this brought by the vibrations, thus it was clear than these last were responsible of the force present between the two disks. So the demonstration was done that, in a fluid, a vibration was able

figure 4-5 : La Nature, May 27, 1899

to give birth to a static force. The authors of the two experiments thought this way orientating the researches of the physicists specialised in electric and magnetic fields towards a not yet formulated property of ether, this of being the siege of vibrations. Around the same time, only a few years later, the industrialist C.L. Weyher proposed another explanation way for the action process of electromagnetic forces, with the help of devices made by him and using aerial currents, produced and directionally controlled by both complicated and smart structures, of which we can appreciate the ingenuity by looking at figure 4-5. We see here, but it is not the only device imagined by Weyher, a cylinder cut in two parts, the whole of which supposed to represent a magnet, with inside a double serial of very light paddle wheels, which inhale the external air by the two sides of the mounting and eject it by the middle, the movement being, if necessary, completely inverted following the willingness of the operator. Trying all the possible rotating senses combinations, it's possible to make that, presenting two cylinders one to the other with a common axis, similar North and South poles could be created with, following

figure 4-6 : La Nature, February 26, 1887

the relative positions of the elements, an attraction or a repulsion. We have this way another analogy, where the lines of force correspond, no more to vibrations, but to fluid currents closed up onto themselves. But Weyher went a bit farer and tried to quantify the mechanical action of the fluid currents onto the objects. Figure 4-6 shows both the mechanism of another experiment set, complementary to the previous, and the display of its result. As we can see, it's very simple: a multi-paddle helix sucks in the air before it and creates a vortex which acts on a rigid disk fixed onto the tip of the rod C, of which the position may vary between pretty large limits. Onto the other end is attached a small cord which passes on a pulley, and which is kept tight by a weight. When the disk is far from the helix, this weight leads it to the thrust on the left of the picture. When it is too close, it's squarely stuck to the helix, of which the aspiration force is then the strongest. Between, there is an intermediate position, in other respects unstable, that we can find in sliding the rod by hand, and where the two forces acting on the disk are in balance. We have at that moment a direct measurement of the attraction force of the disk by the turbine: it's the value of the weight. Making this weight vary and noting each time the equilibrium position, we can draw, which is automatically done by the equipment, a variation curve of the attraction force as a function of the distance. And we see then that this force varies in inverse proportion of the distance, which formerly was a property that we considered as specific of the universal attraction and the Coulomb's laws.

All that begins to be too much: so many efforts ignored till today to try finding simple explanations to invisible phenomena; so many concordance signs to see in the whirling laws the hidden meaning of the magnetic and gravitational forces; so many unfruitful attempts, from genuine scientists or simple amateurs passionately fond of physics, to try to draw the attention of the elite on the capacities of the rational physics, the other physics, the other vision on the world. How is it possible that nobody, among the ruling scientific class, succeeds in initiating any movement which would have as the mission of exploring all the possible ways? And specially to question the acquired knowledge, from time to time, according to the last discoveries? In any case, concerning the attraction forces, one might find in this paragraph all that is necessary to eradicate from healthy minds the poison that mathematicians have, all along the time, conscientiously instilled. The simplification made in putting into equation the phenomena, brought by the creation of the fictive attraction forces, both in the very large and the very small domains, cannot suppress the injurious and pernicious effect they have caused in the statement and the comprehension of physics, through the fact that step by step, in the course of time, mathematical abstractions have become realities which are taught as such, in colleges and Faculties. Vectors, potentials, fields, have become the bases of a physics which has forgotten, since the fifth-form, that it is an experimental science, though exact. But if most of the theoretical concepts of physics are justified by their obvious efficaciousness, there are some among them which lead the reasoning to dangerous ways. Attraction forces belong to this category, and it's absolutely necessary, like for medicaments, to stick onto them a well visible label, in a red cartridge, where is mentioned: danger! When two bodies have a tendency to go one toward the other, we have without delay to take the habit and the reflex, to be transmitted to pupils and students, of first wondering what is the force which pushes one toward the other. This cannot be conceived, in a general way, without making intervene a certain fluid, having particular properties. The right attitude of the beginner in physics, as well as the confirmed scientist, will be then to well take consciousness of the reality of this fluid, to insert it into the data, and to parameter it in the limits of the possible. Only then he will be authorized, with full knowledge of the case, if he cannot do otherwise, to use the subterfuge of the attraction force, considered as a fictive equivalent, to make possible both modelling and putting into equation the studied phenomenon. But it's imperative to indefinitely shelve the pernicious idea that attraction forces really exist. Believe in them is like believe in Santa Claus: we see a number of him at the end of the year, but the prototype doesn't exist.

4-5 : Celerity of the light

For Descartes, light was like a pressure instantaneously transmitted, gradually, by the mean of small solid spheres of which he has made the intimate constitution of the ether. At least, that's what we are told, but we must not forget that the books by Descartes were translated from Latin by the abbot Picot, to whom he had given the full latitude for interpreting the original text, with the assurance of the final approbation of the author. Thus it's not sure at all that he (Descartes) had thought that light was instantaneously propagating, but instead, like Galileo, that it was so fast that it could be considered as instantaneous. That's not the same at all, and if in that time it was not important, it's different to-day, knowing what we know on the topic. Let's say, to resume, that by the middle of the 17th century the word “infinite”, used to qualify the light speed, was more synonym of “incredible” than wearing a rigorous epistemological meaning. It's Römer and Cassini that we owe the first reliable estimations of a finite light propagation velocity. In 1676 Römer, who was studying the period and the occultation time of Io, the first satellite of Jupiter, produced a notice that triggered all the following. The astronomers, to refine the precision of their measurements, have got into the habit of extracting them from an average value, made on the biggest possible number of observations, practically several hundreds at the minimum. After what they deduce prevision tables which allow them to know by advance when the event, in the occurrence the submersion and emerging of Io behind Jupiter, will happen. It's precisely there, concerning the Jupiter's satellite, that things are going wrong. Indeed, the previsions from the tables were more or less correct all along the year, but if they were globally in accordance, sometimes even exactly, most of the time they were either early or late. After two successive measurement campaigns, these time differences were measured by Römer himself at 11 minutes first, and further 7. Looking more carefully at the phenomenon, Römer noted that the biggest delay was corresponding to the larger distance between Earth and Jupiter, and conversely the smallest to the shortest, and that the whole period of the phenomenon was exactly of a year. And thus the evidence came, first in Römer's mind, and then in this of Cassini and the other astronomers, that the only explanation to give was that the light was propagating at a finite speed, although very fast and incommensurable with all that was known and imaginable at that time. The principle being acquired, there was only to quantify.

Taking an average value of 18 minutes for the light to cross the earth orbit diameter, which is about 300 million km, they arrived to 278 000 km/s. In fact, at that time, the estimation of the size of the orbit diameters was different, and comparing this value to the to-day's official value has no much signification. Far more serious are the calculations of Delambre, director of the observatory of Paris, a century after Cassini, and by whom the determination of light speed is related in the volume 4 bis of the "oeuvres de Verdet" (p 654). Delambre took back all the measurements on the occultation of the first Jupiter's satellite for 140 years, that is a pretty thousand, the biggest number of them having been made by Bradley, and remade all the calculations, using the new astronomical data. This way he could, not only more precisely establish to 8 minutes and 13 seconds the time put by light to cross the Earth orbit diameter, but also give a new value of its propagation speed, that he estimated as being 71 000 lieues per second, and assort that result with an error calculus which gave a precision of plus or minus 2 000 lieues. It now remains, to be able to compare his results to the present data, to convert these values in IS units, which obliges us to know what was a "lieue" in that time. But there were two lieues: the 20 per degree lieue and the 25 per degree lieue. Behind this obsolete appellation are hidden two of the distance units which were used in France during the 17th and 18th centuries for the long distance measurements, and of course in the astronomical domain in particular. The 25 per degree lieue is in fact the length of an equatorial circumference bow, determined by a centre-angle of 1/25 degree. The equator length being by definition 40 000km, we deduce:

$$\text{1 lieue 20 per degree} = \frac{40000}{20 \times 360} = 5,55\ldots\text{km}$$

and the same:

$$\text{1 lieue 25 au degré} = \frac{40000}{25 \times 360} = 4,44\ldots\text{km}$$

So the light celerity computed by Delambre, the conversion being done, is thus 315 556 km/s, plus or minus 8889 km/s. If we write this result as a bracket, that means that the light speed in the interstellar vacuum would be comprised between 306 667 km/s and 324 445 km/s. This result is so perturbing that, on one hand it is never mentioned to-day, and on the other hand it is considered as wrong, with the main argument that at that time there was an error on the estimation of the Earth-Jupiter distance, and in the same way on the Earth orbit diameter. That's very easy to say, nobody in the general public has the necessary data to state for or against, and nobody, in the observatories where the archives can contain the necessary documents, will only move the little finger to unearth a document, even fascinating, which would bring the wind of the revolt in a so peaceful place. Unthinkable. Let's remind that the official value is 299 792 458 m/s (AFNOR 1988), and you are prayed not to smile and not to make any notice on the measurement precision: this last being asserted with 9 significant numbers, that means that the light speed would be known to within one m/s! Let's repeat it once more: a good measurement, a precise absolute measurement in physics, is a measurement within the thousandth. An excellent measure is within the tenth of a thousand. A measurement within 10^{-9}, that's a swindle, except if we have decided before that the measured quantity was a universal constant, which afterwards allows any fantasy. And this will remain as long as the published result will not be supported by the measurement conditions: location, altitude, etc, etc... The summit of the unconsciousness, from the scientific staff in general, is to have now defined the standard of length, the metre, previously fundamental unit, from the speed of the light. To-day, the metre is the distance covered by the light for a time of $1/c_0$, c_0 being given in m/s. The madness of the relativists has enforced us of this absurdity that a length would depend on a time. Thus the metre is a variable unit of length, because, let's never forget it, the light celerity is eminently variable itself.

That's not all. There was another method, here-called astronomical, for measuring the light speed, made by the English Molyneux and Bradley, and using the phenomenon of the aberration. This phenomenon consists in and angular variance between the apparent position of a star said "fixed" and its real direction, and comes from the fact that the observer, being on the Earth, moves with it at a speed of about 30 km/s in the Sun reference axis. To use the pedagogic comparisons by Einstein, with his train and his embankment, let's suppose that there are two observers, one in the train and the other on the embankment, and that both let fall a stone on the ground at the precise moment they are face to face. The one who is on the embankment will see the other's stone draw a parabolic arc, this of the train will see instead, relative to a referential bound to the train, the stone of the embankment draw a straight line, but not vertical. The angle made by this straight line with the vertical is the aberration angle, which will be all the more open that the train speed will be high. In the astronomy, its measurement and its calculation make intervene the ratio of the two speeds, this of the Earth and this of the light, thus we can calculate this last if we know the value of the aberration. That's a method still less precise than this of Römer, but like this last it has the property of being made in the interstellar space. They are, so far, the only two method which have this particularity. Now, the method using the aberration leads also to a value which is above the legal one: 306 408 km/s (Verdet, volume 4bis, p 684), with, to tell the truth, a precision of 1,5 % which now puts again the official value in the bracket. But as we know, since Vallée's works, that the light speed is linked to the gravity field, and that, for example, in a field like this of the Earth gravity, it must be less than in the free space, far from any mass, we feel obliged to take this confirmation into account. The wet-blanket relativists will probably say that the measures by Römer and Bradley are wrong, according to a more recent evaluation of the distances used in the calculus, but we may say, in the same way and with the same good faith, that this new evaluation has been made on the Einstein's basis that the light speed is a universal constant, and that all has been made to make the results coherent with that hypothesis.

This point of view merits to be more elaborate, and makes necessary to come back to the events that have marked the steps of the determination of the light speed. After the results of Römer were published, scientists have quickly looked for a mean of getting rid of the astronomical measurements, which had for unpleasant consequence, for some of them, to be obliged to use observatories, their heavy equipments and also the astronomers. For that purpose, they endeavoured to design laboratory devices easier to handle and possibly more precise, in fact of a precision without comparison with this of Delambre. However, it was necessary to wait 1849 and Fizeau to see the first attempt, which by the way was only half a success, since it led to a result of 71 000 lieues 25 per degree, that is about 315 000 km/s. It was exactly the same result as this found by Delambre. The method consisted in sending a light ray on a well away mirror, through the notches of a toothed wheel rotating at a controllable speed that could be progressively increased. When there is extinction, the knowledge of the distance to the mirror and of the rotation speed allows computing c_0. As emphasizes Verdet, neither the experiment details nor the error calculus have been published, so that we consider today that, apart from the historical interest, the very measure, too close to this of Delambre not to be suspected and too far of the value presently agreed, is not available. To go in this direction, the provisional mounting he had imagined, too ill-assorted to be reliable enough, with a toothed wheel and a mirror distant of more than 8 km, mechanically suffered of a lack of cohesion. Moreover, the measurement precision of the distance between the wheel and the mirror didn't put him, from this point of view, in better conditions than in an observatory, behind an astronomical telescope. However, the same measurement principle was used afterwards by experimenters may be more gifted, or simply more careful, we don't know, specially by Cornu in 1874, to give results close to 300 000 km/s. But any way, it's Foucault who, improving

Authors	Date	Résults (km/s)
Fizeau	1849	315 000
Foucault	1862	298 000
Cornu	1874-76	298 500 300 400
Michelson	1879	299 910
Young et Forbes	1882	301 400
Michelson	1885	299 853
Perrotin	1902	299 880
Michelson	1902 1926	299 910 299 796
Michelson, Pearce, Pearson	1933	299 774

Table 4-7

the principle of the Wheatstone's rotating mirror in 1862, was the first who found a light speed under this symbolic barrier, with a value of 280 000km/s.

All these experiments and measurements have in common an exceptional complexity and a drastic exigence of care, because the point is not to find or check a physical law, but to realise a fundamental measurement, of a precision never demanded before, and on which are going to depend so important interpretations concerning the constitution of the Universe. Their mishaps are well resumed in the book supervised by the CNRS[14] and pub-

14 Centre National de Recherche Scientifique

lished by Vrin on Römer and the light speed, to which we refer those who want details (see biblio). Table 4-7 is content with giving a partial view, but significant enough, of the attempts diversity, and the most prominent results. We can note that, seeing this chronological presentation, starting from 1885 the results are more closed and converge towards the presently official value. We thus may interpret this apparent coherence as a normal progression of the measurement precision, linked to a more and more elaborated technicality, more and more matched to the target to reach. We'd rather be a bit more cautious, and not run away. First of all, we have to let come into our mind that, the light speed being assumed by all the physicists, before the Synergetic Theory, as a universal constant, no one among them had doubts on the principle that its precise determination was only a question of technology and that, the more time passing, the more the gap with the searched precise value will decrease. The idea that it would be a variable quantity, depending on other parameters, never crossed their mind. When a physicist confirms a result found by one of his colleagues, it's not forbidden to be mistrustful, because the history of sciences is full of examples where a scientist, either of good faith or at the contrary influenced by the reputation of another one, has corrected a measurement that he found too different from what has been found so far, simply by the fear of being ridiculous, by the lack of confidence in himself, by the lack of personality or something in the same vein, bound to the human nature more than to the very physics. We may ask ourselves, for example, why Fizeau, responsible and initializer of the first terrestrial measurement of the light speed, has felt obliged to assume a result identical to this of Römer, when all those who have used after him the toothed wheel method have found a completely different value. But the problem is not there.

If we go back to the methodology of the measurement of the light speed, we can see, according to what is written before, that there are two groups of experiments totally distinct: first the measurements made from an observatory, which estimate propagation times and interstellar distances, and secondly the terrestrial measurements, that we may consider as local measurements made in laboratory. The advancements made in the precision of the last ones have progressively led the scientists to stock the previous ones in the archives, and then to only recognize for them a historical interest, while suppressing them any experimental value, asserting a wrong knowledge of the true dimensions of the solar system. We can accept, but we'd like to know if someone has taken pains to re-actualise the calculations of Delambre, for example, and make us taking advantage of his investigations. But there is a passage, in the book on Römer hereabove cited (page 267), which at the contrary gives rise, instead of clarifying the situation, to supplementary interrogations:

"*It's thus a shrewd use of two of the most spectacular results of the physics of his time which allowed the fruitful exploitation, for the measurement of c_0, of the Römer's discovery following which the period variation of Io's eclipses, during half a year, could be interpreted by the play of the successive delays brought by the propagation time of the light.*

This method has been used with more and more refined means and more and more meticulous theories until 1909. At this date, R.A. Sampson perfects, at Harvard, a photometric proceeding to determine the happening of Io's eclipses, and exploiting to-day a posteriori his results with a more exact value of "a" -half of the larger axis of earth ellipse-, the error on c_0 had become as low as $\pm$ 60 km.s^{-1}. ".

This last sentence is of a considerable importance, because it should have opened the door to a re-visitation of the results obtained by Römer, Cassini and Bradley, and to the possibility of confirming or invalidating the fact that the light speed is not the same, either at the level of the earth's crust or between Jupiter and Earth. But alas there is no following in this article, no waited development, no supplementary precision, no revelation on what could be a new start on a key-question of the physics. We remain hungry, and we don't dare wondering why the author of the article suddenly stops at the more exciting moment, when the doubt is swept away at once, and when the blinding clearness of the truth splashes you. But no, absolutely nothing. Once more science refuses us its support at the precise moment it has all the cards in hands for stating, but may be it's indeed here that the problem is: all the truths are not good to say, and the formula "woe betide them, those by whom scandal comes" has no better echo than in the scientific sphere. The fact remains that the mystery remains unsolved, and we can measure on this occasion the fantastic inertia of the scientific society, unable to shake its torpor and to face the reality, even and principally when it is question of objectively

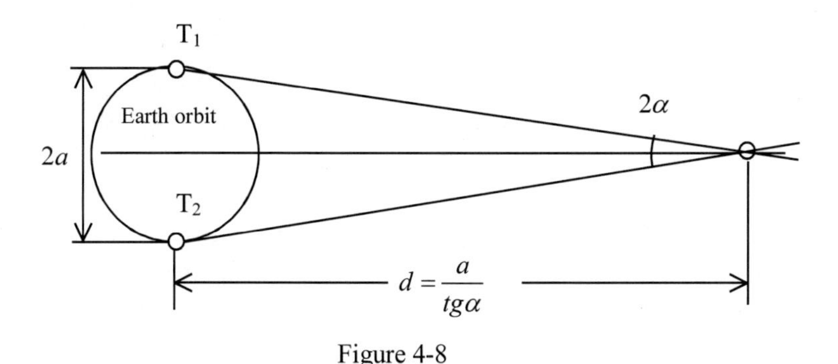

Figure 4-8

taking bearings, doing an inventory of fixtures, which could allow a salutary calling into question of the acquired knowledge, in order to eventually make physics move.

Whatever the managing scientific class thinks, the light speed is not a constant, it varies with the gravity field, and may be with other parameters. However, we must see well that to-day, all the distance measurement devices which use EM waves, from the elementary laser of the bricklayer, or the quantity surveyor, to the astronomical radar, include a microprocessor in the memory of which is registered the official value of the light celerity. Any other distance measurement device cannot be authorized if it waives this principle: thus the system is perfectly closed up, the law of the Trade Code has been inserted in that of physics, and it is well evident, in these conditions, that nobody will ever find the variation laws of c_0 if things remain in this state. But how could they change, since it has been **decided** that c_0 is a constant? Nevertheless, this doesn't thwart relativists to explain us, we poor deadbeats of the suburbs, that the black holes are so dense stars, with such an enormous density, that they forbid light to get out from them. In other words, the bastards who drum into us that the light speed is a fundamental constant simultaneously acknowledge, when it's providing some advantage to them or when they are short of arguments, that not only it varies, but it even can be nullified! That's too strong. We really are nothing in front of this people, who handle contradiction with so much talent, natural and elegance! Except if, more simply, they take us as imbeciles?... No, let's quickly repel this improbable and discourteous idea, and let's go back to serious things.

All the astronomical measurements, for the mean distances, are made on the basis of the parallax. It is a question of the angular difference which appears when you shot the same fixed star, successively from the two diametrically opposed positions of the major axis of the Earth orbit (figure 4-8). Trigonometry allows then to calculate the distance between the star and the Earth, if we know the length of this major axis. The two measurements are thus made with a time delay of half a year. On the point of view of the principle only, this seems to be very simple, and the method, in itself, has been moreover used very early at the surface of the Globe to measure the distance with an inaccessible point: the other side of a river, the middle of a marshland, an island,... To achieve that, you build a triangle, isosceles if possible, of which the base is constituted by the segment between the two measuring points that you have chosen. Then you measure the two angles of the triangle base and, knowing the length of this last, you deduce this of the height or the median, and it's done: geometry has allowed man knowing the distance to an object that he cannot physically reach. The big problem, in astronomy, is to know how was acquired the data of the length of the major axis of the earth orbit. Astronomers have not an infinity of means to determine the distance of remote stars: they measure times and angles, and first, when they are measuring an angle, they must compute the correction to be brought because of the rotation of the Earth, which moves and on which they are bolted. In other terms, they measure the Universe from a turning manège, and this is why measuring and defining time and distance standards are so important for them. Knowing durations is for them a basis of their profession, like weighing was for the chemists, and defining a time unit is for them a necessity still more drastic than for the general people.

The first time unit of mankind, at the dawn of its presence on Earth, was the day, quickly complemented with the lunar month, and then the solar year. When it was necessary to be more precise, the day has been first shared into phases: morning, evening, night. It seems that it should be the science and specially astronomy which has, for its own needs, introduced the measurement of time intervals by counting periodic phenomena. Human inventiveness then began to list all what the nature puts at his disposal, and matched it to his level and his technological possibilities of the moment: sundial, water clock, pendulum clock, and so on, until a correct definition of the second of time, in a reproducible and transportable way, thanks to the invention of the watch, this technological jewel, guardian of so many secrets of the mechanics. Until about 1950, that is yesterday, the second was defined by the astronomers as the 86 400th fraction of the mean solar day. It was in that period that this conception of time was abandoned to be relayed by totally terrestrial phenomena and related to the intimacy of the matter, the atom being considered as an indestructible reference structure, of a perfect stability.

It was first a multiple of the cadmium red ray period, of a wavelength of 0.64384696 micron, before being replaced by the radiation that accompanies a certain energy level change in an atom of caesium 133, of which we have to count something like one billion periods to make a legal second. We easily imagine the complexity of the equipment which allows this kind of result, but does this fantastic technology, which uses the most complete and modern arsenal of high frequencies, give more warranties on the time standard? The ambition of those who have designed the atomic clock is insane in its pretension: no shift of more than a fraction of microsecond per century! The relative precision of the clock, this which is announced, is to-day of 10^{-14}, waiting for better because the ideas people are not satisfied yet. We have there a supplementary illustration, if proof were needed, of the human madness which has definitely reached and putrefied the hermetic circle of the fundamental research. Indeed, to cast off the limited precision of the traditional astronomical methods, it has been decided that the only intimacy of the matter, considered as the siege of the basic natural phenomena of physics, was authorized to provide men with the absolute references he's desperately looking for. This matter is thus credited with an inviolable and total stability it has never had, but in which physicists believe like in a god, with the same bias. It is then considered as a so perfect mechanism that it would be inconceivable that it wouldn't keep in itself the keys of the World mysteries. It was thus decided that atomic and nuclear phenomena were all of them quantified in an absolute proceeding, and that it was there, in the calibrated flea jumps of the electrons and not in the interplanetary space, that we had to look for the time standard.

No engineer who has passed at least ten years in a laboratory, where each day he makes pointed measurements in electronics, can believe that some quantity can be determined with a precision of 10^{-14}. We can make a differential measurement with that precision, but in this case the word “precision” must be replaced by “sensitivity”. This means that, using this sensitivity, you only can quantify the difference between what you are measuring and another object of the same nature that you have chosen as a reference, that is something that you arbitrarily consider, for a while, as exactly known. It's a decision, a postulation, a choice, all that you want, except the truth and the perfection expected. This continuous ever-increasing quest of the zero error has something pathetic. It only can be the fact of people who have completely lost the sense of the reality and the reasonable. The atom is submitted, like the rest of the matter, to an environment of which we don't know what is the exact influence on it, and there is no available reason to assert that this influence is null: the gravitation field, the Earth magnetic field, the temperature, the humidity, the cosmic rays, all the possible and imaginable environmental factors are susceptible of acting, directly or not, on the energetic levels of the atom. Is this action detectable with the usual means? We can easily come to an agreement on that: it's no. But say that it doesn't intervene at 10^{-14}, it's really too difficult to ingest, as it is obviously impossible to annihilate totally and rigorously all the causes. For example, when we are told that an experiment is currently going on in a regulated oven, so as to eliminate the influence of the temperature, we must know that this kind of regulation consists in limiting the temperature variation between two pre-determined values, as close as possible but inevitably different. Thus the temperature is constantly varying between two close values, with a variation law in the form of a sawtooth, even if it is stated "stabilized", word which is not synonymous of constant. This is only an example, chosen among the simplest ones. In what concerns the real precision of a measurement, the error we make can be hidden by the very choice of the process, as well as the choice of the standard. This last is obligatorily, by definition, con-

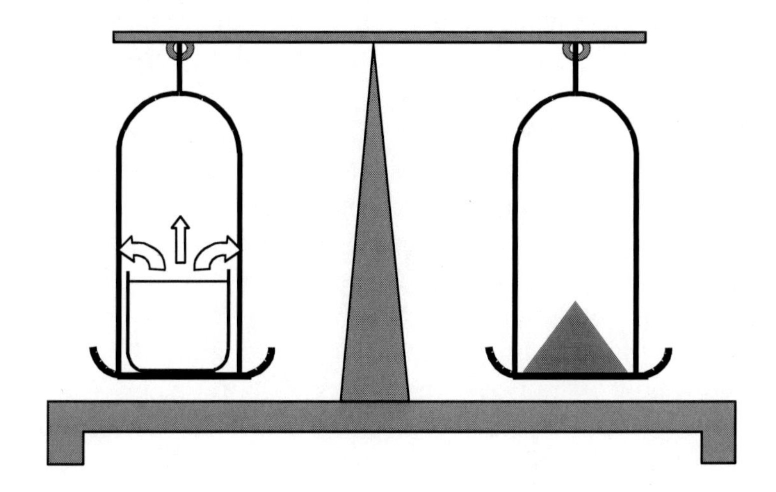

Figure 4-9

sidered as a perfect reference, even if we pertinently know that it is not. But even if we have doubts on its legitimacy, a standard remains a standard, and at all events we are obliged, at a moment or another, to refer to something. That's the human conditions of a laboratory.

To show well at what point the questions of metrology, often of an evident appearance, are in fact intricate and overlapped one into the other, let's take the following example: we want to know, by weighing, the quantitative law of water evaporation in the normal conditions. For that, we'll put on the plate of an assay balance a container, that we'll fill up flush with water, and we'll counterbalance it with a tare of fine sand on the other plate (figure 4-9), at the time t_0. Water evaporates, and progressively the left plate looses its mass and goes up. At the moment t that we have chosen to make a measurement, we re-establish the initial equilibrium by adding besides the container a corresponding weight Δp, which thus is the mass of the evaporated water. We can do then the same experiment several times and obtain a curve as a function of time. What will be the precision of the measurement points? May we conceive that it could be 10^{-14}? Of course no, and for this example, close enough to the current life and to the average knowledge of the same-named citizen, nobody will contest this evidence. However, let's try to analyse the operation more in detail. It must be borne in mind, in this significant example, that the mounting allows the access to two fundamentally different measurements: one is an absolute measurement, this of the weight of water which was originally in the container, the other is a relative or differential measurement, and determines the proportion of evaporated water in regards to the initial mass. But we immediately may retort that the evaporated mass of water has been compensated on the weighing machine by the addition of known and certified weights, and that it had been indeed intrinsically determined. As for the initial mass of water, it can be known also by weighing the sand tare, with the difference that it is here a double weighing by the method of Borda, which eliminates the dis-symmetry of the equipment, and which results in that it is a more precise method than directly weighing the evaporated water. But as it concerns an incomparably more important quantity, the absolute error is finally greater too. But... etc...etc... We can endlessly discuss and hold forth about these double-barrelled notices, we are obliged to note that finally, the precision with which the law of water evaporation has been established depends on a chronometer and a standard of weight, or mass, which is kept somewhere at Sevres[15], or elsewhere, and that this standard, which also evaporates like any matter, even less quickly, will never be determined at 10^{-14}. Finally, we can turn over and over the problem in all senses and burn our neurons in trying to find a way towards certitude,

15 Location of the French standards

the vocation of this last seems to refuse itself to man, as Karl Popper so smartly demonstrated it. Very imperfect in its demonstration, the here-above example has no ambition but showing that the simplest experiment can become very intricate, see inextricable, if we ask too much from it. Let's repeat that an absolute measurement with a precision of 10^{-4} is an excellent measurement, and that engineers are generally satisfied with less, which doesn't thwart them from being efficacious.

Let's now come back to the speed of the light. All the astronomical measuring methods have been abandoned for terrestrial ones, of which the sensitivity has never ceased being improved. In parallel the radar, the radio-telescopes, allow knowing the distance of a solar system element with a resolution less than a metre, but assuming a light speed considered as a constant and stated to 299 792 458 m/s, following terrestrial measurements. Thus it's not now possible to find another result, except by still increasing the sensitivity of the method used, and consequently the quantity of significant numbers, and this wastefully since the process is not well-founded. The system is therefore completely locked up, like a microcomputer which is blocked and that we cannot use again if you don't make it restart by a reset to zero. Who will make endeavouring a new measurement campaign of c_0, now that we, rational physicists, know that it varies with the gravitation field? Who will do this reset? What unconscious hero will dare defying the scientific Nomenklatura and, facing an old flabby in charge of a public research establishment, will tell him: “Sir, all of us are wrong, we should make a U-turn!”? The world of the research is so full of comfortable cushy jobs, where the excess graduates of the High-schools and Universities take life easily, waiting for a retirement at 55 years, that there is very few likelihood to see appearing this sort of Don Quixotic before long. And even if he would exist, what mandarin would pay attention to him? What high responsible will dare risking his career to suddenly remember that the most important part of his task is to promote physics, to never be satisfied with the ordinary, to every day try to learn a bit more than the day before, to guide his herd in the so risky, but so gratifying way of the unknown? There's however no big risk, but at the contrary the hope and the perspective of exceptional satisfaction, in arranging in a part of the budget, in the organisation of the teams, in the planning, a little place for a special study group which would play the role of the devil's advocate, which would systematically call into question every knowledge considered as acquired, which would hungrily explore the rejected ways, and principally which wouldn't make any hypothesis without basing it on the existence of the ether. But apart from these considerations, of a general and ethic character, there is something more serious: if the light speed is not constant, it's absolutely necessary to hurry up onto the problem and analyse it with all the necessary rigour, and not ignore it with pride and contempt. Since the moment a theory, even only one, explains with clearness the relation which exists between c_0 and gravity, any physicist worthy of the name should have the moral and professional obligation to get aware of it, to thoroughly study it and make a statement on its well-founding. And if this last is confirmed, it's worth publicising it and draw all the conclusions which go with. So, who begins?

4-6 : The atom and the quantum physics

The atom, which in Greek means “indivisible”, was considered in the Antiquity as the ultimate component of the matter. Its present conception, with a nucleus surrounded by a cloud of electrons, is hardly only one century old. It's probably the discovery of the radioactivity which has led to the idea that the atom could have several constituent parts, among which the electrons were able to spontaneously leave the edifice, previously considered as inviolable and chemically characteristic of a given element. The first work was to identify these electrons, becoming well aware, through more and more precise experiments, that matter could emit, in certain circumstances, strange electric particles. These particles were always the same, with a certain energy and a certain mass, which were not directly measurable, but which could be deduced from dynamic considerations, for example by studying their deviation by an electric or a magnetic field. We moreover have here the main problem and the specificity of the physics of the particles, that is being only made of models, models of a reality definitely out of reach for a direct observation. One could of course immediately retort that it's there the general problem of physics, may be it's true but not at this point: when we are wondering on what is the mass of an object that we have in hand or in front of us, for example, it is well visible, we can touch it, weigh it, divide it, sample it, while taking for that all the necessary time. In atomic physics, instead, we have first to build a representation, necessarily wrong, of particles already not easily identifiable among themselves, always quickly moving, and then cause an alteration of their trajectories by means of an external force, in order to deduce afterwards, by calculation, their static characteristics. That's to say what uncertainty can exist on the validity of any representation of the atom, dynamic system but of a static appearance though, and the complexity of which increases with time, as we discover or guess new components. By comparison with the indivisible unit of Democrite, the 21th century atom reveals a growing complexity where the nucleus, which for Rutherford, a century before, had replaced the concept of the ultimate unit, is nowadays seen as a world in itself, a world where all is to be discovered, but of which the components are more and more hypothetical. What brings doubts on the nuclear physics, such as the physicists present it to-day, is precisely this complexity which inflates along years and goes against the idea, may be too simple but which was the hope of the first modern atomic physicists, that things would have gone clearer and clearer following the technical progresses of the investigation means.

As it has just been said, we must go back to the end of the 19th century and the discovery of radioactivity by Henri Becquerel to approximately place the origin of the present conception of the atom. Noticing that electrons or other particles were able to leave it made incontestable the idea that there was, inside, a mechanism far well complicated than what could have been imagined before. Making up one's mind on the way the particles were stocked, before being ejected, was not so simple. It first might take into account the stability of the matter in general, and on the other hand produce a model where the electrons would be already present, while remaining prisoners inside a relatively sturdy edifice, but close enough to the surface, either, in order that an energetic event could eject them, and only them, prior to anything else. It's to Ernest Rutherford that was attributed the planetary model, which consisted in seeing the atom as a solar system in miniature, the nucleus having the role of the Sun and the electrons being the planets. In his model the electrons were revolving, like the planets, in a same plane, but his meeting with Niels Bohr gave birth to another model, more sophisticated, which afforded the last discoveries and introduced the fundamental idea of quantification: the peripheral electrons are no more in a same plane, but move in separate concentric layers, in which they have at each moment random positions, but corresponding each to a precisely defined energy level. That means that they may migrate from a level to another one, from a layer to another one, only if they receive or lose a certain quantity of energy, always the same. Although Rutherford had made an image of the atom copied from this, well visible, of the solar system, we can see that there is an enormous difference between the true planetary system and its miniaturized copy: in a genuine solar system, planets may occupy any position, on the simple point of view of their distance to the sun. We are in the habit of seeing them at such a position, but they could as well obey the Kepler's laws on any different orbit, presently empty. Inside the atom, it's different. Following the quantum mechanics, there are movement zones that are authorized and others which are not. The electron may not do what it wants, it is given the authorization of turning, but not anywhere. It's in order to mathematically justify this particularity that a plenty of well renown scientists, among whom the best-known were Einstein, Schrödinger, Heisenberg, Dirac and others, hurried on the Bohr's model to, each of them, apply to it his own way of thinking and his personal techniques by interpreting, by the mean of the wave equations, this refusal of continuity that the nature seemed to have until now taken as a rule, at the sub-microscopic scale.

This “Copenhagen school”, represented by the photograph of the 5th Solvay congress in 1927, gathered all the renamed scientists of that time, including those named here-above, and flaunted, as their profession of faith, the renouncement to understand what really happened inside the atom, for the benefit of an important number of new concepts, made for the aim of creating an efficacious model, that wouldn't have been not predictive but simply coherent with the reality. Nothing new under the sun, this model, which couldn't have been something else than sheerly mathematical, was only the climax of what can propose the theoretical physics, in front of this atom which suddenly unveiled its heterogeneous aspect and its hidden complexity, without offering an

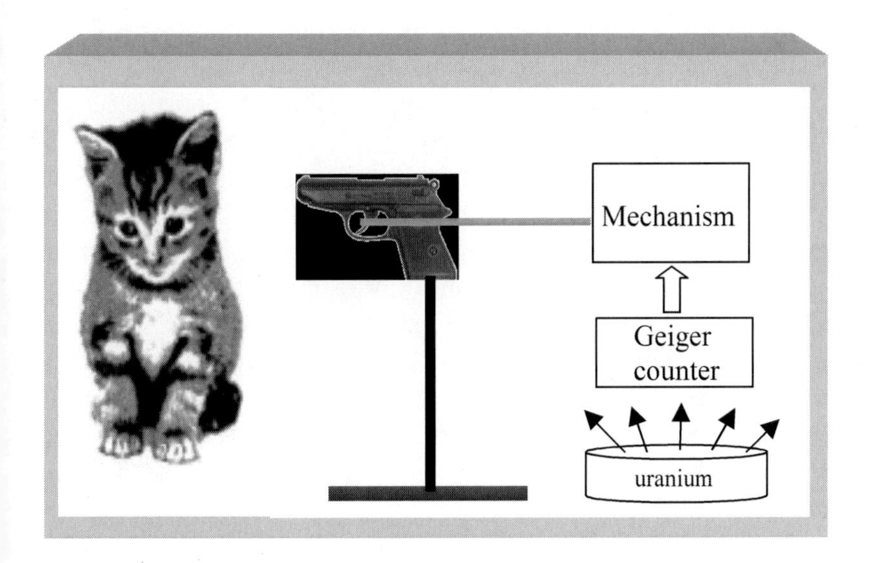

figure 4-10 : the Schrödinger's martyr-cat

evident key to step into. Paradoxically, the impossibility of getting rid of this enormous scale difference between the observer and the object observed, a bit similar to this of astronomers, but in the other sense, made also possible all the interpretations. The imagination was necessarily taking advantage on the observation, and physicists were delighted in doing that, especially since the relativist context had largely opened the way to all the kinds of all wild imaginings: it was practically allowed to say anything, to emit the most way-out thesis without daring to be ridiculous, and the scientists connected to the atomic universe didn't need to be told twice, and went frantically this way. All the barriers patiently erected by the determinism were suddenly attained by a sort of fragility, and then the classical conception of the trajectories, where the position of a mobile is perfectly defined as a function of the time, was abandoned to let place to the introduction of statistic and probabilities notions, in which the electrons were losing their individualities, similarly to the molecules of a gas that only the Maxwell's daemon in able to count one by one. Einstein took advantage of the golden opportunity to make everyone informed that the Relativity had foreseen all that the quantum mechanics has demonstrated, and the thousands of his faithful and discipline epigones still keep on, a century later.

The theories by Bohr, Heisenberg, Schrödinger and the tens of others implicated in the case have reached such an abstraction level that this little world of extraterrestrial people, not to look too pretentious and above the rest of the World, felt obliged to lay simple images, not to say simplistic, to reach and if possible interest the general public. Among these, one of the best-known, only by its name because so much abstruse, is this of the Schrödinger's cat, supposedly imagined by him -Schrödinger- to show the dangers of applying to the macroscopic phenomena the principles of the quantum physics, and to highlight to what sort of absurdities led the fact of infringing it. In this physics the electron, prisoner of its atom, is no more a particle but becomes a probabilist wave function, called orbital, of which the instant existence is defined only statistically. Besides, the question is put to know if the observer, through the simple fact that he is observing, doesn't perturb the phenomenon observed. We see very well, through these so pertinent points of view, the twisted and elitist side of the quantum physics. But let's come back to the cat. The poor beast is prisoner in a special cage, besides a lethal mechanism which will kill it if, in a one minute while, a sample of radioactive matter will have emitted an electron which, once being detected, will act the system (figure 4-10). The problem is, for the observer and without looking inside the cage, to guess if the cat is alive or not. However, the existence of the cat is bound to this of the atom, which can be disintegrated during the one minute duration of the experiment, and in particular to the electron which can leave it. This last, before the observation, can be or not to be, like Hamlet, but this simultaneously, thus the cat can be both alive or dead. Do you understand? Neither we do. What is sure is that the tom cat, it, perfectly knows if it is alive, and in this case we can imagine that the only preoccupation of the poor animal is to save its skin, and that its immediate thought, if we may suppose it has one, might resemble something like: "*Help! Let me go out!*".

Let Schrödinger forgive us this tendentious presentation of his fake-demonstration, but it is both necessary and salutary to react and be confident on one's good sense, when we are taken as imbeciles. This pseudo-demonstration is of the same nature than this of the Langevin's twins, with one of them who travels at light speed -what a realism!- and the other -lucky guy!- who remains with us on our good old Earth. The first one sees his relativist clock slowing, thus he's ageing less quickly than his brother, but this last sees the other moving away with the same relative speed, in the contrary sense, and he ages less quickly too, which is contradictory, but the one who is travelling must make a U-turn to ascertain the age difference, thus he's twice, instead of once, in a Galilean reference system, so he's the winner and ages less quickly than the other.....must I go on? By the way, where did he make his U-turn, and how did he manage at that speed? Let's be serious, or let's try at least: all these esoteric fables by mad scientists have in common to have unrealistic and perfectly unverifiable starting points, as there are in the infinitely large and the infinitely small, there where we'll never go otherwise than by thought. We may say that there is a deterministic and pragmatic physics, which quite well describes the phenomena of the world at our scale, and a venturesome, unrealistic and weird physics, which has, as in the theatre, worlds that we are only able to imagine. In this last all is allowed, the most insane hypothesis can be emitted, as long as they don't too visibly shock the common logics, and we must agree that the human species is very gifted for that sort of exercise.

The quantum physics takes out its name from the hypothesis, in other respects acknowledged, that the energies attributed to the electronic orbits varies by steps from one to the other, and that there are forbidden values inside the atom. Indeed, if we want the electrons prevent themselves to collide, we must suppose that they must move in distinct layers, and it seems quite logical to approve this condition. But this notion of quantification of the energetic levels has quickly been expanded to the energy itself, and from the whole of the Copenhagen interpretation and the Relativity, now indissociable, was progressively born the hypothesis, implicatively or explicatively expressed, that this energy could have the shape of "packs" or "quanta", and therefore vary discontinuously. Energy is equivalent to a machine work, in the meaning given by the physics to this word, and a mechanical energy is always the mathematical product of a force by a distance. Its dimensional equation is thus expressed by the formula:

$$W = M.L^2.T^{-2},$$

where it appears that it is the product of three continuous quantities. Consequently, it cannot be non-continuous in any case. The formula here-above may be read like this: energy is proportional to the mass, to the length squared, and to the inverse squared of a time. Colliard, in "les deux éthers"[16], gives a very pedagogical image to illustrate this salutary and necessary recall to the sources:

"*As for the quantum notion by Planck, it reveals itself useless and wrong. There is no, it cannot be, a discontinuity of the energy, because the quadratic speed, which is one of the elements of this energy, may vary in a continuous way. This doesn't necessarily implicates that any phenomenon varies through the continuous increase of its energy. Linked to an engine working regularly, we can put a gear box of which the clutch will give to the stirring -wheel discontinuous speeds. We may not conclude of it that there are speed quanta*".

The denial of the ether by relativists has made them pass besides mechanical explanations of a so great simplicity that it seemed incredible to them, in the sheerest meaning of the term. In what concerns more specially the quantum physics and the wave mechanics of De Broglie, it's good to remember that it exists in physics, a phenomenon, and only one, known in all its branches except precisely these ones, where the quantification exists: that is the phenomenon of resonance. The resonance is a phenomenon of

16 The two ethers

amplification which occurs each time an incident wave, in a determined volume, finds a way where, after reflection against an obstacle, in general a partition or an extremity, is in phase with itself at the input. There is then a quasi-instant accumulation of energy in the volume where the resonance is arising, until a maximum which only depends on the losses: caloric losses when the resonance is mechanical, ohmic losses when it is electromagnetic. When a volume is resonating, all is happening as if the vibratory energy, coming from outside, was accumulating inside, without any possibility of going back, and this until a maximum is reached, as if the volume in question possessed an energetic "capacity" or "content". For example the resonant cavities are, in high frequency electromagnetism, of a current use in the technical domain of the high selectivity filtering. A cavity of this type shows, at its resonance frequency, particular properties which constitutes a vibration "mode". It is this mode which is quantified by essence, as it depends on the **geometry** of the volume where it is acting, and on a phase condition. There will be as many modes, that is as many resonance frequencies, as there will be directions corresponding to a certain multiple of the associated quarter wavelength, added with the even and odd harmonics, following the way of excitation. This well-known phenomenon is the only one, both in rational physics and classical physics, this last which is taught in elementary and secondary schools, where appears this discontinuity between the stocked energy levels in a given volume. The physicists of the beginning of the 20th century have forgotten it, and for that reason they found them obliged to invent the quanta of energy and all this new physics which allows their use, but however neglecting the possibility of another model of the atom than this of Bohr-Rutherford.

That the atom is the siege of stable periodic movements, source of extremely high frequencies waves discovered by Compton, everyone agrees. That these movements are orbital, that's in return a convention that, despite of the fact that it has been lasting for more than a century, we have the right to contest, under the condition, of course, of proposing another model which would have, in comparison with the previous one, incontestable advantages: it's the physicist's law. Well, if there is a model to be tried, it's obviously this of the resonant atom, which doesn't need that quantum physics exists, but which in return leads to far more seducing interpretations of the atomic phenomena. However, it cannot be conceived without the presence of an ether, and this probably is the main reason of its non-existence, being given the systematic denial that the scientific authorities oppose against this eventual-

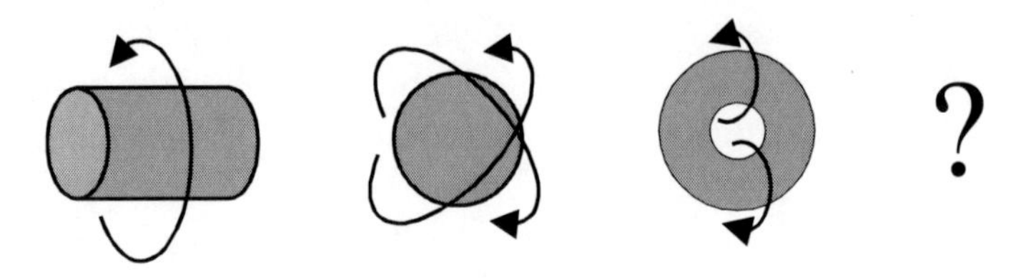

figure 4-11
What are elementary particles really like?

ity. The Synergetic Theory has made us know the creation of particles by a phenomenon of cavitation, qualified of electromagnetic by Vallée, but that we prefer calling "etheric", because it cannot happen elsewhere than in a fluid, and also because the previous adjective is not explicative enough.

The best-known effect of cavitation is, in mechanics, the formation of bubbles when a propeller churns up water to move a boat, when a certain speed is reached. The attack of the water by the paddles, when it is vigorous enough, can become stronger than the inertia of the liquid, which means that this last doesn't remain able to follow the too fast movement, and that there is creation of a momentous vacuum. This local vacuum, in water, is immediately traduced by the formation of bubbles. We must see in the same way the creation of matter in the ether: when, at a certain place of the never and nowhere empty space, an instant and sufficient energy is acting, the ether cavites and then appears locally the equivalent of the bubble in the water, with the possibility of other shapes like this of a smoke ring, of which the divers master perfectly the fabrication, or something else (figure 4-11). Each time any way, there is the creation of an autonomic stable particle, which definitely keeps a certain part, or the totality when it is an elementary one, of the energy that has been necessary for its creation. When we say that the ether cavites, this means that, at the place where this is happening, the fluid splits, gets separated of itself inside a tiny portion of space, where a sufficient while enormous energy density appears, and that, where the ether was a continuous fluid, locally appears...vacuum!

We'll see later that our senses and our habits of thinking lead us to see the things like through glasses which invert the reality: our vacuum is fullness, our matter is vacuum, the black is not nothing but all, mass is everywhere, the matter has no mass, etc, etc... And to finish with, the rational physics, which recommends us to

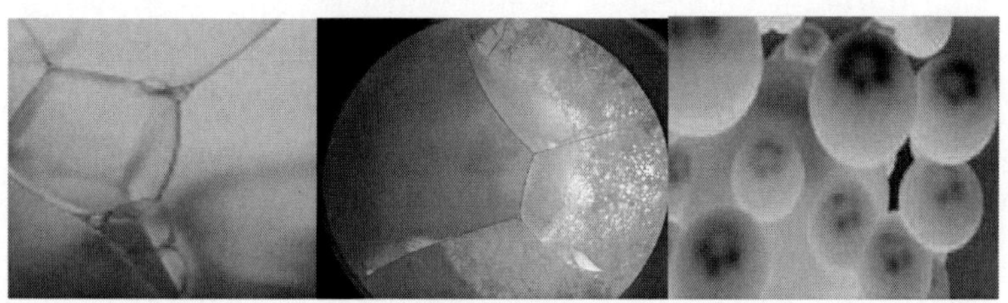

Internet free pictures
figure 4-12

What does the the resonant atom resemble? Conglomerate soap bubbles completely filling space? Simply gathered? All is to be imagined and discovered.

largely use analogy as a reasoning tool, leads us to see the creation of particles in the ether as the quasi-identical reproduction of the creation of bubbles in water, when water is shaken enough to momentously be obliged to abandon its continuity. The image it proposes us of the reality, this reality that we see in total inversion, is indeed more solid than the deceitful appearance on which scientists work since physics exists, but which however constitutes our world, this that we feel and believe true. It's well evident that this way of seeing, which in particular leads us to a total revisitation of our conception of the atom, is not compatible with the too simple images of this which have been the bases and the starting point of the wave mechanics, and then of the quantum mechanics. Considering the atom nucleus as a sun and the peripheral electrons as little planets revolving around it, like butterflies around a buddleia[17], is perhaps a charming and bucolic scene, but also completely unrealistic. It simply proves that mathematics and theoretical physics don't need a representation of the reality as close as possible to the truth. It's enough for them that their model, which any way is indispensable, doesn't bear too obvious incoherences. We must agree that the fact of replacing the representation of the atom closer to the truth doesn't automatically mean a simplification of its manipulation: we are now too much prisoner of so classical images, that are inserted so early in our small brains, and it's possible to get free of them only at the price of a long-duration contesting, which must be upheld only upon the other option of the metaphysics of space, that is the existence and the evidence of the ether.

Figure 4-12 proposes a short display of the samples of ether "bubbles" that our imagination will have to build, to illustrate this new innovative physics, but it is well sure that we are stepping here into an unknown territory, and that any proposed model must be in accordance with the verdict of the laboratory rules, in other words it might be compatible with the acquired knowledge. But we'll ask a bit more to these models: they will have to offer a mechanical explanation, at the meaning of William Thomson and the British school of the 19th century, of the quantum theories. The small cylinder, on the left, in fast rotation on itself and quite well representing the "gyrostat" of Le Bon, is also the barrel-shaped model of the Vallée's electron. We never find these shapes in the bubbles of the aqueous cavitation, which have been for us a representative image and a simple analogical model, but ether is not water, and the fact of being a perfect fluid, that is in particular without internal frictions, probably gives it particular supplementary properties. The advantage of this model is to quite easily lead to an explanation, already developed in the Synergetic Theory, of the electron spin and its moment, but it is not the sole vision of the elementary particle: the dynamic sphere and the smoke ring are also conceivable shapes, and the interrogation point on the right of the three drawings contains all the other hypothesis to come on their geometry. May the tooth-wheels of the Maxwell's

17 Appealing tree for butterflies

ether take place in this family? The decision will belong to physicists who will accept to launch themselves in that way, but it's possible that they would become useless and in desuetude: nobody can guess it in the present state of our knowledge.

Another task of the coming explorers will be to build a resonant model of the atom, edifice still quantified but then in a classical manner, constituted by new-look particles that will be as many resonant domains. This model is waited by an increasing number of specialists who, without confessing it because they dread possible reprisals, want to get rid of a model of which all the resources have been exploited, which doesn't give anything more and which, any way, has never provided something remarkable. It's sure that visually representing this new concept is, at least presently, a difficult exercise, but figure 4-12, where we see something similar to a static collection of soap bubbles, could open the way to something approaching: a complex edifice, resulting from a gathering of basic spherical or ring-shaped forms, each of them being the siege of a circulation of waves, trapped by themselves and resonating in a perfectly defined volume, immutable as long as some shock with another particle or some sudden flush of vibratory energy will not make a change in its dimensions, and thus on its resonance frequency. In that schema, an electron coming from outside doesn’t complete an under-peopled orbit, it inserts in a complex system of which it locally modifies the shape, while becoming a supplementary component. When we'll say that an electron is ejected, for any reason, that's not a particle which leaves its orbit, but an all new particle, anonymous, which has just been created, and of which the type, the resonant frequency and the dimensions are related to the quantity of energy which has been the cause of its ejection. It's not sure at all that this model is presently simpler to manipulate than this of the quantum physics, at the beginning, but the question is to know what we want: either make mathematics, or know the truth. Electron which revolves around a nucleus, that's a fable, it is not not at all what really goes on.

4-7 : Conclusion of chapter 4

Science, carried on by the human intelligence and the genius of physicists and philosophers, patiently, inexorably, tirelessly goes forward, building stone after stone a unique and magnificent construction which sees its dimensions and its sturdiness each year, each day, may be each minute increase, and leads us following a limpid way to the knowledge of the World, supreme award for our methodical and organized efforts. That's what some people wants us to believe at all events, it's this way they present it to us, but it's completely wrong. Science hesitates, stammers, mistakes, goes back, loses its way, and the moments when it really progresses are rare. Most of the time, under deceitful appearances of a swarm of ideas, which indeed are only flights of fancy, pathetic and disordered efforts, it's staying, hidden behind its technological shop window that is often dressed of the name of progress. And each time it's going into a blind way, it needs more than a century to go backward and take then the right direction. But as a matter of fact, it doesn't do what we are really expecting from it: allow man to emancipate, to improve himself, to acquire step by step the moral, intellectual and social qualities of which our grand-fathers and grand-mothers have made the ideal to reach. The consumer society has swept away all this on behalf of pleasure, power, immediate enjoyment and avidity, this last about which Marcel Boll[18] had said, with great reason, that it was the thing the best shared over the world, and not the good sense, as it is commonly said. Of course, we cannot say that there has not been any progress, specially in what concerns social justice, protection of the individual, acknowledgement of his elementary rights, peace in Europe, but conversely we may not affirm that the average intellectual level has been enhanced. But it should have been: the progressive discovery of the structure of the Universe, of the physical laws and their fantastic and impressing logics, the timid but more and more secured domestication of the Nature, the technological progress, all that should have had a positive effect on the human evolution. And nevertheless, reading Aristotle or his mentor Plato doesn't give us the impression that, 2500 years ago, men were less intelli-

18 French physicist and philosopher of years 1900

gent, less capable of reflection than to-day. So what the hell is going on?

We have got into the habit, since the dawn of time, to consider that science is a separate entity, something that man manipulates or surveys as he can, with his means, something about which he confusedly feels that he must know it, like a distant territory that he only guess, but that his atavistic believes prevent him to consider it as a friend. It's good to say that the great traditional manipulators of the gregarious masses do all the necessary for that, and do it well. Stuck between commerce, finance, politics and religion, the human spirit has not yet succeeded in getting free and bringing its intellectual efforts, deliberately and with priority, toward the knowledge of its cosmic environment. Science is not only a simple activity, a muddler of technology, it's also an intellectual quest of what Malebranche[19] called Truth -but thinking then about God-, and which is in fact, for agnostic people, the knowledge without prejudice of the structure of the World. None of us has dodged, one day, as an individual, the fundamental interrogations we have on the Universe and the sense of our life, and science only is able to make us go forward, in a serious and reasonable way, in this personal quest, guiding us by its method and its rigour. Thus it is both natural and logic to behave so that it keeps this character, and when it mistakes, when it doesn't see more promising signs of progress being drawn in front of it, when it appears that it is losing its way and that it is more and more important for it to make a salutary balance sheet, as well as questioning the acquired knowledge, it must not hesitate to take its courage in hands, show humility and acknowledge for its errors. So the aim of this chapter is clear: having a critical look onto the whole of our knowledge, try to find out errors and incoherences, press where it's painful, shake the coconut tree and denounce the passivity of all the Research in front of the stagnation of its discoveries. It is not normal that the repeated snags in the rush to the controlled fusion, to take that example only, shouldn't have led the scientists in charge of that branch to change their tack, after forty years of negative attempts. Besides, when there is a Vallée in the

19 Disciple of Descartes

country, it is unforgivable not to listen to him, and still more to prevent him from speaking.

That's not all, either: all the physics is ill. Ill from its myths, its methods, from the socio-economical ascendancy, from its structure, and more generally by the isolation in which it has closed itself along years. The language it uses, so far from this which is practically used elsewhere, had since a long time isolated it from the rest of the world. Its allegiance to the industrial world and to the politic class, the last at the orders of the former, had cancelled its creativity, indispensable condition for its efficaciousness, as well as the serenity which is necessary to a long-duration work. Instead of what, it's the permanent stress, first at the level of the managements, which restlessly struggle to acquire the budget for the coming year, and also at the level of searchers, who feel being observed and of whom the initiative is reduced to the minimum, by lack of sufficient credit and liberty of action. The future doctors of university, to-day, doesn't defend free thesis, but thesis which are proposed, either on the own capital of the University they are studying in, or by an industrial company which will give them a job later, to realise the product that was the theme of the thesis, according that it will be question of a patent on this product, planned to be developed and sold. But there is also, besides, the fact not necessarily wanted of shutting oneself, in a rigid frame of preconceived ideas and abstruse theories which paralyse brains, maintaining them by force in a dogmatic yoke, of which the former pages have drawn some main lines. Let's add to that the overweighting and paralysing plethora of searchers, working in various disciplines but having no relation with exact sciences, that an organism like the CNRS, which at its beginning was only concerned by physics and chemistry, welcomes to-day to offer a shelter and a salary to supernumerary members of universities, of whom the utility can look hypothetical. But that is another aspect of things, and it's better keeping them apart for the moment, if we don't want to cause some bloody polemics. In order to quickly come back to our beloved physics, the day its supporters will join up in a salutary revolt, which will repel in block the Relativity and its unbearable incoherences, to at last agree on the existence of the ether, giving back the importance that they denied it since now, let's bet that the time of the great discoveries will come back very quickly. It will not be a simple revolt, but a revolution. Waiting for that, we may dream, it's not expensive.

Chapter 5: the Ether (3)

5-1 : Impressions and reality

From renouncement may come complication. Descartes, Maxwell, Le Bon, Tommasinna and Vallée have opened the way toward a System of the World depending on the existence of the ether, but without defining it with precision enough, so that the other physicists were not tempted to abandon their mistrust about the hypothetical fluid, and to follow them in a more systematic study on it. Instead, not knowing how to manage a concept congealed by the lack of obvious clues, and wanting in a way to cut the Gordian knot, Einstein and his theory took along the quasi-totality of the corporation into the relativist phantasmagoria, more flashy and more esoteric: people always enjoys mystery and fuzziness. The contesters were reduced to silence, those who hesitated were carried away by the force of a current still going on, despite progressively losing its power, in comparison with its initial strength. But what an inertia, this of the scientific society! What could we do to make them move and change their way? Whatever it can be, we have lost a century in looking for the truth where it was not. This stubborn obstinacy to clutch at a whole of ideas which has gone bankrupt, because, as a sheer invention, it didn't lead to any major discovery, is dramatic. There are of course some reasons not to be interested any more by the ether, may be by discouragement, and all of them can be resumed in a major one: it cannot be perceived. It escapes all our senses, thus our body tells us it doesn't exist, that vacuum is everywhere out of the matter, and following this process our brain registers and classes. The sixth sense only, this of reflection and logics, this that invents mathematics when it cannot do otherwise, only this one is able to discover the hidden Universe, the authentic one, the real world that our normal constitution makes us see reversely.

If we had to qualify in block our five senses, considering only the whole of their functions, we could say in short that they are the "hunter's senses", those which are indispensable to man to exert his activity of survival in the environment he is in. We could also say that they are, in an atavistic manner, exactly matched to the terrestrial life, and nothing else. If to-day, after an accident following a catastrophe caused or not by him, man was suddenly finding back himself in the same conditions than when he appeared on Earth, and that he should have to survive and only that, he would know how to do because, just as millions years before, his sight, his sense of smell, his sense of hearing, would allow him hunting and fishing, because his hand would give him the necessary skill and sensations to create weapons, to know what is cold and what is hot, because his sense of taste would tell him what he may eat or not eat. The sixth sense only, that we may call either logics or intelligence, allows going farer than the necessary and discovering reflection, human specificity as it is told with so much confidence.

The complementarity of the five senses is absolutely remarkable when we class them by their ranges:

- The sense of taste, first, which acts inside the body. Its use is to appreciate and recognize, in a manner that belongs to it, what we bring to the mouth, and which is not necessarily edible. It brings both pleasure and security.

- The sense of touching, then, which finds its sensitivity at the surface of the body, but remains attached to it. It is located at the limit between the inside and the outside, it's the permanent agent which gives us information, among others, on the temperature and the presence of objects in the obscurity.

- The sense of smell comes after, less developed for men than for the rest of the animal world, but which is the first sense about which we may use the word of range: by it we are informed on something which is remote, we smell at a distance. This last is not so big, it generally is counted in metres in the normal use that we do with, but can be expanded to several kilometres if the source is powerful enough, for example a volcano or a forest fire.

- After the sense of smell, in terms of range, comes the sense of hearing, which is able to detect a weak noise in the vicinity, but also another, stronger, at tens of kilometres, and which adjusts its sensitivity following the ambient noise, so as to constantly be a little more above this last. (3dB indeed: in electronics, we call that an automatic gain control).

- Then at last comes the sight, which theoretically has no limits, since with the help of optical instruments we are capable of seeing at billions years of light. But already, to the naked eye and without assistance, it makes us touch the stars, that we'll never reach otherwise than by it.

So we better understand the expression "hunter's senses", which defines these ones in their whole, because they provide us with all that is necessary to move, find his food and if possible survive at the surface of the Earth. But moreover, what is particularly remarkable is their complementarity, their regular and harmonious arrangement as a function of their range, from the inside of us to the infinity. Among the five senses, the dominating one is obviously this of which the perception has no limits, that is the sight. If we compare the eye to the ear, physic-chemical marvel, or the other captors, it is not the most sophisticated. To-day, the simplest digital camera does better than our retina, but it is nevertheless the one of which nobody would want to be deprived, if we had to make a choice, so primordial it is for us. However, the fineness of the details that it communicates to the brain also makes the weakness of anyone who trusts in it too much: there is too much information in a picture, and we cannot completely digest such a flow of information. A mental selection, either wanted or automatic, is obligatory, and unfortunately what is repelled is sometimes the most important. A lot of parasitic professions have understood that since a long time: cheaters, conjurers, swindlers of all sorts, but also the dream machine of Hollywood and all that is gathered under the smoky vocable of audiovisual, industrial activity which uses in abundance of the incapacity of our cortex to treat quickly enough so many data, which push one another in the optical nerve and make larks of us. To come back to physics, let's apply ourselves to a particular aspect which will give us a supplementary key for the awareness of the existence of the ether.

Figure 5-1 shows two aspects of a same subject, in the occurrence the group photograph of the Solvay congress in 1911, one in positive and the other in negative. Who would be able to recog-

figure 5-1 : the Solvay congress, 1911
positive and negative

nize Marie Curie among the others on the negative? Without cheating, of course, for example by using the fact that she is the only one not to have a beard. However, from a technical point of view only, the transformation which allows to pass from one picture to the other is such that it is difficult to imagine something simpler. Indeed it's enough to invert the level of the grey from one to the other, on the same pixel, that is on the point which has the same coordinates on the two prints.

In the still recent time of the silver-salt photography, the negative was the first step in the picture fabrication, once the snapshot was done. The negative image thus contains, neither more nor less, all the informations of the so-called positive image, at the only difference that inverting the levels of the grey makes the personages indiscernible on one and perfectly clear on the other. This proves well, but physiologists know that since a long time, that it is the brain which sees, and not the eye, and taking this fact into account is fundamental for what is following. This process of the vision, which induces our perception of the external world, and also this of the well-founded of the apparent reality, has been demonstrated by experiments only known by specialists, and of which one of the most amazing was to equip a human guinea pig with special glasses, which inverted the sight up and down. Completely lost at the beginning of the experiment, and unable to find his way without the help of his hands, like a blind, the patient found back a normal vision after a few days. His brain had reprogrammed itself so as to put in conformity its experience and the new optic through which he was then obliged to see the external world. Once the inverting glasses were removed, the same time was necessary to the volunteer, or more exactly his brain, to find back the normal conditions, but he did that with no more efforts than for the first travel. This experiment, and some others in the same vein, makes us ask the question of the foundation of what we call "true". We may suppose that we could, to-day, do again this experiment with other glasses which, instead of inverting up and down, would transform the positive into the negative, with the hope of a similar result.

This approach of the reality, more exactly of what can be hidden behind this too simple word, is a philosophical problem that many thinkers have already endeavoured, but generally without being supported by complementary scientific data. But the problem of the ether and of its nature leads the physicist, convinced of its existence, to totally modify the way he tackles that question, and overturns his habits of thinking. Rational physics, as we are speaking about it, needs to precise, the most and the best possible, what does involve, in the particularity of its method, the difference between what our senses give or suggest to us, and what really happens. We may not, with impunity, declare that we are in a massive and omnipresent ether, and do as if it didn't exist, or say like a number of sceptics that whether it exist or not has no importance. The question of knowing if what we are seeing is the reality is indeed of prime importance. It's even, paradoxically, as much important as it is never posed. Ordinarily nobody indeed, scientific or not, has the least doubt on the fact that it's only the reality that his eyes can see, that when he sees nothing between him and some object, that means that there is effectively nothing, that transparency means vacuum, that this last is everywhere where there is no matter. This certainty of every day and of the theoretical physicists, upholds on a word which, for most of us if not the whole mankind, resumes all, prevents from any reflection and creates the consensus: “evidence”.

Evidence is, for anybody, the expression of a fact, so clearly felt that it is perfectly useless to question on its validity or its reality. We commonly say: that's evident, full stop. This so-called evidence is assorted and accompanied of ready-made formulas, which withstand time and reinforce the gregarious mysticism denounced by Marcel Boll, like for example: “if it was not true, it would be known”, peak of stupidity, or whether that we have made Saint-Thomas to assert, “I only believe in what I see”, formula not really stupid but terribly incautious. This is to forget that evidence is a personal and temporary feeling. Personal because something evident for someone can be not for another, which is frequent, and temporary because what is evident for you at a moment can be not a moment after, which is as frequent.

Thus we must be on our guard, like for the pest or the sleeping water, in front of the evidence, this strong and incoercible feeling, cement of the certitude and source of an incalculable number of conflicts and errors. The rational physicist, in particular, must exclude it from his vocabulary, and get the reflex, when he is tempted to use it, of looking for “the other explanation” of what can be “evident” at a first sight. The statement of the chapter 3 has laid the bases of a new look on the relative position and structure of man and space, specially on the location of the mass, which is all in the ether. And we can see, at this occasion again and like by chance, that the whole of our perceptions gives us an inverted image of the reality, a reality that the rational physics only can make us discover, thanks to our sixth sense: the reasoning.

Matter is imponderable, it doesn't possess an inherent mass. Its apparent mass, either inert or weighing, is this of the ether which occupies its volume, just like the weight of a wet sponge is the weight of the water inside. By the way a short problem, apparently simple, will illustrate the fragility of our certitudes. It's the following: what is the best method to use if you want to know the weight of a dry sponge? Most of the interrogated people will probably answer that it's better weigh it as it is, after having got sure that it is really well dry. Here is precisely the problem: what exactly is a dry sponge? An atmosphere asserted to be dry, let's say at an hygrometry of 40°, which must be close to the Saharan air, contains water any way. Similarly, a sponge supposed dry, and which thus is not, at least completely, has a weight which is partly due to the presence of a certain quantity of water, however small. Hence this consequence which can seem paradoxical at a first sight, but which in fact brings an answer through a rational analysis: to get the exact weight of a dry sponge, weigh it in water! It's not worth insisting on the provocative aspect of the formula, its message is far more important, and opens the way to an analogical representation of the ether which should convince everyone. Technically speaking, first, it is evident that, for a weighing in water with weights normally used to weigh in air, it will be necessary to bring the correction of the Archimedes' pressure, which will find expression either in rectifying the crude weighing by a mathematical formula, or using special weights, having a different volume that the "normal" weights and calibrated for a submarine use. We discover, by the same occasion, that a weighing in air, which wants to be of a special precision, should also be added of an analogue correction, because air is also a weighing fluid which causes an Archimedes' pressure, proportional to the volume occupied, pressure which must be subtracted to the crude weight. We also can see, and it's not the least important, that in the case of the aquatic weighing, the volumetric mass of the water doesn't intervene in first approximation.

This being assumed, the parallel made between the sponge full of water and our body full of ether is direct and obvious enough so that it's not necessary to insist any more: it's this way that we must now understand the reality, as it really is and not as it appears to us. The upper paragraph, and also those of chapter 3 which treat the subject of the mass, have only the aim of destroy-

Figure 5-2 : a tree such as we see it, in (a), and such as it must be in reality, in (b), the hole in the clouds letting see the black mass of the ether, however transparent to our eyes.

ing the pernicious idea, which prevails since Descartes and fully comes from our sensory habits, that the ether is something infinitely light and infinitely rigid, the word "infinitely" being to be taken in an all relative meaning. If we mind enough, we find normal that we couldn't see the ether, since it is everywhere: if we saw it, we'd only see it, that is nothing else. In revenge, and this is absolutely disturbing although true, it is the indispensable propagation medium which, completely occulting its own existence, allows the light waves to transmit to our brain the information on which this last will build the familiar image of "our" reality. But also, instead of being the intangible fluid of which the apparent subtlety led men to give it the name we know, synonymous of lightness, the ether has got a volumetric mass of which we only know, without we could exactly evaluate it, that it is so enormous that it is able to carry away stars and planets. Figure 5-2, which cannot be exact since we are not able to know the "true" reality, except if we imagine it, is simply here to help representing things as they are, in a physical world where all is inverted in comparison to what we feel. In front of a structure of space which escapes to them, the eye and the brain do together what they can, with the elements and the capacities they have, to provide us the essential indications we need for our existence. With the help of the other senses, they allow a rational use of the space, while ignoring its nature, and they make us adapted to the terrestrial life without giving us the keys to understand what it really is, and what it is made of. And it is obligatory to reach a certain level of reflection, both physic and metaphysic, to at last lift the veil of the great mystery.

The question of the sensory perception is crucial since the moment we want to take into account, and understand, the fundamental role -we have a tendency to forget it-, of the observation in physics. Observation is at the base of any knowledge, but more particularly in some sciences where it constitutes the starting point of the future hypothesis. A badly driven observation leads to wrong hypothesis and to prejudicial losses of time. But we cannot correctly observe if we don't know, with a sufficient precision, the tools we use, and our senses not only take part of these, but are their quintessence. The microscope and the astronomical telescope are only additive equipments which simply extend the range of our eye, and for the other senses we also have equipments which improve them, often considerably, but which don't replace them. Thus is it necessary to well master them and, to do that, it's worth knowing enough their characteristics, their qualities and principally their weaknesses. But before all, we must never forget that a given sense is made of the association of a captor (the eye, the ear, etc.…) with a cervical zone, where is built the sensation which gives it its name (the sight, the sense of hearing, etc...). There is nothing, in this principle, which could be able to warrant the absolute well-founded of what we are perceiving: the two composing parts are fallible. If the imperfections of the first are known, those of the second come under a personal appreciation which is rarely rational. The sight and the sense of hearing, to only cite the two major senses, have actions about which we have *a priori* no question to ask: we use them, they are right here, attached to us, being part of us, and those of us who believe in the existence of a creator have got the conviction that the work of the Great Engineer can only be nothing but perfect, and consequently that the capacities we have been given when we were born, all of us, cannot be suspected of being insufficient, mediocre or of a bad conception. But even if we are not a believer, nothing can awake in our mind the least suspicion on the authenticity of our perceptions, as well as the least doubt on the fact that they transmit us the reality of the things, so dazzling with evidence they are. And nevertheless...

Is black a colour? For the physicists, it's rather the absence of colours. But if you ask for a pot of black coating at the drug store, you'll not hear that it doesn't exist, and similarly if you have got a black car, you'll not say that it has no colour, OK? White is also a colour, if we go this way, but however the physicists will tell you that it is the composition of all the colours or, when it suits them, of three fundamental or complementary colours. Black and white thus have not much to see with each other, but we have got the habit of both gathering and opposing them, as two limits of a suite of colours, spread along a spectrum, moreover without knowing where exactly locating them: one close to the other, very far at the contrary, above, under? Whatever we think about these questions, which are at the limit between science and philosophy, it happens a curious phenomenon for what more specially concerns the sight and the perception of the black. When we ask someone, child or adult, what he sees when he closes his eyes, the answer is immediate and final: "I don't see anything". Making him conscious that he's seeing something can thus appear as an odd, fancy and meaningless enterprise. Here once again, the notion of evidence appears with its habitual strength, and cuts short any attempt of reflection: when we close our eyes or when it's night, we see nothing, full stop. To show what fundamental error is hiding in this affirmation, even so indisputable a priori, it seems necessary to use a more understandable intermediary, a short story invented for the sake of the problem. So let's suppose that, in a cold morning of the autumn, somewhere in the Sologne or in the Cevennes[1], or elsewhere, two ramblers find themselves trapped in the fog, a good fog that you could cut with a knife. One asks the other: "do you see something?". And the other answers: "no, I don't". In fact the two individuals don't see any object, no precise shape having a contour, or a colour, or anything else which could emerge from the unvarying background. But this last has an existence, incontestable and prehensile: it's fog, and fog is matter. And if the first questioner makes his neighbour notice that he actually sees something, in the occurrence the fog, the other will probably agree and, this way, will do a pretty short step toward the discovery of the critical mind and the hidden reality. In return, if the same scene is happening in a cellar without light, or outside during a deep night, without moonlight, the black that we then see will be qualified of "nothing", because the reflex, which makes us say that, has been acquired since the prime youth, with the agreement and the encouragement of the adults, victims of the same ancestral habits, that they transmit with natural to their descendants, on behalf of the evidence. And if by chance this black was not "nothing", but "something"?

In the last few decades, let's say since the beginning of the 21th century, the astrophysicists, which share their life between the Earth and the far galaxies, and who put the evolution and the movement of these last into equations, have noted that the coherence of their results needed the hypothesis of a new parameter, that they called the "hidden black mass". Only the name they gave it shows the disarray of these fragile brains, in front of a phenomenon which escapes them, and on which they however cannot have a doubt: this mass, which is indispensable to them, is black, and more it's hiding! But that's not all. The strongest is that, still following their calculations, this hidden black mass would represent by itself a so big proportion of the total mass of the Universe that it is quite impossible to believe it: 80% for some people, 90% for others, see 98% for still others! And all these high-level specialists, which represent the cream of the brains, are searching for this fantastic accumulation of mass which, if we trust in what is said about the black holes, should attract in its

1 French provinces, specially foggy in autumn.

gravitational precipice all that exists in the Universe. But, fortunately or not, all is quiet, we don't notice anywhere, apart from a hypothetical expansion, perfectly contestable and any way quite difficult to put into evidence, a point of cosmic aspiration where should be ingested, first the environing stars, then the rest of the Universe, by continuity. Where the hell can be such a mass, if it is not confined? Because this enormous mass exists, and the fact that it had appeared, as a necessity in the calculi of the theoretical physics, is a subject of meditation that nobody may refuse. Here is indeed something miraculous and marvellous: two radically different conceptions in physics lead, by opposite ways, to the same conclusion that the Universe is provided with an enormous mass, or even is constituted by this mass. But if the theoretical physics doesn't succeed in getting rid of a new parameter which suddenly appears, like a dog in a set of skittles, springing out from the equations, and which doesn't find a place in its edifice, the rational physics, at the contrary, greets the acknowledgement, by the theoreticians, of a truth it already knows since a long time. Where can just be located a so important mass if it is either not localized, or localized in the far nebula, at a distance beyond our understanding, at more than ten billions years of light? “damn and blast it, that's sure!”, would say Monsieur de Lapalice[2], if it is not localised, that means that it is distributed everywhere! And subsequently the hidden black mass doesn't represent 80, 90 or 98% of the total mass of the Universe, but exactly 100%: **it is** the Universe!

But then, if the ether is everywhere, why do we have no consciousness of it? If it was true, it would be known, say the theoretical physicists and the imbeciles, using one of their favourite formula of omniscient people. But precisely, it's because the ether is everywhere that it is not detectable. We only see it when we close our eyes, when these last are not at work with light radiations, but then we declare that the black we are seeing is “nothing”. We cannot seize it by hand because we are, as beings of matter, essentially composed of vacuum, and because the universal fluid lets it being crossed by our moving body, our moving hand, just like a wide-meshed net does when pushed in water. It has no smell, it

2 French celebrity, specialist of evidences.

doesn't transmit any sound, because the only vibration it propagates are the EM waves, and among these, only the thermal waves may excite our skin, but this last doesn’t think. There is another logics, still more elementary, which stops us seeing the ether: being given that it is everywhere, if we saw it, we couldn't see something else, and we 'd be led to the same condition than blinds, who see the ether permanently but don't know it.

We all bear in ourselves a patrimony of innate knowledges, coming to a great extent from the experience of our ancestors, and of which certain are not real knowledges, but more habits of thinking which have been wrongly reinforced, all along the existence of the species. As long as the questions to deal with were the terrestrial survival and the following day uncertainty, the human race has not suffered much prejudice from that. But when men began to have free time, when they discovered their ability for reasoning and logics, when they invented science, the wrong ideas were already so solidly, so strongly implanted in their deep memory, that now we never wonder about them, as natural and out of suspicion they seem. In the visual domain, it's the case of the black, but it's also the case of the white, which like the black is both a colour and an absence of colour, but that the physicists have defined as he whole of the colours, as demonstrated by the prism properties, which effectively realise the operation in the two senses, either addition or partition. We don't go further, here, than the systematic errors of the vision, but all that concerns mass and inertia, for example, which involve in a more confuse manner the whole of the senses, is also the object of regrettable but incoercible habits, dating back millenniums ago, and make so that we often understand things inversely to what they actually are.

We have the impression that the Nature protects itself from the too curious people we are, by presenting us the phenomena through a distorting glass, and that it is impossible to the human species to lift the veil of the great mystery, before it had reached a certain intellectual level. Obviously it's not the case yet, but to get it, it will be necessary for us to master that sort of extraordinary filter which inverts all the images, visual, audible, tactile, that our brain gives us of the external world. To get back to the black and the white, and to animate the fragility of the perception we have of them, let's take the example of a cloud, a big nimbus well fat, ready to burst, and seen from the ground as a black mass stopping the quasi-totality of the sun light: this same cloud would appear perfectly white to the passenger of a plane which flies over. This shows us that colour, and there are many other ways to be conscious of that, is not a physical characteristic, but the simple description of an aspect. Colour has no physical value, which doesn't prevent it, minute space in the electromagnetic spectrum, from being for us of a vital importance.

Whatever it may be, the hidden black mass, will we be pleased or not, is here, around us, in us, everywhere, and its name is the ether. The astrophysicists, who only see it through the evolutions of the far nebula, are like this mushroom searcher, sad and alone in the autumnal forest, his basket desperately empty and his face crestfallen, who doesn't see that the two trunks between which he is are in fact the feet of two gigantic four meter high boletus. To find something you are looking for, in physics, it's sure that mathematics can help, but nothing can replace a directional idea. And that population has none.

5-2 : Matter and vacuum

In chapter 3 can be found the introduction of a new concept of the mass, seen as the quantity of the ether that imbibes a certain volume of matter, and gives it this way the classical characteristics known under the name of weight and inertia. This vision of the things would implicate that two equal volumes of matter, though having different measured volumetric masses, would contain the same quantity of ether. Thus they should have the same mass. This is an affirmation it shouldn't be cautious to assert, so violently denied it is by the facts, but which besides poses a certain number of basic problems on a notion which is one of the foundation of physics, and which has not so far been treated, neither seriously nor correctly. All the doubts we may have on this subject have been expressed by a good number of physicists who, in despair over their efforts to correctly define the mass, ended up seeing in it only a handy coefficient, expressing the relation of proportionality between a force and the acceleration it causes.

The ether, by the fact of carrying away the planets, must have a considerable volumetric mass, of which we'll try further to estimate its order of magnitude. If we have supposed to it, for such a long time, an extreme lightness and rigidity -apart from Malebranche[3], who had understood it might be "heavy"-, it's because all, in our way of perceiving the world, suggests us with so much strength that space is empty, that we might fight against our habits and our reflexes to dare imagining something else. Besides, the acquired knowledge of science is so impressing that, even if we have a tendency, from time to time, to send some critics, even if the great theories regularly give rise to polemics, it is incontestable that an impression of solidity and of constant progression ra-

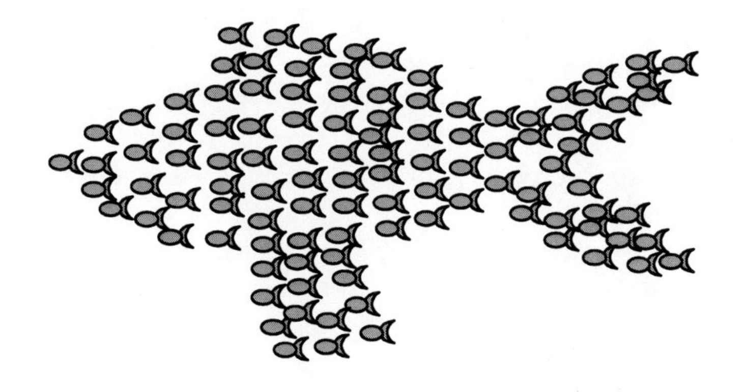

figure 5-3 : the « mass-fish ».

diates from that so particular activity, which differentiates men from animals.

All the pages before, nevertheless, show that a particular attention, and a minimum of curiosity, lead to think that there is much to say on what we are taught under the name of "studies", and that contesting the vision of the World, given us by the science, is very largely founded, so much this last sometimes shocks our logics. Of course, we can decide to be confident, to tell oneself that

3 Disciple of Descartes

the scientific community cannot mistake in its whole, nonetheless the big discoveries were, all of them, born from the dispute, at one moment, of an individual against the dogmatic ideas. Imagining an ether, omnipresent, heavy, massive and black challenges our understanding. Nevertheless this idea not only gets the job done, but it appears to be the answer to the incoherence of the 21th century physics. Its logics is not common, it stands up against this which usually is ours, but it finally makes a more solid whole than the torn coat of the modern physics, which stuffs in its scatter, its too large complexity and its contradictions. But it's also obvious that the intellectual effort, necessary to accept the idea that the real world is practically the contrary of what we see, needs time and a considerable willingness. Losing his habits is something that man has much difficulty to accept, it's the reason for which it is necessary to multiply images, analogies, comparisons and all the possible techniques of persuasion to make this iconoclast theory penetrate minds.

We first must insist on the basic questions bound to the matter, of which a new reading has been initiated in chapter 3, and of which the true nature must absolutely be well assimilated, so that people could be able to digest the replacement of the colourless vacuum by the ether, both black and nevertheless transparent.

Some helicopter pilots, evolving over a particularly limpid sea, were one day surprised to see, in the water, big dark masses, vaguely spherical but undulating and resembling nothing known. The beautiful film on the oceans, by the actor-producer Jacques Perrin, shows an example of this phenomenon, seen in the water under the name of "ball of chinchards[4]" -see Internet-, but it took, to the first aerial spectators of this amazing show, a certain time and some discussions with other witnesses, to realize that what they had seen were only shoals of fish, so dense and so big that, at a certain distance, they appeared as a strange entity, both massive, unitary and opaque, made of a heavy matter. A diver swimming in the vicinity would have at once seen what this matter was really, and at the same time he would immediately have had the visual and physical certitude that the actual volume of the fishes was negligible in comparison with this of the water containing

---

4 Fishes which move in group, like sardines.

them. What he would have remembered of the living show would have been for him something very casual, despite beautiful, with no relation, *a priori*, with the description given by the pilot in the helicopter. It's however question of the same object, but seen at another distance, through another way, and in the same way the image we have of the matter, whenever living or inert, is also subject to the conditions of the observation: when this last is done at a long distance, referring to the details of the thing observed, matter looks continuous. When the observation distance becomes of the same order of magnitude than the unit element of its construction, discontinuities appear, and if we get still closer we discover that there is vacuum between the "bricks". Let's now suppose that

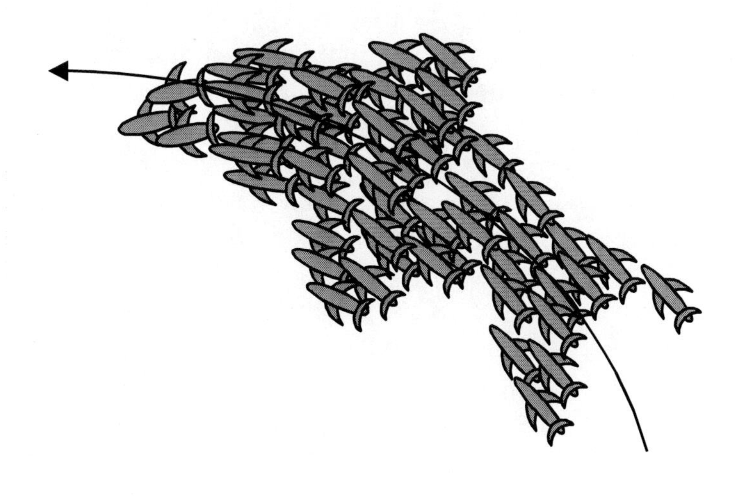

figure 5-4 : Analysis of a bend in group.
Deformation of the combined structure of the mass-fish.

the shoal of fish has itself the shape of a fish, that we'll call "mass-fish" (figure 5-3), and that this enormous and multiple fish moves straight forward. Seen from far and high, no doubt he would have been reported as a unique being and considered as the fabulous discovery of an unknown specimen, but it doesn't matter. The important is to note that it has a shape, even changing, and that we have the right to globally consider it as a real and independent entity. Doing that we are, in a way, giving an autonomic life to a shoal of fish, because we have given it an autonomous behaviour. Let's now suppose that we could command each individual fish to brusquely turn left: what would happen and what would be the response time? The answer is simple: the mass-fish, without changing its shape, would pass from an axial translation to a lateral translation, the time for each elementary fish to do a 90° turn, but without bending its trajectory like would do a normal fish, because this wouldn't be possible. If the mass-fish collectively wants to behave like an element-fish, in other words not transgressing the inertial laws, it might do like this last: take support on the water, more on one side than on the other, and use its softness, its extensibility and its compressibility to slightly change its shape, but without losing it the time it's turning, and find back its normal shape once it has chosen its new cape (figure 5-4).

We'll unfortunately see that this example, imagined to reinforce an idea, this of a porous matter which only takes its mass from the fluid it contains, poses in fact more questions than it gives answers. This means that all the problems linked to the mass and to inertia, its other aspect, have been until now, admittedly, more metaphysic reflections that scientific studies, in the absence of a good starting hypothesis. If we opt for this of a massive ether, and if we assign it all that was previously attributed to the matter as something belonging to it, all the phenomena related to the mass become of a puzzling simplicity, revealing themselves as being the manifestations of a permanent interaction between the ether and the matter, completely analogue to this which exists between the water and the mass-fish. This mass-fish is fascinating, as it gives us the key of the explanation of inertia, while besides posing other problems that the theoretical physics, working in the vacuum as it does, cannot treat. Nothing prevents the individual fishes to break with the group at any moment, and as a matter of fact it's what they do when they are attacked by a predator, in the case of a real shoal. But apart from this exceptional circumstance, they don't, and apparently follow a leader, like racing cyclists do when they are grouped in a bunch. There is here a general phenomenon which concerns all the groups of individuals and which is the rational base of any study on the crowds psychology -read Gustave Le Bon-. Therefore, it opens the way to a lot of investigations, full of promises and probably exciting, but which don't take part in the subject treated right here. So let's come back to our mass-fish and its physical properties.

Since the moment we consider the mass-fish like a traditional fish, simply bigger then the others, by doing so we assume that there exists, between its constituting elements, a binding force which is the cause of the dynamic permanence of its shape, let's say its very existence as a living being, unique and identified. For the ordinary matter, living or not, this force is the MBL pressure of the dynamic ether, which pushes the molecules one against the other, and that the theoretical physics sees as the action of intermolecular attraction forces. For the mass-fish, it's something else. To precise it, let's suppose that it is constituted by a shoal of some tens of thousands sardines, that it has got the shape of a cachalot, and that it is swimming besides a true cachalot of the same dimensions. When the true one moves and makes a bend, to make it able to do that, a part of its body contracts, this which is the closest to the axis of the bend, and the opposite part simultaneously gets longer, so that the whole takes the right shape to follow the curved trajectory. The mass-fish must do the same, thus the sardines which are nearer the gyration centre get closer to the neighbours, while the others do the contrary, and the whole changes its shape just like the true cachalot does, when we see it moving at some cables from there. There is however an enormous difference between the two creatures: one is made of a matter which seems to be continuous, the other of distinct elements, having each an autonomous existence and merging without a physical phenomenon intervenes. Nevertheless, to speak physically, it's difficult to imagine that the cohesion of an organized and identified body shouldn't make intervene external binding forces which explain its mechanical properties: extendibility, compressibility, hardness, elasticity, etc... There are thus necessarily forces which maintain the individual fishes at a constant distance one from the other, in order to build the mass-fish, but which forces are they?

This is a problem susceptible to embarrass the theoretical physics. Its tactic, in the presence of a new effect, is to create a new force, to quickly give it a name, a matriculation, a serial number, and to stick it in the general catalogue, so as to be able, in case of need, to take it out conveniently.

But here, how could we define that force which prevents the mass-fish to dissolve in the ocean, in a more and more vaporous cloud of libertarian fishes? It's neither an attraction force, nor a repulsive one, since the individuals move in company, but it's also both one and the other at the same time, actions which balance at each moment and determine, stating the average distance between participants, the volumetric mass of the whole. Is the origin of this force only in the brain of the fishes, like we imagine it's the same case in a bunch of bike-racers, or is it here somewhat the link between the material and the spiritual? And then, at all events, how to name this thing, this power, this action, which maintains living objects at a given distance one to the other and gives their whole a certain configuration, otherwise than a force? There is here an extraordinary opening of the physics toward a domain only exploited, so far, by the charlatans of the "psychic sciences", but where the rare attempts of serious studies have been crushed down by the Establishment -still it again- which has excommunicated and struck with ostracism the irresponsible persons. Among them Yves Rocard, renamed professor at the prestigious ENS[5] but unlucky author of "le signal du sourcier"[6], who had dared adventuring in unknown and forbidden territories. But we are here putting a foot, just touching a new possible study able to suit with the rational physics, in a country made dangerous by the mystic pollution, which keeps it among the deceptions and the most confused activities of the para-science. So let's go back, after this brief glance through a strange half-open window, to the question of the mass, equally ambiguous but more academic.

What is the mass of the mass-fish? Is it the sum of the individual masses or the mass of water contained in the volume occupied by the shoal? Is it bigger, or less, than this of the neighbouring cachalot which has the same volume? Does it depend on the

5 Ecole Normale Supérieure, highest school in the French tuition branch.
6 The signal of the water diviner.

number of individuals? How is its inertia defined? The most impressing, in these sometimes gigantic shoals of fish, are the flashy accelerations they suddenly take, when for example a predator is approaching, and which don't at all correspond to dimensions which are instinctively associated to an enormous mass. These masses, composed of tens of thousands individuals, move from a place to another with a vivacity which shocks experience, because we expect an inertia, a response time, proportional to the volume, when at the contrary they react with the speed of one element only. A school of sardine or similar, weighing several tons, can at once rush like a volute of smoke carried away by a burst. It's that way we have to see the movement of the matter in the ether. All those questions, and all the others we can imagine in the same order of ideas, have no evident and peremptory answers, but show the complexity of a notion which is nevertheless one of the original bases of physics, and that however nobody is able to define by a simple word, let's say by a short, well clear and unequivocal sentence. "*What is well designed is clearly stated, and words come easily to say it*"[7], said Boileau in "l'art poétique". That's a citation we can hurl like a javelin to any member of the theoretical physics, of which the language causes such a scattering, among the possible listeners, that we are well obliged to speak of an insane elitism, to define in one term only a physics which, like all the dictatorships, desperately fastens itself to the power by the coercion. There is however few to do, a small concession to the logics, to give it again splendour and legitimacy: to agree on the existence of the ether, and specially give it its indispensable attributes, mass and energy, while clearly précising that it is the origin of all the mass and all the energy of the World.

It's also true that we may dodge the problem and consider the mass only as a useful mathematical coefficient which relates the force to the acceleration, as long as we have not found an available definition, and this is right now what has been officially done. The big problem of the theoretical physics, which goes this way, is that all these postponed notions begin to be overstocked in an infernal backlog, a sort of dustbin where you can find, higgledy-piggledy, a plenty of new formulas and concepts which

7 Free translation of a French well-known poet.

hide the true definitions, these which should be the bases of the physics tuition, and of which the study is unceasingly deferred until later, by lack of ideas but also of willingness: mass, vacuum, ether, matter, atom, quantification, neutrino, speed of the light, etc., etc... No student, to-day, endeavouring superior studies, has got the advantage of having a preceding logical formation, solidly supported by well plain principles, apart from a dominating mathematical culture which makes him believe that it is only by that way that we may discover the mechanisms of the Universe. We are however very close to the truth, we touch it. It only should have been enough that, when we were children, during our primary education, at the so-called low school, teachers had began to warn us against appearances, made us capable of correctly using our senses and teach us how to master our perceptions. This is a physic-physiologic education which is totally missing right now, but which is absolutely indispensable to be installed if we want to develop the rational physics. In the same way it is necessary, previously, that its existence and its well-founded might be accepted, and that the little people of the science finds it interesting, which is quite a business.

What would it take to convince? The examples and the arguments proposed all along the pages before should normally incite some ones to revisit their course in physics, to shake their torpor, to react against the notions to which they don't adhere, while teaching them reluctantly, but this is without taking into account the fantastic inertia of the system. Let's agree that this word "system" is quite practical, by its confused and undecided side, but it is precisely used here because it points to a whole that it is very difficult to split. It's question, in fact, of the whole education and also the practice of the physics in its totality, including all its aspects, its social role, its interest, its appeal and all that makes in us what we think about it, the idea we have of it, students, professors and general public. Physics is stagnating because it refuses to put itself in question, because it refuses to imagine, one second only, that it could be mistaking, and moreover because it has constantly neglected and ignored the major problem of the physics, this of the ether. The re-education, at the strongest meaning of this term, will thus begin in this way: what could we do so that, one day, people would be taught since the earliest age that the world appears to us just like the windows of the so-named system by Microsoft, that a picture generally hides another, which just needs a skilled click to make it appearing, and that we are constitutively massless, the mass we believe to have being this of the ether which is inside us?

When we can discuss about the problem of the ether with some University members, they generally listen to you and accept the idea, because they are not idiot, that it must exist. But we are surprised to often hear, from them, this amazing question: "what interest?". Even when you make them being conscious that the ether is full of energy, and that consequently it is a source that we possibly could exploit, this idea seems to them so fancy, and in a word so incredible that they never consider it at all. This very frequent reaction, even by those who accept to do the effort of listening to a speech different of these they have already heard so far, is very revealing and symptomatic of the cultural isolation of the teaching people. All of them are teachers, we must not forget that, and above all we must not laugh at it, because they are educated, obliging, helpful and gentle people, who will be in charge of the education of our children and will put in their brain, by a phenomenon of continuity, all the ineptitudes that they were themselves obliged to ingest for their studies, without any possibility of criticism or protestation. It's this way that we may explain the intellectual inertia of the society, and what superhuman efforts must be engaged in order to try changing the attitudes and the habits, when we endeavour the promotion of a new idea.

If there is no invention, or no prototype, or something obvious to make the novelty in question incontestable, and prove its interest, it's not worth trying to publicise it. An inventor must absolutely establish the well-founded of his invention by physically realising one of its possible applications, if he wants to be paid attention to. The only exception to this rule is found in the theoretical physics, where high level searchers have acquired the right to say no matter what, as nobody understands them and so cannot contest them in a well-argued speech. For this little world of initiates, who know so many of these so much extraordinary things that we, ordinary people, cannot even conceive, the fact of existing or not for the ether has absolutely no importance. Let it be energetic, moreover, so that it possibly represents the final solution to our energetic problems, doesn't worry them at all, doesn't rise in them any particular interest, any willingness or run-up toward a new track where could flourish their bright intelligence: they are happy in their soft bubble full of mathematics and don't want to get out, may be afraid of catching a cold, unless they dread getting diseases at the contact of ordinary people. That ambiance is really unbearable.

5-3: The new deal

It's time, now, to go beyond criticisms, though it's always something pleasant to do, and definitely settle the bases of a new way of thinking, that physics will not be authorized to ignore, whatever happens.

Before all, it's indispensable to insist on the probable reasons for which the question of the ether has always been so tiresome that it has discouraged generations of physicists, to finally come to the drastic strong line taken by Albert Einstein, and the advent of the disastrous Relativity. It's absolutely necessary, if we want to restart on the right foot, to analyse the process and try to understand why this problem has been so badly approached, and particularly how it must be tackled so as to be more efficaciously treated. All the evil comes, as it has been said in the former paragraph, from our conditioning to our immediate perception of the World, coming from our senses. These ones, let's repeat it once more, are perfectly matched to the terrestrial life, but nothing else. They allow us to evolute in our small daily universe without too many problems. The fact of seeing, hearing, smelling and touching gives us a sufficient information on our close vicinity to give us the feeling that we don't need more. It's available for the casual life, it's not enough since the moment we try to understand how the Universe works, that is when we do physics.

Our eyes and our gaze enforce us, in a strong and instinctive way, the notion of vacuum. Our number one sense is so powerful, so dominating, that it leads our brain to this evidence that we only can accept, without any condition: between an object which is at a certain distance and ourselves, there's nothing but the air that we are breathing. The transparency of space irresistibly builds in us the notion of vacuum, as strongly as our weight and our inertia make us believe that this mass is something which belongs to us. Only the reflection of the physicist, as well as a well led reasoning logics, can cast doubts on this apparent reality, and allow him to look for the scientific truth where it really is, hidden behind the so clear but indeed confusing aspects of the things.

This would still require that this physicist could have had the possibility of escaping the common fate, and beneficed, at a moment of his life, of a liberty of reflection that he hasn't got, probably, if he has followed the normal and classical course. In our modern society indeed, we may not both be formed at a certain level, high enough to have at least the sensation of well knowing our subject, and avoid the dogmas of the tuition. This last has the softness and the handiness of a road roller. It is designed so as to enforce into the trusting brains all that the humanity considers, wrongly or rightly, as an indisputable acquired knowledge, on the validity of which we couldn't cast doubts. All the more so that students, to whom no possibility of critical reflection is allowed, show a thirst of learning which sweeps off all the doubts they could have on the well-founded of the theories they are taught. They are well helped in that by overloaded schedules and programs, yet hardly sufficient to make them ingest a macro-physics dehumanised and drowned in the mathematics. To change all that, it would probably take almost nothing, may be a simple sudden awareness, a questioning about our real intellectual possibilities, a pause in the Research, all that together either, and may be something more... Is it, all simply, that it is not foreseeable because owing to the human nature? Nevertheless, the history of the sciences teaches us that they have seen, in their progression, long periods of stagnation, brusquely shaken by the earthquake of an incredible theory or invention. But the big problem of our limited brains is to be unable to see the difference between genius brainwaves, like Maxwell's equations or the discovery of EM waves, and bullshits like the Relativity or the Big Bang. So, what could we do, in order that physicists might be interested again by the ether, and why aren't they? What exactly is that corporation which seems to be so excited on what happens -or more exactly happened- at 15 billions years of light, and which has been squelching for forty years in the quest of the fusion energy? Which is happy only when it painfully pounds a new equation, out of mind for the average people? Which tells us fables on the structure and the history of the Universe, like if it was in front of low-school pupils? What can we do to convince a fistful of these peoples to leave their crazy ideas for a while, and have a glance on these of this book, for example, either to take them into consideration, or at the contrary to explain to us why they are wrong, if by chance they are?

The ideas and the concepts previously asserted on the perception of the black are the starting point of any attempt to get conscious of the existence of the ether. From the moment when, shutting your eyes, you know that what you are seeing is not "nothing", but the hidden black mass that astronomers locate away, when instead it is yet under our nose, you may say that half of the way is done. It's a very long process, it's an idea which goes its way so slowly, to take place in a standard head, that we have to be patient and let it the necessary time to be installed. Afterwards, it's a question of willingness. But it's also a question of interest, in the intellectual meaning of the word: most of the searchers-teachers are to-day content with the physics as it is, as they learned it and as they teach it. The edifice of the science is apparently so solid, it needs so many years of studies to be assimilated, that nobody has the time, or simply the idea, to let germinate in his mind the least dispute, the least doubt. When a too abstract notion generates trouble into us, we generally try to understand, by habit or by obligation, why we feel so uneasy in front of it. May be it's by lack of confidence in our capacities, instead of wondering if, by chance, it couldn't be too twisted to be honest. "*Teachers teach*", they say over the Channel, "*only students can learn*". Whatever it is, it's that way that things are going on, and there are only two possibilities to make them change. First possibility, a creative inventor, using the ways that a well understood notion of the ether leads to, offers to the mankind and its industry a disrupting discovery, like for example an electric generator which seems to work without any apparent source of energy, or an anti-gravitational device, let's say a science-fiction object in a way. Second possibility, the scientific progress knows a so long period of stagnation that the physicists feel obliged to leave their computers aside, in desperation, and then try to make going back to work what they have kept of their grey matter. The first hypothesis will be actually realised, one day or another, the second is probably going on and will come helping the first. Waiting for that, we must wait with patience. Only the Bogdannof brothers[8] and their field kitchen find that physics is exciting to-day.

We are now coming to the clearest justification of the distinction that has been done, so far, between theoretical physics and rational physics, and of the necessary existence of this last. What could have seamed for sceptics, at the beginning, a play of writing or a hair-splitting is now revealed in its depth and its pertinence. We cannot progress in physics if we don't simultaneously or alternately use these two approaches, similar, by their complementarity, to the two cerebral hemispheres. It is strongly recommended to use them in couple, and the present evil, in sciences, precisely comes from that the theoretical physics has completely stuffed its sister, which officially doesn't exist at all. Another evidence springs in the same time, reinforcing the former one: if we concede to the theoretical physics that it is necessary to the tuition of this major science, as its mathematical language is the only one bringing with it the indispensable rigour for this exercise, the rational physics is still more indispensable, as it is the one which makes it progress, thanks to the inventiveness given by the innate and the intuition, which leads to the new ideas and the true discoveries. Thus each of the two physics has got its particularities, is totally different from the other at the point of view of the reasoning and the logics, while being complementary and unable to exist one without the other. Used together, they improve the efficaciousness of the searcher, who cannot be complete if he forgets half of his patrimony on the way. And that is precisely what is done when we let to the mathematicians the mastery and the control of the education, while forgetting the sheer reflection. It's absolutely necessary to use both of the two physics, and never accept to be intoxicated by the flaming side of the mathematical lo-

8 French TV presenters of a scientific show

gics, of which the implacable rigour can lead to madness, if we make of it our lover or our credo. Cantor died mad.

The essential bases of the rational physics are, on one hand the existence of the ether, on the other hand the principle of its permanent interaction with matter. All the acquired knowledge of the theoretical physics must now be reinterpreted in accordance with those original verities, and each phenomenon must be examined as being going on in the presence of the ether. This last can play a more or less important role in its progress, from almost nothing to almost all, but it is always everywhere, unmissable, quiet, immutable, indispensable. Taking it into account in the observation and the description of the facts leads, almost always, to a revelation. Let's take as an example one of the phenomena which regularly plunges us in dread and amazement each time they happen: tornadoes.

This impressing event, well known by the inhabitants of south USA, is regularly the subject of reports, which are always the same, apart few exceptions, but with images we are never tired to watch. We see a central upward whirlwind surrounded at its base, making the progressive link between the intake column and the ground, by what the technicians of the weather forecast call "feeding zone" by, where the air whirls more and more fast as it's going nearer to the centre. It's thus question of a sort of vortex, different from the astronomical vortex, described in chapter 3, because it happens, contrarily to the previous, in a very compressible fluid, hence giving it more approximative shapes. In return, it quite resembles a hydraulic Rankine's vortex, which would have been simply vertically inverted. It's in the feeding zone, so called because it "feeds" the central column, that we can see the transportation of an incredible number of objects, pulled up by the extraordinary power of the meteor. We can see, floating like pieces of cork on the water, all that the whirling mass has found on its way, without any distinction. As far as it is question of iron sheets, boards, branches of trees, tools, we are not so surprised. But when we see motorbikes, cars, trucks, may be tractors sliding in the air like balloons, when we see these heavy objects moving like sponges carried away by a submarine flow, like if they suddenly had lost their mass, there's something not normal, that overtakes understanding, which doesn't fit well to what we have been taught. Are we well conscious of the size and the power of the suction device that we would have to design to make a tractor leave the ground?

The wind which blows in the feeding zone, where are floating the objects pulled up from the ground, can reach a speed of 500 km/s. Let's take a reasonable and minimal value of 150 km/s, to settle a practical basis. An average tractor weighs about 3 tons, and we'll say that its wind surface area is something like 3 squared metres. If we suppose that all the kinetic energy of the air is converted into a pressure, this last should be of the order of one ton per squared metre to make the tractor leave the ground, and furthermore should have a vertical upward direction, which is in contradiction with the observation. Besides, the central chimney being inhaling upward, the slope of the air flow in the feeding zone, far of the centre, should be downward. But if the heavy objects are pulled up from the ground and transported to a certain altitude, we must conclude that, at least at a certain distance from the centre, this slope is at the contrary upward. Here too, it's paradoxical, but we cannot deny the facts. Let's go on, any way. Let's imagine we are inside this lifting zone, just to do a short calculation, only to precise the orders of magnitude. For that, we'll suppose that the levitation force exerted on the tractor is only due to a vertical upward wind, blowing with a speed of 150 m/s (horizontal component of the real speed) on a rigid surface of 3 m^2 and with a leading angle of 10°. The kinetic energy of the fluid, which is sensed constituting the force which lifts the tractor, generates a pressure $p = \frac{\rho V^2}{2}$, where ρ is the volumetric mass of the air, that is 1,3 kg/m^3, and V its speed. In the present case we thus have $p = 0,5 \times 1,3 \times 22500 = 14600$ Newtons/m^2. Being acting onto a surface of less than 3m^2, this represents a lifting force of an order of 5 tons, which effectively could lift our tractor, but under the condition that this power exerts vertically. To take into account the fact that the effort is not vertical, but oblique, and that, seeing the reports, the tractor is more floating than going upward, we must multiply this value, like for a plane wing, by a coefficient in the order of a few %, which makes us go back to about 50 kg. A force of 50 kg to lift 5 tons? That's a joke! Either the witnesses have exaggerated and never seen a tractor flying in a tornado, or there is something hidden behind the visible to explain a power that a light fluid cannot develop by itself. And there is effectively something else, that we now are beginning to be well aware of.

The interaction between the ether and matter is effective and available for any moving body, let they be light or heavy. The ether carries away the planets, but at the surface of any of them, any moving body carries away a part of the ether, and this quantity is a function of the mass and the speed of this mobile, either in translation or in rotation. Though the air is light, the mass of a rotating tornado, being assumed its dimensions, is considerable. If we consider an "average" tornado of which the feeding zone, supposed to be a cylinder, has let's say a diameter of 1 km and a height of 500 m, we have a whirling volume of about 160 000 000 m^3 and a mass of air in movement of 200 000 tons. With an average speed of 100 m/s, we have here a kinetic energy of 2 billions joules, totally convertible into a pressure: that's not nothing, and we may think that, in that zone, the driving of the ether is consequent and well fitted to explain the destroying phenomena that we observe, and which cannot be attributed to the only action of the air. When we see a tornado, or a waterspout which is a small one, or still a typhoon which is a big one, we may imagine, hidden behind the gaseous appearance of the whirling fury, the presence of another fluid, immensely heavier than the air and of which the minute quantity in movement, because the driving phenomenon is only partial, is enough to communicate to the air a supplementary mass far bigger than its own mass, and responsible, this one, of the results and the havocs that go with those exceptional phenomena. The air only doesn't explain all what happens, because it is too light. But the colossal number of its molecules, in fast movement as it is the case in a tornado, is well suited to partially drive the ether, which will give to the aerial fluid a dynamic density which has nothing more to do with what we can statically measure in a laboratory. And it -the ether- will thus lift houses, cars and tractors.

It's well evident that this interpretation of the power of the whirling phenomena from the weather forecast completely changes the approach of their study, but this example is only one of the most striking, among all the others that the direct observation provides us.

You shouldn't believe that this is here only speculation: the existence of the ether has been put into evidence far more directly and peremptorily, in a completely different field. It's question of advances in the technology of the high frequency passive filters, used in the base stations of the mobile telephone networks, a domain which has been the speciality of the author for more than twenty years. The modelling of these filters, made possible by taking into account the existence of a propagation medium for EM waves, medium which in this case is limited by a well determined volume, and to which we thus may give geometrical characteristics, leads to a sheerly mechanical equivalent, which indeed is also the real mechanism. This model has allowed for an evident and unprecedented solution of certain difficulties in the matching of these filters, which are not explainable by the traditional electromagnetism.

5-4: The volumetric mass of the ether

This critical parameter of the ether has constituted the blocking point of the Synergetic Theory, as Vallée didn't succeed in making it looming out from his equations. Was it possible, knowing the method used? It might be necessary, to be able to answer, that a mathematician of the physics take again the subject in hand, but unfortunately we aren't sure to find someone both interested, volunteer and specially competent for this venture. It should however be fundamental to find at least an order of magnitude of a quantity which, once established, would bring a new light on most of the phenomena that we usually consider as known.

Before going further, let's make a parenthesis on the acceptability of this iconoclastic hypothesis made in the previous chapters, and which consists in asserting that the ether is not an extremely light fluid, as it was supposed to be so far, before Vallée, but at the contrary extremely dense. This belief in a light ether comes from a rudimentary reasoning, lacking of a sufficient reflection, which would want that the value we know of the light celerity couldn't be explained otherwise than by extreme properties of its propagation medium, that is to say an infinite lightness, suggested by the transparency of the emptiness, and an infinite rigidity, the word "infinite" being to be taken at the imaging meaning of the term. Let's suppose, to illustrate this note, that we wouldn't know the speed of the sound in water, but that we try to estimate it, starting from the knowledge of its value found in air, supposed known and measured about 340 m/s. Following the same sort of reasoning than mentioned before, knowing that the volumetric mass of the water is 770 times this of the air, we might deduce, only taking into account this difference of masses, that the sound speed in water must be $\sqrt{770}$ times smaller than this in the air. Bad luck, instead of finding it 28 times lower, we would measure it 5 times more. The fault from the compressibility, much smaller, and the very different values of the specific heats: a liquid is not a gas, it's another state of the matter, as everyone knows. This to say that the hypothesis of a very big volumetric mass is not incompatible with a propagation speed very high too, if we suppose in parallel that the compressibility of the propagation medium is quasi-null. This property has been more or less implied before, specially when we were talking about the solar system and the whirling drive of the planets.

The principle being now correctly expressed and enhanced, the task which owes us is to find an approach or, if we can't exactly calculate the volumetric mass of the ether, at least give it an upper and a lower limits, or an approximate average value, let's say an order of magnitude. When, in rational physics, radioactivity is explained by an instability of the matter, due to a too strong interaction with the ether, it's because this matter is seen as being submitted to the ceaseless harassing of the radiations, that shakes and transforms it, above a certain density, into a sort of boiler at work. In other words, matter can absorb vibratory energy, but only up to a certain limit, a certain volumetric mass beyond which it cannot be stable any longer, the edifice becoming too dense and unable to contain an energetic supply too important for it. According to this way of thinking, we may imagine that the value that we are looking for might be much higher than the density of the heaviest known stable elements, density which is about 100, and which corresponds, by chance, to the radioactive elements. Is it 10 times, 100 times, 1000 times or still more? It will not be surprising if we find a volumetric mass bigger than 1000 tons per m^3, but it's even better to find a method of approach which allows quantifying, through what we already know in physics but arranging things differently, and come to a result more or less in accordance with the classical formulas, simply reinterpreted.

The electromagnetism, this so complicated and so improved science, which wouldn't exist without the mathematics, was born from the fact that the physicists, one day, have renounced to know what is the Universe constituted of and how it works, to sever in chapels of stubborn specialists, each of them attached to explore his branch while ignoring the others. It's this way that mechanicians have progressively left the electricians, the chemistry having plaid the pioneers in this behaviour. So that, in certain cases, there are to-day several manners, each of them autonomic and assorted of "its" mathematics, to tackle a given problem. May be it's, by some sides, a regrettable situation, but probably inevitable so much the specialisation, bond to the knowledge progression and the corresponding increase of what is necessary to ingest before being authorized to wear the name of searcher, has become indispensable to our too small brains. However, it's precisely this state of affairs which will allows us, paradoxically, to find a way to grasp more closely the mystery of the volumetric mass of the ether.

In rational physics indeed, electromagnetism is only considered as the mechanics of the ether, the EM waves being nothing else than sheer mechanical vibrations, absolutely identical to these of the sound propagating in water, and which cross us without perturbing us since we are included in the propagation medium, being imbibed by it. This doesn't mean that we have then to repudiate all the classical physics, and throw the Maxwell's works, and the other's as well, to the dustbin. It wouldn't be reasonable, or even conceivable yet, to do without a science so confirmed by the experiments, and without such a number of formulas of which the utility is not to be demonstrated anymore. Let's rather say that it's now possible to justify them, by supporting them with a physical model that now represents correctly the reality. This being agreed, the way is open to a connection which was not so evident, but which becomes to be so, between acoustics and electromagnetism, specially in what concerns waves propagation. We now must try to write in two different languages, in these two distinct domains of the general dynamics of the vibrations, the expressions of a same phenomenon when we make intervene the mass of the medium, and the mathematical identification of the two expressions should lead us to the result we are looking for. It will be question, in the occurrence, but there probably are a plenty of possible ways, to estimate a same quantity, well known both in the two disciplines, and that we'll choose as being the characteristic impedance. The impedance is a parameter which, in electricity, expresses the proportionality between the voltage and the current, or in mechanics between the pressure and the speed. When it's question of the propagation of a wave, this quantity becomes one of the characteristics of the medium, hence its name of characteristic impedance. We already talked, before, about the electro-mechanic analogy, in a restricted meaning, that is simply seen as a way of introducing an alternative proposal in the explanation of a phenomenon, or in its interpretation if we prefer. Now, that notion will be generalized by showing it, not like a simple exercise of style, but as the unification of two different theories in only one, more general and containing the two former.

Let's consider a plane wave, either sound or EM, which propagates in a given medium. In the first case, in acoustics, the characteristic impedance is $Z = \rho c$, ρ being the volumetric mass and c the celerity of the waves. The pressure and the speed are linked together by the relation $P = \rho c v = Z v$, analogue to the formula $V = Z I$ which relates the voltage to the current.

In the second case, in electromagnetism, if we are in the classical case of a propagation in the vacuum, the characteristic impedance has for value:

$$Z = \sqrt{\frac{\mu_0}{\varepsilon_0}} = 377\Omega ,$$

where μ_0 and ε_0 are respectively the absolute permeability and permittivity of the vacuum. But, following the thesis which is championed here, we have under our eyes two equivalent relations which concern the same kind of fluid, one in rational physics, which consider the EM waves as mechanical vibrations, the other in classical physics which sees these waves as particular and only obeying the electromagnetic laws. These two relations are, in fact, relative to the same propagation medium and, as the characteristic impedance is unique and well determined, they are necessarily identical. We thus may write:

$$Z = \rho c_0 = \sqrt{\frac{\mu_0}{\varepsilon_0}} \text{, from what } \rho = \frac{1}{c_0}\sqrt{\frac{\mu_0}{\varepsilon_0}}.$$

But we know $c_0 = \dfrac{1}{\sqrt{\mu_0 \varepsilon_0}}$, from where, bringing it in the previous equality:

$$\boxed{\rho = \mu_0} \qquad (5\text{-}1)$$

In front of this equality, surprising by its simplicity and which suddenly unveils the true signification of the magnetic permeability, we feel a bit like someone who has just broken through an open door, and by the way like a hen which has found a knife[9]. Indeed we are looking for a volumetric mass, which is expressed in kg/m^3 in the International System, and here we find it equal to a quantity of which the unit, in the same System but in the electromagnetic section, is the Henry per metre. If we want to express this magnitude through the fundamental units, we only find in the cross tables the relation:

$$\mathbf{1\ H/m = 1\ Wb/(A.m) = 1\ m.kg.s^{-2}.A^{-2}}$$

And so we find ourselves completely disillusioned, because of the **A** which means ampere, which is the independent -or fundamental if we prefer, it's the same- unit for the electric current, which stops us in our attempt to express the permeability of the vacuum with sheer mechanical units: we are going round in a circle. It was foreseeable according to what we know about the birth and the history of the electromagnetism, which is an auto-

9 Typically French expressions.

nomic discipline. We have thus to re-write the relation (5-1) this way:

$$\boxed{\rho = k\,\mu_0} \qquad (5\text{-}2),$$

where k is the numerical coefficient which can allow us to pass from the mechanical units to the electromagnetic ones, but of which we unfortunately don't know the value. It thus seems that it is impossible to determine the volumetric mass of the ether, and that we might limit us to the hypothesis made before on its probable order of magnitude. Unless...

However, all this is full of insights and calls for two very important notices. The first one concerns the units system(s), since it's apparently one of the blocking factor against our endeavour. Adding the ampere as a fundamental unit, with the Kelvin and the Candela, was made in 1946 with the recognition of the MKSA system, also called “Giorgi rationalized”. At that time, there were so many available units systems that one of the physics exercises to which the future bachelors were trained consisted in passing from one to another, as it was legal to use anyone at the free choice of the user: CGS, MKS, MKpS, MKSA. It's only in 1960 that the last cited became the International System, with the supplementary addition of the Mole. We thus see that the progressive fractioning of the physics in several distinct and autonomic disciplines, consequence of botching the study of the ether, and finally renouncing to lead it to its term, has caused a multiplication of the fundamental units, before gathering them artificially and trying to make them coherent. It's this way that the ampere, unit of electric current, came cohabiting with the other traditional basic mechanical units -length, mass and time- to try to join again two disciplines become effectively independent, though they are so close since the moment we accept to make the ether intervene. We are at this point right now, and nothing, since Maxwell, Vallée excepted, with the result we know, has been endeavoured to make a change.

The second notice is that identifying volumetric mass of the ether and magnetic permeability is founded, under the condition of well making allowances. The expressions (5-1) and (5-2) show that it is obviously the same magnitude, the same property of the space, but also that it is impossible to quantitatively determine it because of the lack of footbridge between mechanical and electrical units, and we presently know why: the unit Ampere has been added to the mechanical system as an independent one, with the only goal of bringing more softness to the whole. We thus must read the first relation only as the simple formulation of the physical identity of two parameters, and the second as a more rigorous mathematical expression, the only one to be considered if we want to obtain the numerical value of ρ, impossible to reach if we don't know k. Moreover, and this is the most important, implying that the magnetic permeability, the μ as they say in the jargon, represents a form of inertia, will not shock many people in the secluded world of teachers. Which physics professor has not practised, when the time of the course on the Lenz's law has come, the little pedagogic play of the electromechanical analogy between a balloon full of water, which splits when it crashes onto the ground, and the spark of the disrupting extra-current? Or of the supplementary effort that we have to exert on a hand dynamo, at the precise moment we connect it to a bulb? Or still what happens when we insert an iron rod in a coil? All this with this little unpretentious attitude, carefully polished up, this knowingly half-smiling of the master who wants to suggest to his students that there is more to know than apparently, but that unfortunately he may not tell them? But that it will be possible further, when they will be learning in the uppermost university cycle? Almost all of them do that, but no one goes further in the conclusions, except at the occasion of personal reflections, about which we can see, when discussing with them, that they are not as rare as we could think. The teaching profession is to-day mature to bite into these new-old ideas on the true nature of space, under the condition that they would be re-looked as much as necessary, and made presentable. It would be enough that a few of them wake up, create study groups and act by constituting themselves in enterprises. So, why wait longer? Go on, profs, full speed ahead!

Waiting for that, we still have not, so far, would it be an approach only, the least clue, on the value of the volumetric mass of the ether. Its identification to the permeability of the vacuum, although fundamental, having not given anything from the quantitative point of view, we must try another method. But before that, a very important notice is to be done, which immediately ensues the context, as soon as we postulate the existence of the ether: contrarily to the relativist formula, the mass doesn't indefinitely increase when its speed is approaching this of the light. When an object with a volume v is submitted to a constant acceleration, its mass increases up to an asymptotic limit -that is a limit of which it will be closer and closer without never reach it- which corresponds to a total drive, by the mobile, of the ether inside it, and which thus would be equal to $v\rho$, ρ being the volumetric mass we are looking for. It will thus be necessary to complete, while complicating it a bit, the relation $m = m_0 \frac{1}{\sqrt{1-\frac{v^2}{c^2}}}$, which is only available in first approximation, the same way that the Van der Waals's law went précising this, too simple, of Mariotte.

This being assumed, let's go back to the determination of this volumetric mass. The principle chosen to reach its numerical value is to compare the same phenomenon, judiciously chosen, through two usual well-known presentations, let's say under the aspect of two mathematical formulas, one in mechanics and the other in electromagnetism, and then identify the two expressions. That's not all: we have previously seen that the unit of electric current had been added to the mechanical units of the MKSA rationalized system, become the International System, as a fundamental unit, thus independent from the others. So we'll have to, either avoid using it, instead of what we'll be again in an already explored blind way, or establish its hidden relation with the mechanical units. This being agreed, one of the rare ways which seems allow us to build a bridge between mechanics and electromagnetism is this of the energy, which in these two domains is manifesting by identical thermal effects, that we can both calculate and measure with the same units, and which consequently are absolutely comparable. Another one would be to appraise the power of the vibrations, transmitted into the ether by a given source and considered under its two possible aspects, one as an antenna, the other as a mechanical pulsating sphere, both being omnidirectional and with the same power (figure 5-5). We know this power, that we are able to evaluate in the two cases. For example, the

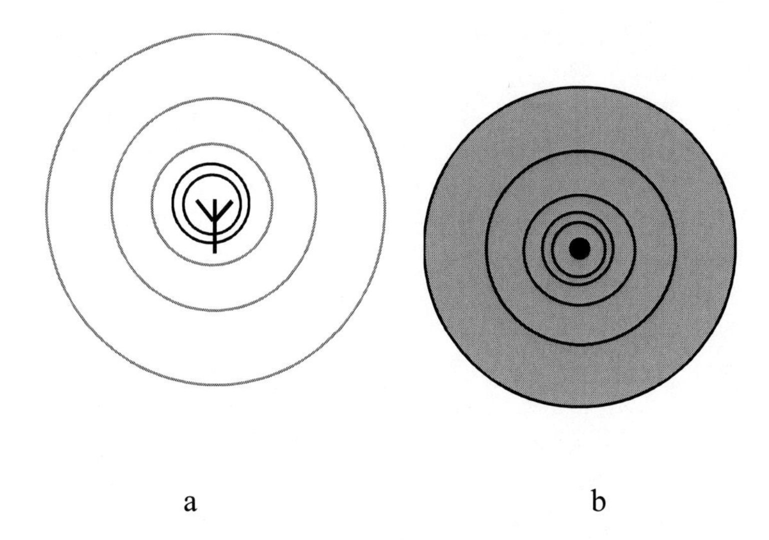

figure 5-5: the two sides of an omnidirectional EM source. On the left the antenna which emits EM waves in the vacuum, on the right a pulsing sphere, vibrating in the ether following the same laws than in acoustics.

power radiated by a dipole of small dimension, let's say in the order of a tenth of the wavelength, is:

$$P_1 = \sqrt{\frac{\mu}{\varepsilon}} \frac{\beta^2 I^2 L^2}{12\pi},$$

where $\beta = \frac{2\pi}{\lambda} = I$ is the average current and L the length of the dipole (Antennas, Kraus, Mc-Graw & Hill, p216).

In what concerns the equivalent pulsating sphere, we find the value of the radiated power in Rocard's (Dynamique générale des vibrations, Masson, p344):

$$P_2 = \rho c \frac{8\pi^2 a^4}{\lambda^2} U_0^2,$$

where ρ is the volumetric mass we are looking for, $a \prec \lambda$ the radius of the sphere and U_0 the amplitude of the vibrations. Supposing reasonably that $L = 2a$, it's enough making $P_1 = P_2$ to make the value of ρ get out. Unfortunately, and it was absolutely necessary to join this point to well see what is wrong in the enclosure of the theoretical physics, we see that ρ remains a function of I, expressed in amperes, and thus without any relationship with the other fundamental units. Indeed, the ampere has been defined in the International System as the intensity which, when it crosses two parallel conductors placed at 1 m one of the other, generates between them a force, repulsive or attracting following the respective senses of the currents, of 2.10^{-7} N/m. A force being homogeneous to a mass multiplied by an acceleration, we might think first that there is here the footbridge we are looking for, but no: this definition is effectively based on a physical phenomenon, well-known and incontestable, but it is totally conventional and purely **numerical**. In other words we cannot reach neither by this method, nor by any other which would make intervene classical mathematical relations, a value of ρ expressed in kg/m^3.

It will thus be necessary to find another method, radically new and non-conventional. To do that, it's once more worth opposing the theoretical physics and the rational one, now in their different vision of a phenomenon, so well-known and so mysterious in the same time, that we call electric current. Let's first notice that calling it "current" already evokes an idea of a fluid in movement, against what no physicist has ever protested or opposed. We moreover suppose that, the free electrons being the only mobile elements in a conductor, the current can be attributed only to them. We'll thus consider that they are, in their whole, comparable to an incompressible gas which moves in block, as soon as a tension is applied to the circuit in which they are. We have the impression, at first sight, that this circulation is done at a high speed. In fact, the measurements made by the physicists show, or seems to show, that the own speed of an electron which moves in a conductor is extremely weak, in the order of some centimetres or tens of centimetres per second. However, the establishment of the current in a circuit appears as a phenomenon which has a speed of the order of this of the light. To explain these two aspects, contradictory at the first sight, two speeds have been defined, both measurable *a priori*: the first one is the speed of an electron, seen isolated, the second one, strangely named "information speed", is this of the whole cortège considered as an incompressible cylinder, where the movement is transmitted instantaneously, like water in a garden hose: when we open the knob, the water flows im-

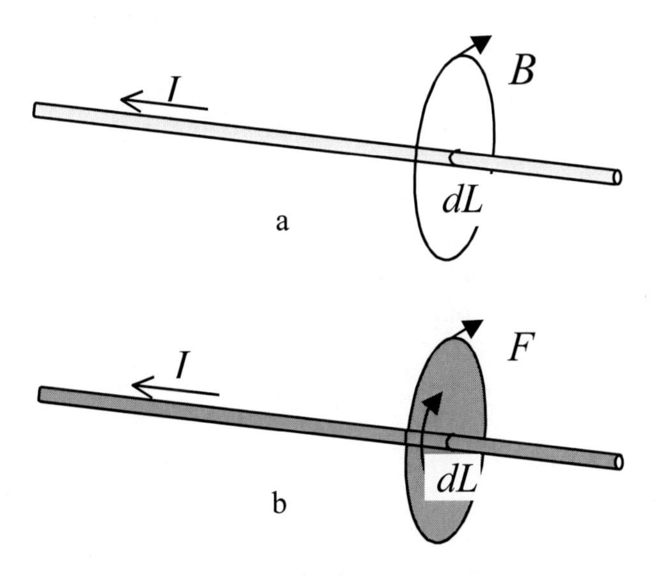

figure 5-6: the two aspects of the electric current. In (a), the classical presentation with an electron flow which creates a magnetic field B perpendicular to the conductor. In (b), the same current becomes the fibre of a Rankine's vortex, of which the sheet drives a magnetic body.

mediately at the extremity of the hose, but the molecules which get out are not those which have just come in: these last will have to wait several seconds to reach the output. This is obvious if the hose is first empty, but the water flows instantaneously if it is already full. In the case of the conductor and its free electrons, the hose (the conductor) is always full. Figure 5-6 shows the two visions we can have of the phenomenon. The upper part (a) is the classical version of electromagnetism, with the electrons which go their way in the conductor, and a magnetic field, able to orientate a compass, which is created in the vicinity. The part (b) is the reality of the rational physics, with in each element *dl* of the conductor a Rankine's vortex, of which the chimney, guided by the metal, constitutes the true current, and of which the perpendicular sheet drives mechanically and selectively the magnetic bodies and orientates them. The fact that the electrons are supposed to move slowly in the conductor would probably have much surprised a number of physicists of the last century who, like Planté or Le Bon in France, have put into evidence the violence of high intensity electric phenomena. All the more so between the two extreme speeds cited here-above, we are proposed a third, this of the electronic currents in the vacuum of the radio lamps, which would be about 15 km/s, but that is how it is: we must trust. This being said, let's go back to our topic.

The two pictures and the two interpretations of figure 5-6 are fundamentally different, but we'll say that each of them obeys the laws of the universal logics, and that they are not in contradiction, but simply independent one from the other. Consequently, we'll pose as an hypothesis that, in any of the two conceptions of the electric current, the moving mass is the same. This mass is constituted on one side (5-6 a) by the whole of the electrons, the only mobile entity able to represent a current, and on the other side (5-6 b) by the vortex trickle which flows at a high speed in the electric cable, supposed to be in copper with a length of one metre and a section of 1 mm^2. But we have, from the classical information sources, all the numerical data necessary for the quantification:

- 1mm^3 of copper contains $8{,}6.10^{19}$ atoms, and as there is one free electron per atom, this number is also the number of the free electrons. 1 metre of copper wire thus contains $8{,}6.10^{22}$ such electrons.

- We know the mass of the electron: $9{,}1.10^{-31}$ kg. The total mass of all the electrons in our portion of conductor is thus: $8{,}6.10^{22} \times 9{,}1.10^{-31} = 7{,}83.10^{-8}$ kg.

- We know the "information speed" in the copper, that is 273 000 km/s. We'll now suppose -we have no other choice- that this is the flowing speed of the fillet of ether in the conductor. We cannot imagine, indeed, that a jet of fluid without friction and incompressible could move inside itself at only a few cm/s. But the physics of the conducting matters proposes us only the two possible values mentioned above. The mass flow is consequently, in our copper wire of a 1mm^2 section, 7,83.10^{-8} × 2,73.10^8 = 21,4 kg/s. So it comes here a mass of 21,4 kg which is moving in a volume of 10^{-6} m^3. The volumetric mass being the ratio between the mass and the volume it occupies, we finally have:

$$\rho = \frac{21,4}{10^{-6}} = 21,4.10^6 \, kg / m^3$$

$$\boxed{\rho = 21400 \, tonnes / m^3}$$

So we are, at last, in front of the first efficacious attempt to quantify the volumetric mass of the ether, that is of the Universe. Is this result available? Are the hypothesis made to reach it receivable? It will be now to the inspired opponents to state on it, as much as they will be keen to consent losing a bit of their precious time to have a glance to this amateur work. For what we are concerned, we consider that our task is completed. If we had found 10 times, see 100 times less, we would still have been satisfied, the essential being to arrive to a value large enough to explain, first the driving of the planets by a consequent whirling mass, and for example the limit of density for the stable elements at a value in the order of 100 (i.e. 100 t/m^3). The ether, indeed, is unceasingly crossed by energetic vibrations of any frequencies, qualified of EM waves in classical physics but, let's remind it once more, sheerly mechanical and perfectly analogous to sound waves. We may thus suppose that, since the moment a nucleus presents a dense structure, with a sufficiently large number of elementary particles, it's full up of vibratory energy and cannot any more resist the internal pressure, which at the end takes the advantage on the MBL pressure. It's probably this way that we should imagine the phenomenon of the radioactivity. It's similar to a boiler which cannot contain its steam any longer.

We are thus arrived, at last, to the end of our efforts to try to find a value, whatever it is in a first time, for the mysterious volumetric mass of the ether. Is this value exact? Nothing is less sure. But at least, it's not a value which doesn't destroy all the hypothesis made so far, and specially this of the eminently whirling character of the Kepler's laws. At the contrary, it's going to comfort the major idea of this modern theory of the ether, which takes the opposite course of the physicists habits of thinking, and which denies the idea of a light fluid, suggested by our senses, in which we trust too much and too systematically, with all the contradictions that come with. Even if the result is wrong (is it?), it opens the way, by the method used, to a reflection that the future opponents will be obliged to do. Because it exists now, it would be advisable to find the arguments, either to show it's wrong, or to rectify it, or even, why not, to confirm it. Of course the starting hypothesis are bold, but they are emerging out from a desert landscape, where there is no real competition and where new ideas pain growing.

We'll talk again on that later, in the last paragraph.

5-5: the muscular force

Up to this paragraph, apart from some necessary digressions on the five senses of man and their limits, it has been only question of physics, in the most traditional meaning of the term: mass, gravitation, astronomy, electromagnetism, thermodynamics, etc., without wondering if the ether had its role to play in other disciplines.

We must however note that, when physicists come to speaking about forces, of which they generally enumerate a quantity that is now limited to four categories, they always forget another one, that they leave with respect and cautiousness in the investigation field of their colleagues, physiologists and doctors, specialists in the puzzling living matter: it's the mysterious muscular force. Why mysterious? Simply because nobody, whatever his speciality, has succeeded in dismounting its mechanism. Those of us who have tried to find, in the specialized libraries around the Medicine Faculties, some work clear and well explained on the myology, know what it is about: it's a quasi-total emptiness, as if no specialist had found good to write for the general public. And no opportunity for having the least clue: if by chance you can capture the presence of a representative of the medical profession, quite easily noticeable with a minimum of seasoning, you are at once in front of a reaction of a surprising modesty, coming for a so much educated people, and which gives this kind of answer:"*Ho, you know, it's not at all my speciality*", or something similar. But no advice, no recommendation, not the least orientation but, as a bonus, a dodge in the form of an escape, as if something particular, apparently explaining this systematic withdrawal, was attached to this speciality, however of a considerable importance. In fact the truth is of a luminous simplicity: nobody really knows how the muscle works!

For the obstinate who, in spite of that, wants to know more, the salvation will come partly from the hospital libraries, those which have got a myology department, and partly from another parallel profession, more modest but as useful, where the authors do a laudable effort to make their texts accessible to a general public. If we indeed persevere and ferret in the counters, you finally can find a consequent source of information from the physiotherapists, source with which the amateur will be content, and which indeed gives us all that we are looking for. You'll never find anything more, concerning the real mechanism of the muscular force, but a good anatomic description, nevertheless detailed enough so that we could elaborate new hypothesis.

We are often amazed at the athletic performances of the small animals, like fleas or grasshoppers, which easily jump ten or fifteen times their size with the highest facility, when our greatest champions have difficulties to slightly top theirs. If we now pass to the elephant, which can just leave the ground of a few centimetres, we may deduce from these first observations the sketch of a first law: if the force was proportional to the volume of the muscle, the weight/power ratio would be about the same for the grasshopper and the elephant, and this last, under the condition of taking off from a firm ground, should be able to also jump fifteen times its size. We easily imagine the results! But everybody knows that it's not the case, and that the more the beast is big, the less high it jumps. Accordingly to this notice, we have the temptation to think, looking back to the very muscle only, that the force which is available inside is proportional, not to a volume, but to a surface. Which surface, exactly? Let's say its active surface, or may be its efficient surface, that is this which is perpendicular to the direction of the contraction. This being assumed we better understand, if the hypothesis is well matched, that the bigger is the animal, the less forces it has to move its mass. And we'll further

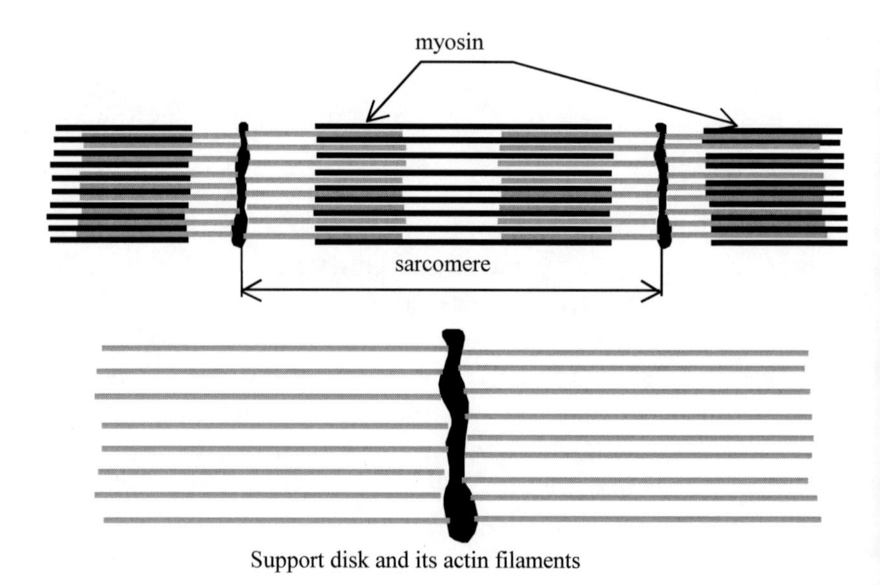

Fig 5-7 detail of the muscular fibre

see that putting now the muscle in the etheric context allows a simple and convenient explanation of its working mechanism, and brings it with a total clearness.

To begin with, it's indispensable to dive a bit in the muscular anatomy, so as to well define the context, and specially the precise object of the study: the muscular fibre. To be simpler, we'll limit the description to this of the best known muscle, the long striated muscle, made of a juxtaposition of muscular fibres, constituting a spindle which goes from one of its connections to a bone to the other. This type of muscles is also called skeleton muscle, because they cause the movement of the different parts of the skeleton, like for example the biceps, which allows bringing the hand closer to the shoulder, through a movement of the fore-arm. These muscles are built like a sort of bundle of elementary fibres, themselves formed with identical motifs placed one behind the other. Their length is approximately constant and defined by the distance between two successive disks, that we can see by a microscopic examination. Each of these motifs is called a sar-comere.

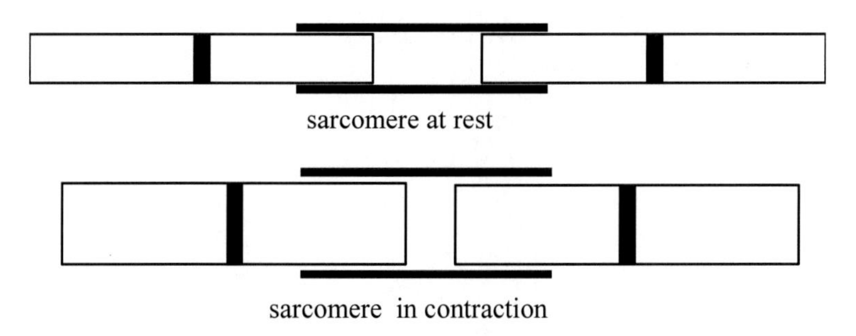

Figure 5-8: mechanical model of the muscular contraction

Figure 5-7 shows, in a wanted extremely simplified way, how is constituted this sarcomere. The disks are in fact the stay of a double filament pencil, perpendicular to these disks, these fila-ments being maintained in fascicles and, in a way, included by other filaments, thicker, of another protein called myosin. The length of all these filaments, either actin or myosin, is constant. Thus the shortening of the muscular fibre during the contraction doesn't correspond to a length variation of these components, but to bringing the disks closer. We thus may consider, if we want to make an equivalent mechanical model, that the fibre in contrac-tion works like a suite of free pistons (disks + actin), which penet-rate more or less in cylinders (myosin), like figure 5-8 animates it. We note on this schema something very important: when the sar-comere shortens, its section simultaneously increases. Everyone of us indeed knows, after having looked at his biceps, that these ones have their diameter increase in the central part when they are requested: they blow up. It's there such a casual phenomenon, so much known, that it could be classified as itself and forgotten, but a physicist, who by definition is more curious than the others, will not miss asking to himself the question of knowing the variation

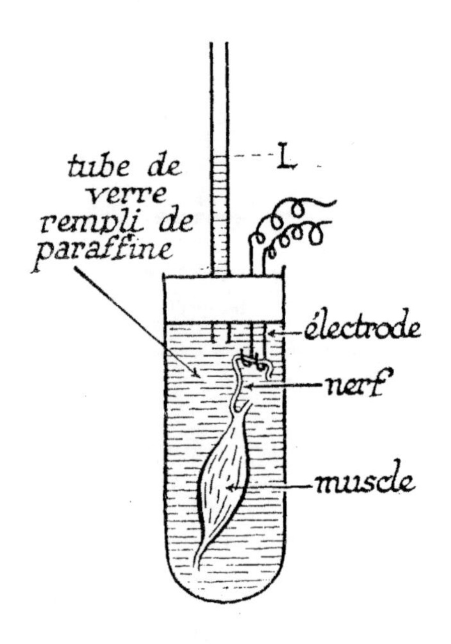

fig 5-9 the muscle contracts at constant volume

law of the diameter, or the section, of a muscular fibre as a function of its linear contraction.

Is this an accessory phenomenon, or is there behind it something more fundamental? So as to be able to answer this question, right now or further, we first must study it in a more methodical way, and see if, by chance, this study couldn't have been already made before. Precisely the answer, or at least one of the answers, can be found in the volume 2 of the book “la science pour tous”[10], by Lancelot Hogben, page 312, where the author relates and comments an experiment, may be pretty aged, but which allows asserting that the muscle works at constant volume (figure 5-9). This experiment consists in installing an animal muscle, living, in

10 The science for all

a transparent jar, sealed and full of liquid, and of which the only communication to outside is a narrow glass chimney, where has been added a few quantity of liquid, so that this last be higher than the jar and show, as if it was the column of a barometer, the potential variations of the internal volume (liquid + muscle). Onto the muscle is connected an electric excitation circuit, which at will can induce the contraction. The result is final: at the moment the influx is sent, the level in the column doesn't vary, though the muscle shape changes. Thus we may say that the muscular contraction works at constant volume.

We may think that this effect, important a priori, is not considered as such by the specialists in myology, science apparently recent and still confidential, because in the schemas that we can find in their rare books, or on the few Internet sites relative to the muscular contraction, there is nowhere any drawing on the transversal variation of the fibre, just like it if didn't exist. The books almost exhaustive on myology are American, the two most recent reference works being "Skeletal Muscle Mechanism", by Walter Herzog (Wiley), and principally the most important "Myology" (Mac Graw & Hill), of which the two volumes of 1800 pages can be considered as being the state of the art in this profession. In fact, in these two books, the part dedicated to the anatomy and how works the muscle is very limited, and treated practically the same way in each: same text and same drawings. The essential of these professional bibles essentially concerns the functional description of the human muscular system and its pathology, since we mustn't forget that the first role of a doctor is to cure his patients. But it's also because nobody has found yet, in the classical and the most general frame of the study of the living, the keys of the mechanical explanation of the movement. It's perhaps in that discipline that taking into account the existence of the ether, and of its energetic characteristics, previously described, will have the most spectacular results, and will lead to definite and determining upheaval in the study of the living. For the moment, the existence of the universal fluid has no more credit in the Medicine Faculties than in these of Sciences.

One of the most dramatic and dangerous consequences of this absence of theory, itself due to an absence of directing ideas, is the well noticeable penetration of the mathematics, in medical books in general, and in myology in particular. The first book cited upper is, from this point of view, frightening. Practically half of the book is covered by formulas of such a complexity that even the specialists must have pain to understand them. But moreover, it's not equations on energetic check-ups which will make the discipline progress: the motor function of the human body is before all a mechanical problem, in which we have first to identify the motive force and the device which creates it. And in this field, even if we remain searching, we don't see in the publications any hypothesis both new, smart and simple, which would allow orientating both researches and reflections. This blocking situation has an explanation which concerns all the medicine.

Since the Middle-Age, the human body has been considered as an autonomic machine which works with only one energy source: food. It's particularly evident -apart from the restrictions already made on this adjective and its relative value- that eating is the *sine qua non* condition to survive: when you stop eating, you die, that is, to speak like physicists, that you loose the totality of your motor function. We thus are well obliged to think that it is through the transformation of the food, inside the body and by the action of diverse cellular processes, grouped under the word of metabolism, that the vital forces were born. It's difficult not to adhere to this simple and peremptory logics, but once the thing is told, once the principle is agreed, explaining the detailed way of working of this indispensable and daily transformation is quite another matter. It's so intricate that biologists, and the medical research in general, have quickly felt the need of going to see if, in the so-called exact theoretical sciences, there wouldn't be some tool available to help them. And as the vital phenomena almost go with thermal exchanges, whichever the scale, macroscopic or cellular, the thermodynamics and its priests were requested.

Doctors my friends, you are lucky and privileged to exert one of the noblest profession that it can be, which brings you the consideration of all the people. What more honourable and meritorious indeed, than giving his life to treat the others and release their pains? Apart from a minority of you, for whom this activity principally means social success and getting well off, it is not thinkable to do your job otherwise than by vocation. Also, the permanent contact with the others, the obligation of trying to understand them in order to better treat them, the necessary dialogue with handicapped people who often see you as their saviour, all that makes that your profession has something particular which doesn't exist in any other. The enormous and indispensable knowledge you must acquire, all these studies of an exceptional length you have to assume to get your grade, and in parallel the fact that, in spite of this, you must talk to casual people, more or less educated, adapting your speech to the level of the patient so as to make him reassured, to communicate him the indispensable confidence which will double the efficiency of the therapy, all that must not be altered by a bad company. Your science and your know-how are situated between the art and the technique, and make of your activity a particular one where the intelligence, to be really efficacious, must be more practical than theoretical. So, don't listen to the sirens of the theoretical physics, remain far from mathematicians and still more from thermodynamicists: they'll bring you only sadness, ruin and desolation. But what's the most, they will slyly attack what you have as the most precious, on the plane of your intellectual faculties, and which is by you of a capital importance: intuition. Beware, that's not here a vague threat, evil is already installed, the worm is in the fruit.

In the preface of a book, written by searchers in biology, you can read these sentences, which sends a cold shiver down your spine:"*So, when it is difficult to speak to-day of a biologic development without appealing to the properties of informative molecules -and also, by the way, to the epigenetic-, the thermodynamic approach takes on an importance that you can't get away, for who takes interest into the cellular functioning.* ***Without its support, the biologist will only be able to describe reactions of exchange, transformations..., he will not really understand their deep economy for all that.***" Followed a bit further by: "*The book,...should find an excellent welcome. Original, it gathers, which is rare, the axiomatic thermodynamics and the thermodynamics applied to the living systems. Useful, it brings to the molecular biologist, to the specialists of the reproduction and the development, of the bioenergetics, or the membrane interactions,* *all the bases necessary to them. Indispensable, it opens, as said before, new ways for the study, how much actual, of the cellular and inter-cellular control.*".

About the very book, we look in vain for an explanatory text, resembling a traditional biology research as we usually imagine it, that is, before all, as an experimental and deductive discipline, in an ocean of mathematics that the prefacer himself confesses not to understand. This doesn't prevent him from citing Schrödinger and Dirac, who, according to him, would have made an attempt of incursion into this domain, normally forbidden to the theoretical physics, and specially to the quantum physics. What the hell Schrödinger and Dirac are to see with the cellular biology? Let's say it clearly: this book, for a simply curious individual who only wants to be aware of the state of the researches in that domain, is absolutely undrinkable and useless. Unfortunately, this is not exceptional in the domains connected to the medicine and to the sciences of the life. The reference books in myology, which have been cited above, don't escape from that tendency, no more and no less than a number of others. All this is in fact the deep disarray of a research that, totally lacking of any directing idea, doesn't see its salvation elsewhere than in a systematic modelling, joining up this way the theoretical physics, about which it thinks, in desperation, that the rigour of its processes could offer it a sheet anchor, through a complete change of methodology. That's dramatic, but unfortunately of an inexorable logics: since the moment you think, and we are obliged to do so in function of the classical hypothesis, that it's feed that, after multiple transformations, is the primary energy from which this of the muscle comes, you are well obliged to look toward these transformations to try to know how a force can come from a food. As, besides, all the vital phenomena, from the digestion to the tiniest explorations of the metabolism, are accompanied with thermal exchanges, it was fatal that searchers orientate their thoughts toward the thermodynamics, ignoring that this last couldn't do anything for them, at least for the very muscular mechanism. Moreover, certain ways of reflection, like these of the transmutations, have been straight out slapdash, and only a few individuals, who in other respects have been immediately marginalized by the scientific community, have led experiments in that domain. These of Kervran, for example, who has showed that a hen, enforced to pick a ground where there was not a pico-gram of limestone, was bravely continuing to make its egg shells in the norms, without waiving the specification notes, are both stupefying, full of outlooks and totally despised by his colleagues.

The thermodynamic way has quickly led to a massive introduction of mathematics in the reference books on myology, with the excess that we have previously mentioned. The complexity and the difficulty of assimilation which go with, because we may add that the sort of mathematics used in these books is particularly off-putting, make that the students who were *a priori* interested by the myology, and they very well have the right to be, quickly turn away from the research, at the profit of pathological studies, more practical, more accessible and less uncertain. What is still more serious is that, when reasoning only on thermal exchanges, check-up and transformations, is that some people have a tendency to forget that, among all these transformations, a certain number of them are not reversible, and that you may not arithmetically manipulate quantities of heat like it is usually done with lengths, weighs or masses. Nevertheless, there are a number who seem not to be worried of that. At the Palais de la Découverte[11], which in France is an enormous living aide-mémoire that many high-grade scientists should more assiduously frequent, would it be only to check their knowledge, you can find workrooms for all the matters, with, in each of them, written recalls for the most important principles. For the thermal, which interest us more particularly, you can read this, for example: "*Energy can spontaneously change to heat, **never the reverse***". To transform heat into work, you need a thermal machine that works in accordance to the Carnot's laws, that is in particular between two sources of heat, a cold and a hot, and besides it is necessary that this machine really exists. Since the moment we have drawn a steam engine, anyone will immediately grasp the causal chaining that, starting from the boiler furnace that heats the water and brings it to the state of steam -what any housewife knows by heart- will cause the translation movement of a piston, that we can transform into a rotating

11 Pedagogic site for the discovery of sciences

movement to make the locomotive move on its rails. In the anatomy of the muscle, despite of the perpetual improvement of the observation instruments, nobody has found a device resembling this one, and nobody will ever find one.

Taking into account the existence of the ether, as it had been described in the chapters before, completely changes the deal. It provides a new working hypothesis by unveiling and external origin, relatively to the human body, of the muscular force, when it was not possible so far to locate it elsewhere than inside the very muscle. Among the two main characteristics of the ether, its volumetric mass first and its vibratory energetic nature secondly, it's evidently the second that comes to play: the primary force that the muscle uses, its first motor, is the pressure that we have called MBL, this which exists everywhere in the space. That the machine could be able to transform this pressure into a motive force, we already know that (figure 5-8), but this new context will give us a completely new dimension and an unprecedented comprehension level. It's question of the sarcomere, of course. Before going further, it's not useless to make an important recall of the first chapters. The multi-frequency vibrations that are responsible of the vibratory pressure in the space can be seen in two ways: in classical physics, they are called electromagnetic waves, that have their own existence. In rational physics, they are mechanical vibrations that propagates in the medium “ether”, and thus have the property to cross us without we realize it, apart from the small quasi-octave of the visible radiations that strike our eyes. But it's now well agreed that they are the same thing, rigorously.

We have modelled the macroscopic structure of the sarcomere, this which is revealed by the microscopic observation, as the concatenation of kinds of small pistons of which the whole, constituting the muscular fibre in all its length, is mechanically able of an irreversible contraction, controlled by the nervous system. In fact, if we want to be more precise, the muscle only know the contraction, the return to the original length, at rest, being done through a release and the simultaneous action of an antagonist muscle. The repositioning of the muscle in the etheric context, that is in an energetic fluid medium that both imbibes and surrounds it, appeals at once for the image of a direct and mechanical action of the

MBL pressure on each of the sarcomeres. The fact that these last simultaneously grow up when there is a contraction is not mysterious, since it had been demonstrated that the phenomenon goes on at constant volume. But this is only an experimental observation, that must hide something more subtle that remains to be discovered. We however can work on this new model, knowing that suddenly discovering the physical origin of the muscular force will not, however, solve the mysteries of the complete process of the contraction, and in particular this, starting from the nervous excitation, that controls the contraction in the longitudinal sense only.

This polarisation, that intervenes at the moment the muscle passes from the resting state to this of work, now that the origin of the primary energy is established, will immediately become the new problem to solve, and announces a huge task. But if we really want to know what really happens, we must be neither discouraged, nor wondering if it is possible that we have worked for nothing so far. It's exactly the same situation than in physics today: the theory of Relativity has led the searchers in a blind-way, from which they can't get out except going backward, like a driver who has lost his way and who has to turn back to the place where he failed. That means that, one day or another, as we told that several times before, it will be necessary to get rid of the myths of the universal constants, and at once agree on the existence of the ether. The same about the myology, it will be worth leaving the idea that the muscular force comes from the transformation of the food, and from the different thermal and electric flows that go with the skeletal movement. These flows incontestably exist, but it is likely that they would be bound to the control process of the muscle, more than to its very mechanical functioning, that is the polarisation of the sarcomere. It's this way that we'll now call the phenomenon associated to its axial contraction, and to the concomitant radial expansion, in other words its activation. This last, very probably, needs a series of physic-chemical commands, that correspond to the forwarding of the "orders" given by the nervous system, and where there are all the possible forms, chemical, thermal, electrical and others, of transmitting the energy of these messages. Information asks for energy, it is energy itself, and we perfectly may see the whole of the energetic flows of all natures, that happens during the muscular contraction, as being totally dedicated to it.

This assumption being made, the explanation of the muscle contraction is totally carried back to the MBL pressure, to which we now attribute the whole of the physical origin of the muscular force. The end for the internal interpretation, the end for the re-

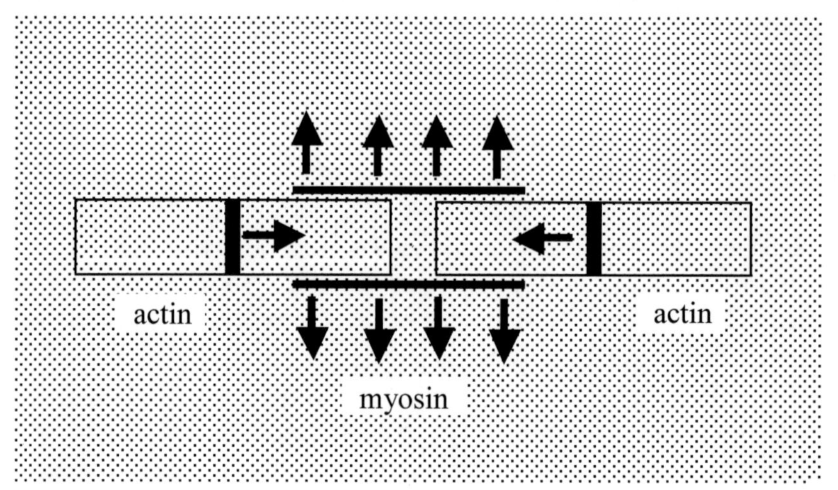

Figure 5-10: the contraction of the sarcomere reviewed in its etheric context

course to thermodynamics. We then must go back to the figure 5-8 to see again how the fibre globally behaves in contraction. Its dark zones correspond to the myosin, of which the length doesn't change, which alternate with light zones, narrower. It's these light zones of actin that narrow when the muscle is working, making that the sleeves of myosin get closer one to the other. The term "sleeve", that suggests that the myosine covers up the actine, is in fact not fully correct, as the actin and the myosin inter-penetrate, like would do the hairs of two brushes pushed one against the other. But in the model we have chosen, which indeed is much simplified but better accounts for the global behaviour of the fibre, we'll do like if the actin would play like a piston inside the cylinder of myosin. Figure 5-10 illustrates this model, represented bathing in the ether that imbibes it with its mass and its vibratory energy, with the two series of force vectors that correspond to the deformation of the sarcomere: the radial expansion forces, which correspond to the growing up of the muscle, and these of contraction, longitudinal. We don't know *a priori* if the ones are the cause of the others, or if the whole of the simultaneous actions have a unique cause, but at once we know from where the motive energy comes. Remains now, and it is not the easiest to imagine, or more precisely to discover, to find how works the process we have called polarisation and which, under the physic-chemical orders coming from the nervous system, modify the different parts of the sarcomere, so as to create an unbalance of the MBL pressure onto its geometry and cause its deformation. If we remember the electromagnetic nature of the ether, there is no doubt that it is here a phenomenon analogue to the gravitation that, remember, is the result of an unbalance of the radiation pressure caused in the space by the presence of a mass, all this transposed to the cellular scale.

It's well evident that, through this way of thinking, all is to be reconsidered. We must restart from zero, and it also seems evident that the myologists will not agree throwing into the dustbin years of work and reflection: it's on this point of view the same problem than with physicists. But they must know what they want, either prolonging a blind work on ill-founded or too intricate hypothesis, or looking for the physical truth, the only correct intellectual attitude for a searcher, whichever his domain. The 300 preceding pages should help him.

This new hypothesis, this new way of thinking, the agreement on ether as an incontestable base of the physics, will have as much difficulty to be asserted in myology as in the other sciences. It's extremely difficult to fight against the habits and the established order: the human species delights stability and viscerally detests changes. The example of the Relativity is fully demonstrative of this state of mind, but let's not insist on what has already been told so many times, and let's rather campaign, all together, for the accreditation of the existence of the ether. This one, on the point of view of phenomena comprehension, changes all, everywhere. Since the moment we give it the consideration it merits, and a minimum of reflection time, instead of repel it without examination, it provides the scientist with all that the aridity of the theoretical physics refused him before: simplicity, intuition, powerful analogies, in fact all that was missing to start again. But taking it into account, even if it brings the satisfaction of a philosophical base that unify all the sciences, exact or not, will be done only when the theoretical physics will have given all that it can give, and will be in a state of sterility. May be it's now the case, but not obviously enough. It's so, we cannot do anything against, that's human nature. Why torturing one's brain with new ideas, specially if they have the outdated and deceitful taste of old ones, even renovated, while the scientific society feels intellectually so well carrying out the studies it has planned, when the credits are available? As long as there will be oil, even expensive, and nuclear, even dangerous, the necessity of the “all electric” will not be resented with so much strength that it would be decided to extract energy from the fantastic and inexhaustible reserve that space constitutes. And the myologists will still not understand how a muscle works.

And any way, no thinking has a big advantage: no risk of headache!

5-6: The electricity

It's the renunciation of the concept of the ether, apparently impossible to master, that led to the birth of the electromagnetism, at the 19th century, and favoured its launching. Even Maxwell, who however was convinced of its existence, never mentioned it in his treatises on electricity and magnetism, though the allusions are so strong, at places, that it is very difficult to ignore them. We have to ferret in the Scientific Papers to find articles that he has dedicated to it, and this Janusian[12] aspect of his work is extraordinary: on one side, for the books of large issues, a rigorous and mathematical account, well in the norms and made not to incense the Establishment, on the other side, in the notes of less important diffusion, short billets treating delicate or fully forbidden subjects, each of them being a treasure of logics and intuition. Those who

12 From Janus, the two-faced God

have read these two sides of his work know that it is because he constantly tried to approach the intimate structure of the ether, that he finally saw too intricate and, therefore, gave up elucidating it. Then, at that time, he found this genius idea of the "displacement current", that was the gate to the unifying equations that bear his name.

If the analogies proposed in this book are not these of Maxwell, they are inspired from a considerable number of attempts, made by the physicists of the former centuries, to make a representation of the mechanisms of the science of the invisible, as it is this way that we must consider electromagnetism: the mechanics of the invisible, of the invisible ether indeed. Seeing a magnetized needle moving when an electric current flows in the vicinity, feeling a magnet, hold in the hand, irresistibly attracted by a piece of iron, all that automatically gives two series of reactions. First, having the feeling that you are in front of a mysterious phenomenon, that is imperative to understand if you are a scientist, then realise that it will be necessary to do precise measurements, varying the parameters, to give body to something intangible, that escapes both from our senses and our innate. That was done. From the measurements were deduced laws, from which scientists tried to mathematically find other laws, and both electricity and magnetism were born, before they saw that they are two aspects of a same thing, that they called electromagnetism. When we say that, genuinely, these sciences, that seem to have each its well determined field of study, well distinct from the others, are only the different manifestations of general properties of the space, even without pronouncing the word ether, that however we'll be obliged to postulate at a given time, we reduce the totality of the physics to the mechanics exclusively, extending this last to all the subdivisions that have been created, by lack of coherent hypothesis, after the abandon of the reflections on the nature of the space. It's this way that were born, as autonomic specialities, apart from the already cited electricity and magnetism, the acoustics, the capillarity, the wave mechanics, the nuclear physics, the quantum physics and others, before the Relativity brought its cortège of phantasms and destroying flights of fancy.

Now that physics is split in a thousand pieces, the work is to rebuild it. It doesn't consist in destroying what exists, but adding the foundations that we have disdained so far, so as to complete the

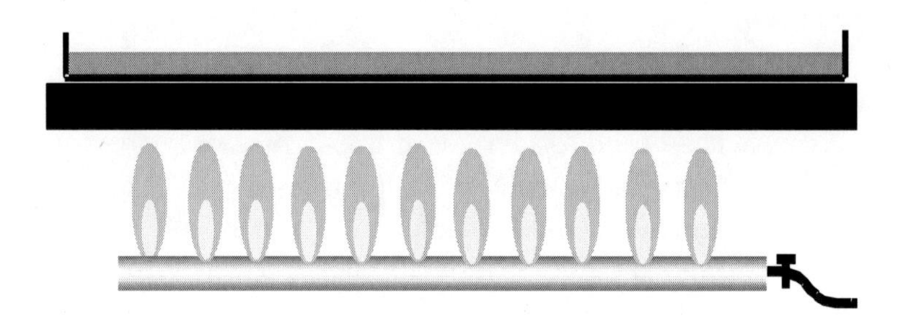

figure 5-11 : appliance of Bénard for the study of the whirling convection in thin layers.

rigour of the theoretical physics, reduced to the mathematical catalogue that we know and that rebuffs so many people. What it lacks, indeed, is that the knowledge is not accompanied by comprehension, great absent from the education despite the constant efforts of the teaching corporation. It was thus evident that, after having cleared up phenomena such as the mechanism of the gravitation or this of the muscular force, *a priori* so different, the etheric hypothesis should have made us have a fresh look at other problems. The parallel between a continuous current and the filament of a vortex having already been established higher (see figure 5-6), we'll limit ourselves here to the electricity, and specially to one of its best-known object, the plane condenser.

It exists in the Notes and Memories of the AFAS (Association Française pour l'Avancement des Sciences)[13], reports on experiments that have fallen into oblivion, and that nobody has tried to re-examine, but that however are of a considerable importance, in the overlooked domain of the electromechanical analogies, the poor cousin of the physics. It's here question, in the occurrence, of

13 Old French publication, commenting the works of the Sciences Academy.

an experiment realised by Henri Bénard, assistant at the Collège de France, related in the 29th session (Paris 1900) of the AFAS and the title of which evokes whirling convection currents -what do you say?...-. The mechanism of the experiment consists in a thick iron plate, of a good surface, that can be heated from below by any mean somewhat regular and uniform (gas, electricity,

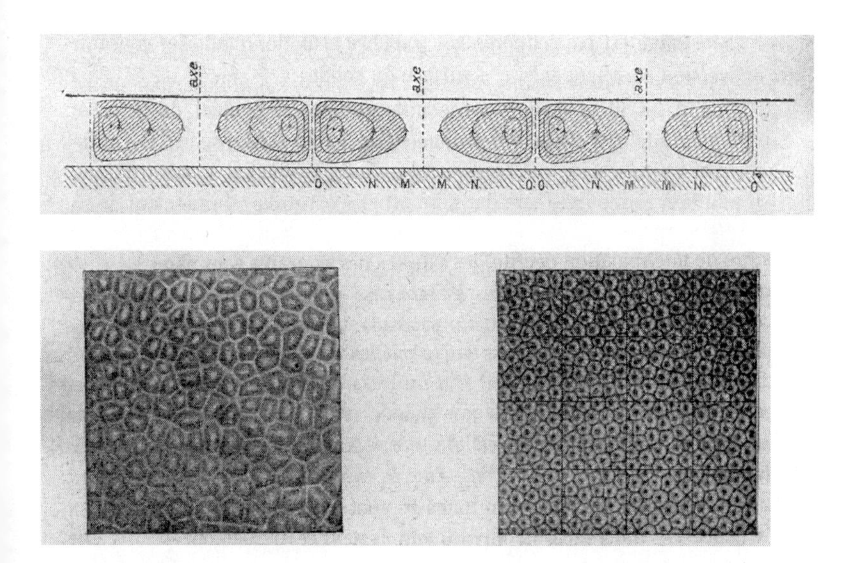

figure 5-12 : the convection whirlpools of Bénard. Model and aspect of the surface after sprinkling.

whatever...), and on which lays a container not too deep and of slightly smaller dimensions (figure 5-11). In this container is a liquid, that has been chosen with a certain viscosity. Bénard will finally use spermaceti[14], just above its fusion temperature (40°). The arrangement is heated, and the experimenter waits the necessary time, for the temperatures of the different parts, to be considered, following adequate measurements, as constant. And then...nothing special, nothing noticeable happens. Seen from far, there's nothing moving, disregarding the heating system that works, but that we can stop for a while, just the time to let come

14 Whale sperm oil

in the room someone who would have seen the device before the beginning of the experiment. If we ask him to say what is different now, he only may say that his eyes don't see any modification. Nevertheless, the precise measurement of the temperatures shows -was it necessary to precise, so much it is evident?- that this of the liquid surface, which is at the contact of the ambient air, is lower than this of the bottom, this last being equal to this of the highest part of the heating plate. In other words, there is in the thin layer of liquid a vertical heat gradient. And still nothing more, at least apparently.

This manipulation, in other respects, is exemplary, because it perfectly illustrates the difference that exists between a physicist and someone simply curious, or a mathematician. For Mr Bénard, physicist, it was not conceivable that the heating of the receptacle and the existence of the thermal gradient didn't manifest in some effect. He had then the idea of sprinkling the liquid surface with a very fine and light sawdust, and at once saw that something was effectively happening inside the liquid. Indeed, after a certain time, he noticed that the dust was leaving some points and grouping in other points, in closed lines firstly irregular, then more obviously polygonal, and then taking progressively the shape of hexagons, not perfect at the beginning but more and more as the time was passing, to finish in a beautiful plane honeycomb structure very similar to those of a beehive. Bénard then improved his observation means and measured at the microscope the different speeds inside the hexagons, that allowed him to exactly determine the shape of the whirlpools revealed by the hexagons and reproduced in the figure 5-12. Afterwards the experiment was done several times, each time changing the density of the powder, this last being first lighter than the fluid and then denser, which allowed him to see what was happening on the bottom. He totally précised the shape of the whirlpools, that we'll call "gyrostats", for reasons that we should have already guessed, and in particular he noted that their diameter was much longer than their hight: these proportions are not irrelevant. Apart from that, there is still an ascending flow at the centre and a descending one on the sides, just like in an ordinary boiler. What is the most interesting in this experiment, in addition to the phenomenon that it emphasizes, is the contrast between the static appearance of the liquid and the frenetic agitation that is going on inside, and that the human eye cannot see.

Let's now wonder which electrical device could present some analogy with this of Bénard. It's here that the rational physics takes advantage over the theoretical physics. Indeed, this last totally ignores the ether, that is a fluid, and its users and supporters have no reason to do an immediate parallel between the liquid that was in the receptacle of figure 5-11 and the space between thee two electrodes of a plane condenser. For the rational physicists, at the contrary, it's obvious: the liquid, in the last case, is the ether, that fills up the volume limited by the dielectric, whichever this last -air, vacuum or any other-, and the voltage between the plates is analogous to the difference of temperature, in the Bénard's experiment, between the bottom and the surface of the liquid. From there, the imagination finds a new field of action where it may frolic, a new degree of freedom where it may venture, under the condition of remaining well disciplined and not lost in too easy phantasms. That's why the thesis that are going to be only suggested further might be taken as simple suggestions, with the cautiousness that goes with, and at the conditional. Their role is not to assert a new truth, but more to show with reserve and shyness a direction not yet explored.

This being précised, let's come back to the object of our reflections: which parallel may we do, and at what level, between the whirlpools of Bénard and the plane condenser? The similarity of the two geometries is evident, it's this one which calls for the analogy. But what do the vortices of ether, stored in the dielectric, represent? Are they the charges? Of course they are! What could they be else? Let's first recall what is a plane condenser: two equal parallel metallic plates separated by a dielectric. This dielectric may be the air, a gas, the vacuum, all that works the same way. But to compare it more easily to the mechanism of Bénard, we'll suppose that the dielectric is a solid and well stuck to the plates, which makes of it a rigid whole, transportable and manipulable. Moreover it allows, under this shape, a remarkable experiment that is related in the old treatises on electricity, and that goes so well in our direction that it is not possible to ignore it, contrarily to what are doing all the modern books on this subject. It's question of the electrophorus of Volta, on which we'll say some words further. When we apply to a condenser a continuous voltage, it gets charged by the way of a current, that exponentially decreases with the time. That means that it stores a certain quantity of electricity, a certain power that could stay indefinitely at

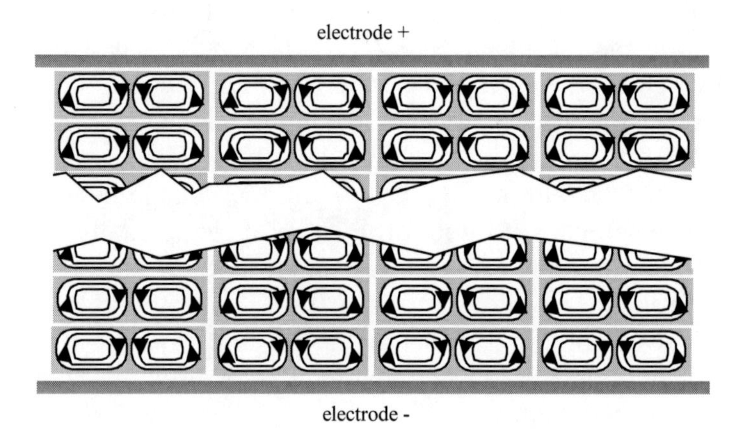

Figure 5-13: the electric charges in the plane condenser, seen as ether vortices

disposal if the dielectric was perfect. The formula that relates the quantity of electricity Q and the charge voltage V is the simplest we could imagine:

$$Q = CV$$

C is called the capacity and depends only on the geometric characteristics (surface S and thickness e), and on the permittivity of the material:

$$C = \frac{\varepsilon_0 \varepsilon_r S}{e}$$

Generally, a current is considered as being a movement of charges, the whole of these being seen by the physicists as a fluid that fill the condenser, seen as a container. We thus cannot see the charged condenser otherwise than a tank containing a certain number of stacked charges that, in the analogy proposed, cannot be something else than vortices of ether, each of them having the shape of a torus, more or less deformed under the action of the vicinity -that is the other charges-. The first consequence of this representation of the captive electricity is that, contrarily to what

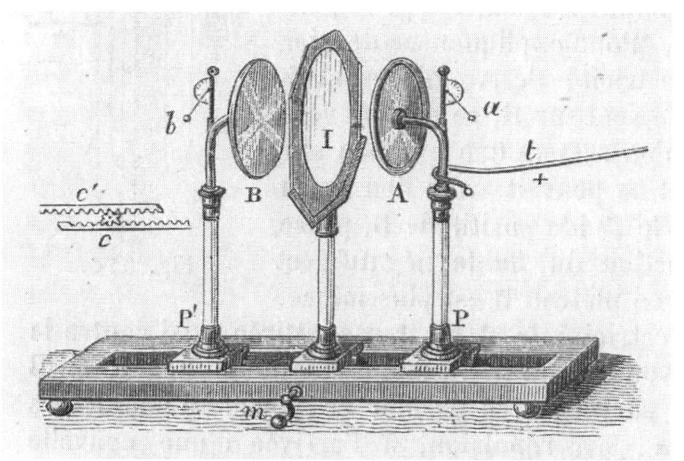

figure 5-14 : condenser of Aepinus

is usually taught, there are no positive or negative charges: there are all identical electric charges, that have a positive or negative polarity following the circumstances, that is the side on their axis from which we look at them. It's the electrodes that put into evidence this difference, as these electrodes are in contact with the two opposite sides of the charges, that we now may call gyrostats, to pay homage to Gustave le Bon. Figure 5-13 illustrates a possible disposition of these gyrostats, directly inspired from the discoveries by Bénard, to which we have to add that, contrarily to the convection whirlpools that don't exist elsewhere than inside their heated liquid, the charges-gyrostats have an individual existence and can organize themselves, both in layers and in stacks.

The illustration shows a geometrical arrangement that corresponds to the condenser completely charged. But what happens when we are in the transitory regime, when the load current *I* has any value comprised between I_{max} and zero? Do the gyrostats have either fixed, immutable dimensions, and in this case do they form a sort of undisciplined cloud in the volume of the dielectric, or are these dimensions constantly in accordance with their number, so as to exactly fill the volume? When we alternately examine the two hypothesis, we very quickly see that the first one poses many problems, when the second, once admitted that the captive charges have variable dimensions, doesn't pose anyone: the charges behave like balloons that we would force into a container, and of whom the individual volume would progressively decrease as their quantity increase. This way, one balloon occupies the totality of the container volume, two half of it each, a hundred the hundredth, etc. etc. This image is moreover extremely vivid, as we can see, in the increase of pressure proportional to the number of balloons, an exact parallel with the increase of the voltage, between the two plates of the condenser, in function of the number of charges, in other words of the quantity of electricity stored.

To provisionally finish with this series of comparisons, let's say a few words on the condenser of Aepinus, announced higher (figure 5-14), that comes to comfort, by the commodity of its experimentation, the representation given above of the charges arrangement in a dielectric. It's question, in a way, of a sort of condenser that can be dismantled, the two plates of which we can separately distance from the insulator, thanks to a groove system, and then put them close again to the contact, as many times as we want and at any distance. The two conducting plates being pressed on the dielectric disk, we begin with charging the so-built condenser, with a generator that is not represented. The elder-balls electrometers, bound to the plates, take an oblique position and indicates that there is a charge on them. We then put them at a distance from the central disk -in glass in the original version-, and we see that they carry the charge with them. We then make the charge collapse by connecting the plates to the ground with the hook: the elder-balls join back the vertical position. Now we make again the contact between the plates and the disk, and we can see that the plates recuperate their initial charge, as indicated by the electro-meters. We may repeat the experiment as many times as we want, we cannot see the least loss of the plates charges. This experiment, and numerous similar variants like this of the Volta's electrophorus, shows us that the charges sent into the condenser are totally located in the insulator, and the schema of the figure 5-13 finds itself both reinforced and justified.

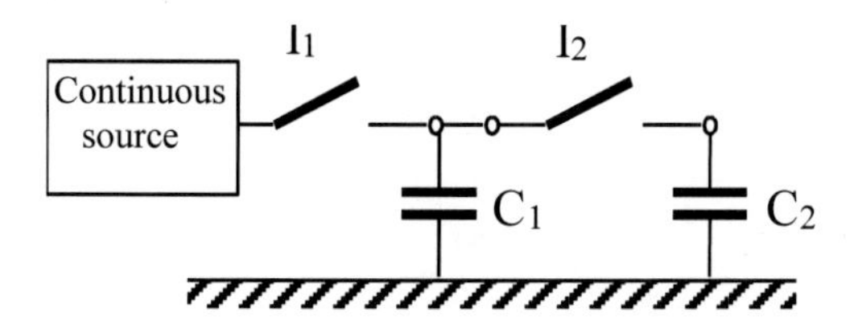

Figure 5-15: Gandillot's experiment

Let's not have the unconsciousness of thinking that all this "proves" the well-founded of a particular electromechanical analogy, chosen among others, left to the choice and the convenience of those who are ready to do the effort of searching one. The final aim was to offer as an example, to those who try to build a representation of the hidden phenomena of the electricity and the magnetism, a simple and pedagogic image of one of them. As a premium, nothing thwarts to believe it's true.

This was a typical example of what the rational physics can bring: clarity of the models, systematic comparison with mechanical phenomena, understandable by everybody, pedagogic simplicity, and also, what is perhaps the most important, a tiny, see a good chance, that things are going on this way in the reality. There are many other mechanical analogies, nowadays forgotten, in the domain of the electromagnetism, an important number of whom owed to the physicists of the end of the 19th century and the beginning of the 20th, and that are sleeping in the old books where it's enough to ferret to find them. Among them, there is one that we may not do no mention of it, so demonstrative it is, and it will be the end of this paragraph. It is owed to Maurice Gandillot, one of the most virulent enemy of the Relativity, therefore already cited in this book. It's the reason for which it is called the "paradox of Gandillot". The experiment he describes, that can be either real or virtual, is extremely simple (figure 5-15): with the help of a continuous current source, controlled by the switch I_1, we can charge two identical condensers C_1 and C_2, having each a capacity C and separated by a second switch I_2. At the beginning, I_2 is

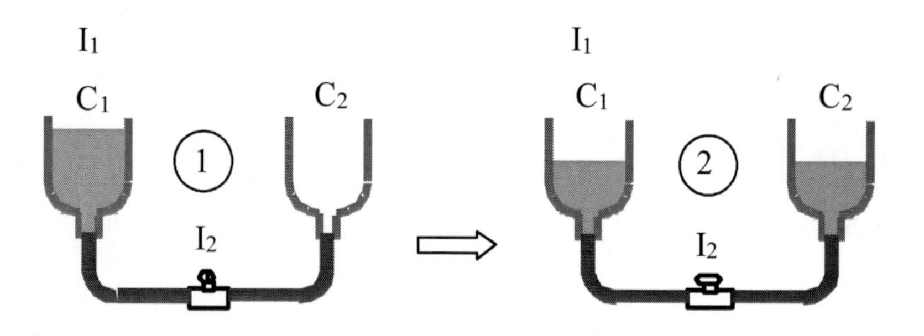

Figure 5-16: the hydraulic analogy

open. We shut I_1 to charge C_1 with the voltage V. The energy stored in C_1 is $W_1 = 1/2\ C_1V^2$. We then open I_1 and we shut I_2. The charge contained in C_1 is thus divided in two halves in each of the condensers, with a common voltage of V/2, and each condenser has now an energy of $1/8\ CV^2$, which gives a total energy, for the whole of the two condensers, of $W_2 = 1/4\ CV_2$. Half of the energy seems to have evaporated, and Gandillot used this result as a proof of the existence of the ether, in which the missing energy would have collapsed. So you can be graduate at Polytechnique[15] and say bullshits. But we already knew that, no?

If Gandillot had read Maxwell, he would have known that the final state of a transformation, on the energetic point of view, is equal to the initial state **minus** the sum of the mechanical works used to pass from one to the other. For those who prefer equations, even short, this means that the apparent result of the experiment here-above is not $W_2 = W_1/2$, but $W_1 = W_2 + W_3$, where $W_3 = W_2 = W_1/2$ precisely represent the invisible operations that happen when the switch I_2 is open, which allows not to bend the principle of the energy conservation. But then the mystery remains total, because we don't see anything going on, like in many elec-

15 The highest French Highschool

trical phenomena, and Gandillot should have been more credible if he had pushed on to there. If you want to prove the existence of the ether, or anything else by the way, you have to take care that the principles, agreed as universal, are not transgressed, and this of the energy conservation is one of the strongest. It's here that the mechanical analogy, and more precisely hydraulic, that respects these principles, will do the demonstration of the amazing easiness with which it can solve a problem, apparently hermetic, in electricity.

Figure 5-16 shows a disposition where tanks will be considered as the equivalent of the condensers. Almost all the physics professors of the secondary classrooms use this image, electricity often behaving like a fluid, and the loading of a condenser being similar to the filling of a tank. In order to reinforce the analogy, we have kept the same notations: the condenser C_1 becomes the vase C_1, the switch I_2 becomes the sluice or knob I_2.The switch I_1 is not drawn, as its opening in the schema 5-15 is equivalent to the filling of the vase C_1 in 5-16. The manipulation in electricity becomes the following in hydraulics: we start with filling C_1, I_2 being close. Then we open I_2, supposing that we can do that instantaneously, and we look what happens, what we cannot do with the experiment 5-15, since we don't see anything. The liquid freed, moved by the gravity, falls through I_2 and, carried on by its inertia, climbs in the vase C_2 to reach the same level it was having in C_1. Then, it is solicited again by the gravity, does the travel backwards and, if there is no friction, the oscillation may go on this way indefinitely. If there is a friction, in other words if we pass from a theoretical mechanism to a true one, the oscillations will damp following a certain logarithmic decrement, let's say more or less quickly if we want to be simpler, and after a while the two vases will be at rest with, in each of them, half of the liquid. We may notice, besides, that the equivalent of the charge voltage is the height of the liquid. The mystery of the Gandillot's hidden energy W_3, that he missed so much, is thus cleared up: it's a dynamic oscillating energy, of which the mode and the process are perfectly limpid, that is transformed, slowly but totally, in heat by the action of the friction.

Now that we exactly know what happens in the experiment described in 5-15, is all this so clear, and must we shelve indefinitely or sweep with the back of the hand a notice that seemed to be a door open toward the unknown? Let's beware. It remains from of all this an experimentally established fact, indeed fundamental, that is also well-known, through other ways, by the high frequencies engineers, that it is not possible to divide the charge of a condenser without losing half of it, and that this lost half is fully transformed into heat. In the same order, these engineers will tell you that you cannot couple two signals without suffering the same sanction. It's then very opportune and too easy to say that the heat produced dissolves in the vicinity, and that at the end it vanishes by splitting by degrees in the other objects, more or less in contact: the truth is not so evident. Whatever we may say, we are in the presence of an energy that disappears we don't know how, unless we take into account, of course, the omnipresent ether? In this case, it becomes evident that things are much clearer, but this is another story...

5-6 : conclusion of chapter 5

This is then the end of the journey, the term of a critical exploration of our real knowledge, in the bushy and complicated universe of the modern physics, and in parallel this of the defence plea for the acceptance of the existence of the ether, as fundamental concept of the science, all the science. We do say "acceptance", and not "demonstration", even if this last, at least we can hope so, is now obvious and doesn't suffer of the least dispute: the ether is here, anywhere, black but transparent, massive and heavy but that the matter, made of vacuum, crosses without any problem, like a fishing net that we drive in the water, and our existence is bound to it just like an embryo is bound to the placenta of its mother. We owe it all that condition our existence: the mass, the energy, the life itself. Why, in these conditions, is there such a disinterestedness, such a mistrust, such a denial, as soon as we try to talk about this topic? The attitude of the scientists in front of new ideas had always been the same and, though it often is demoralizing, it's not the less interesting side of the analysis of the science evolution. In fact, we have too much the bad tendency of seeing in them an exceptional people, with an intelligence above the average, when on this point of view they are exactly like everybody, with preoccupations and worries totally casual and a standard mind. The most perceptive of them know that well, and don't show any feeling of superiority, but all of them are subject, like anyone of us, to the same gregarious reflexes in front of the unknown: mistrust, weight of the habits and horror of change. And for some of them, representing a bunch that pollutes all the professions, we may add as a premium avidity and acting from ulterior motives, which doesn't improve anything but explains certain reactions, conflicting the common interest.

When we have the privilege of capturing the attention of an academic, and when we can manage so as to orientate him, step by step, to questions of general physics, at the limit with the philosophical domain, and concerning, by chance, the propagation medium of the EM waves, something happens. Suddenly, as soon as he sees that we are trying to lead him toward a field where he feels disarmed, his eye becomes suspicious and his attitude changes, being brusquely less cordial. This kind of people has, in general, passed a lot of examinations and degrees, they have assumed answering a lot of questions and, having often been, by necessity, on the defensive, they have kept something of all that inside them. They at once find back their reflexes of stressed students, and it's them who, at their turn, ask questions like this one, that comes regularly and often announces the end of the conversation: why taking interest in that sort of things, about which we know that they lead to endless discussions, when the theoretical physics provides us, to satiety, with the propagation equations that allow us to solve all the casual problems, with an absolute rigour? When one of them, more conciliating than the others, accepts to convene of the possibility that the ether exists, this concession, pulled off in the pain, is in general followed by this final question, asked in many ways but all amounting to the same: "*Ether? OK...but what interest?*". Even when they agree that the ether can be highly energetic, and when they can concede it, if we have been convincing during the discussion with them, they don't see at all what they could do with it. Unless they are members of the species, on the way to disappearing, that has followed the Vallée's courses, the capture of this energy, gigantic and inexhaustible, doesn't seem to come to their mind. That's distressing.

This attitude, unfortunately so common, is deplorable and unnerving. A physicist has not the right of shutting his mind this way. In what more precisely concerns the ether, his role, his mission, his utility, his professional justification, is not to know if it is practical or not, but if it exists. And all that has been written so far has only one aim, that is precisely to show that it exists, while fiercely fighting against the dropping off that has won the whole of the Research, the Education and all those who, as said a Faculty professor, also director of collection at a scientific editor, follow in these domains the politics of the ostrich.

Among the great purposes that, in the preceding pages, have been the subjects of non-conformist thesis, totally competitive with the official ones, let's come back a while on the one that is may be the most important, not only by its philosophical and scientific range, but specially by its accessibility. The fact that anybody is able to discuss about it, in spite of its so simple statement that only depends on the ordinary logics, is a gage of solidity. It's question of the solar system. The demonstration made in this book that the Kepler's laws have been so far wrongly interpreted, and that they are in fact the representation of a whirling system, would indeed make the most limited academics ask themselves questions, no? It's owed to Oscar Wilde the affirmation that only imbeciles don't judge following the appearance. If we apply this iconoclast formula to the aspect that shows the cosmos to anyone, at night, if necessary with the help of an astronomer optical equipment, we can notice that there is a number of spiral galaxies that it is not seriously possible not to see, on behalf of their shape, as the same number of vortices.

We'll note that this word of vortex, or whirlpool, is rarely present in the popularization books on astronomy, and moreover it's not done otherwise, when it is done, than in woolly, innocuous and depth-less comparisons. However, everyone knows that such a geometry, of a shape both so unique and so characteristic, cannot be something else than scattered matter, driven by a fluid moving in phase of vortex. It's of a consensual evidence. And this matter in pieces, in a so characteristic spiralled shape, is nothing but the accurate and visible image of the instant dynamic configuration of the motor fluid, that is invisible itself. A visible vortex always reveals a fluid, like the autumn leaves do with an air vortex, that we couldn't see otherwise. Because it is the fluid that is the motor, and not the mysterious forces associated to the theoretical models. These forces have been created from nowhere by the physicists of the astronomy, so as for them to be able to tackle this sort of phenomena, following the principles of Duhem and the theoretical physics: when you have decided to study a system about which you know, in first analysis, that it cannot be stable, like for example a solar system in the vacuum, you invent and name all the necessary forces to make it stable. Thus the two visions of a solar system, this of the theoretical physics that considers it as a congealed system moving in the vacuum, without initial motor, and this of the rational physics, that replaces the same manège in a fluid that drives it, giving it the structure that we know, are in total opposition. Nevertheless, they both know a way of justifying the appearance of the phenomenon, each of them with its arguments and its methods. That's what is serious. Because a solar system, such as described and introduced by Kepler, may not be stable along the time. When, in the classical presentation, any two planets come into conjunction, they perturb each other by the presumed action of the attraction forces -that, let's remind it, don't exist, see Newton-, but no other force comes to correct this perturbation. However, the two planets gently find back their faithful trajectories, as if nothing had happened: extraordinary, no? Who has once asked an astronomer the why of this miracle? Does it mean, all simply, that nobody has ever asked the question? And what about the other embarrassing questions, that have been already assessed in chapter 3, and to which only the rational physics and its whirling models are able to bring an answer: why are planets in a same plane, for example? Why are their aspects so different?

We thus see well the danger of the theoretical physics, danger on which its furious promoters, like Duhem and Poincaré, have never said a word: when the starting hypothesis are good and the modelling correctly made, it's all right and all the physicists congratulate each other. Except that, the modelling used in that school being made with mathematical objects, with impossible properties, we may expect, sometimes, aberrant results, that has already happened. And when the system fails, when the model denies the ordinary logics, the result can be catastrophic. The solar system is perhaps the best example, but there are many others, like these that have been previously denounced. In head of them, the Relativity, this abstruse theory in front of which the scientists of our time bend like believers in church, but that brings them only mirages and disillusions, with at the same time a useless complexity that they transmit to their students. But try to convince people in the mind of which has been put the idea that they are the guardians of the knowledge!...

It would seem that the pioneers of the nebular theories, who are looking for the “hidden black mass”, are starting to tire. To understand that, we must imagine ourselves in their position: having found, by calculation, by putting into equation the movement of some far galaxies, at the limit of the observation possibilities, that there is, somewhere in the Universe, a mass representing at least 80% of its total mass, and not to be able to say where it is, there is indeed a reason to lose their morale. So let's come back on earth. The engineers who use to work in the high frequencies domain have got the habit of sharing the constants they use in two categories: localised constants and distributed constants. The capacity of a condenser, for example, is localised in the condenser. The capacity of a HF line, at the contrary, is distributed all along the line, and is present and identical to itself at each point of the line. But what is available for the HF constants is also available for a number of other things, it's even a principle that wouldn't have denied Monsieur de La Palisse : what is not distributed is localized, what is not localized is distributed. Let's apply this maxim to the hidden black mass. Either it is localized, and such an enormous mass should have already created, if we only refer to the simple black holes, that have a so gigantic gravity field that they trap the light -let's notice by the way that the relativists who propose us this image, implicitly admit that, locally, the celerity of the light, that is supposed to be a universal constant, can possibly decrease to zero-, an enormous cosmic tsunami that would probably have already engulfed all of us, or it is distributed. In this last hypothesis, things become possible again: if the black mass is distributed, we easily understand that we couldn't be able to locate it. But in this case, at the light -we apologise...- of what have been expounded in the former chapters, we perfectly know what is the point: the hidden black mass is the ether. The most remarkable, in this story, is that the specialists, who are at the origin of this hypothesis of a hidden black mass, have probably rediscovered the truth by their calculi, which would tend to prove, for their credit, that their modellings are correct. Simply, the mass in question doesn't represent 78, or 80, or 90%, or even more, following the sources, of the total mass of the universe, but exactly 100%.

There is thus only a short step to do, for the theoreticians of the cosmos, so that all becomes clear, and that the evidence become obvious to them, but they will be then confronted, like everyone, to the incoercible obstacle of the force of the atavistic habits, the certainties and the weakness of our judgement. The idea that planets could be driven by the whirlpool of a fluid, of an enormous volumetric mass, could be admitted, considering that hypothesis as a mechanical model only. Things can always been introduced this way. All that follows could also be agreed by open minds, receptive to new ideas and knowing reasoning, because the whirling theory of the solar systems doesn't show any impossibility or contradiction, in the same way that the rational theory of the gravitation does. In return, persuading someone, even well-disposed, that we all see inside out, and that we are creatures originally without inertia, moving in a liquid medium called the ether or the mass, this is practically impossible to make him ingest so quickly. If beforehand the candidates to the discovery of the Universe, such as this last is really, are not submitted to a brain-washing, so as to make them getting rid of their fallacious and deceitful reflexes, for example what makes any individual say that he doesn't see anything when he closes his eyes, the mission is impossible, as it has been checked a number of times. When we close our eyes, the corresponding physical action is that our brain is instantaneously deprived of all the stimuli transmitted by the visible frequencies, except what we cannot neither avoid, nor take away: we may call it, following the point of view, the ether or the black mass, suppressing to this last the adjective of hidden, that has no more object.

Mastering the handling of the physicist tools, first reasoning, passes or should pass by this of the observation senses, since physics is, originally and structurally, a science of observation before being a theoretical science. The observation being the basis of its activity, reflection coming **after**, it's hardly conceivable that a physicist could correctly think if he doesn't know well the mechanisms of his captors, and the way his brain builds the global image of the external world. The vision being the main sensation and the eye its primordial element, it should be highly desirable, even indispensable, that knowing precisely its operating mode should be a *sine qua non* condition for practising this noble profession. Long conversations with its eminent representatives, added to the targeted discussions that we can have with them on this subject, show that it's not at all the case. The physics professors, let they be in the primary or third cycle, have no deep knowledge of the physiology of the vision, apart from a minority of them who, more conscientious than the mass of their colleagues, get it by themselves through the interest they have for the neighbouring sciences. As for the other senses, it's even not worth mentioning them. It's in function of that ignorance, transmitted from generation to generation, that these however educated people have no notice to do when someone says, besides them, that we don't see anything when we close our eyes, or when we go downstairs into a cellar. Anyway, it's possible to argument, to make acknowledged that in a thick fog, where someone says by reflex that "*I see nothing*", he indeed sees the fog itself, but this fog has no outline, no shape, and is not perceived as an object, just like it is for the ether and the black. Unfortunately this kind of discussion generally ends with a large smile from the one who listens to, and the following day he has already forgotten.

There is in fact a material proof of the existence of the ether, only one if we consider that the heart of this book is only made of presumptions, even if they are strong. It's question of a discovery directly deduced from a mechanical model of the HF passive filters, those that are installed in the base stations of the mobile telecommunication networks -for the specialists, see *VHF/UHF filters and multi-couplers*, by the same author, ISTE-Wiley-. The technological rise in question concerns the matching of the "combline" filters, of which the adjustment is made by input and output loops in silver-plated straps, called "taps", the surface of them being optimized by mechanical deformation, so as to reach an optimum. It's a very delicate tuning, the surface of the loop must not be either too large or too small, referred to the convenient value, that is unique. If we have to enlarge the passing band of the filter, we must increase the surface of the loop, but we are often limited by the geometry of the filter. But, if we remember the mechanical analogy of figure 5-6, where a magnetic loop is compared to the sheet of a vortex, another idea is to increase the thickness of this last, in order to rise up its effect, or its power if we prefer that term. To do that, the idea came to solder a second loop in parallel onto the ordinary matching loop. In the classical representation, that doesn't change the magnetic crossing flux, the only parameter taken into account and that depends only on the current intensity, that remains the same, whatever the number of loops. So, following the classical laws of the electromagnetism, nothing should happen. Nevertheless, the effect was flashing: we can this way double the matching band of the filter, which considerably increase the softness of use and allows dividing by 2 or 3 the number of the previously necessary models. But what is the most remarkable, from a physical point of view, is that if instead of soldering the second loop parallel besides the first one, we do that in the same plane, which is electrically the same operation, absolutely equivalent, there is no effect at all. It thus seems well that there is here a confirmation of an action on the vortex, in accordance with the new theory. In spite of the fact that this unprecedented proceeding concerns a very narrow industrial niche, quite confidential, its implications in physics are enormous and proves, in an irrefutable way, the validity of the mechanical model used, and consequently the existence of the ether, which is its starting hypothesis. We could think, therefore, that some academics would have had the curiosity of looking closer at that: nuts. They don't care a rat's ass about it, absorbed they are with far more important tasks. May be the diffusion of the book, where all that is related, is too limited to come to their knowledge, but in return it seems to interest the engineers, who see themselves very satisfied of using the process, without necessarily well understanding what they are doing: it works, OK, it's all right! Perhaps that, with the time, a sufficient number of users will be led to mind a bit more on how the system works, and will make things progress, but the hope is weak. For making that coming, for making the whole of the scientific community, searchers, engineers, professors and students, feeling concerned with the problem of the ether, it will be necessary that a pretty more resounding event happens, implicating a bigger number of people, and this in a more essential domain, where practically all the individuals will feel involved.

It indeed seems that there are only two possibilities for an evolution. The first is bound to the stamping of the physics, of which the recent fiascos, such as this of the controlled fusion, will not be tolerated any longer by the providers of the research budgets. These people will fatally be obliged, one day, to enforce an obligation of results, instead of what the credits will be suppressed, or at least be drastically reduced, except that we hardly see how the big machine could really completely stop. Besides, it's not this kind of sanction that will make the Research restart, because its true motor is not money, but ideas. And when all the models of the mathematical physics will have been pressed like lemons, until we'll have extracted all the possible juice, when the directing ideas will be in full shortage, when the physicists will at once be conscious that something essential misses them, to get out from the blind way where led them the decision of the theoretical physics to do without the ether, then and then only we'll probably attend a new start, a rebirth that many demoralized searchers are already waiting for.

The second possibility is that an individual, an independent searcher, rational physicist of course, succeeds in finding the way, so simple that nobody had thought of, to transform the diffuse energy of the space into a form usable by men. May be René-Louis Vallée was too ambitious, or too short of time, when he was trying to directly transform this invisible energy into its electric form. The experiments he had led at the CEA, using the Tokamaks and that are recalled in chapter 2, before their last track was cancelled, were probably one of the promising way to reach this result, what he was close to do. After having destroyed the person and his ideas, the killers decided to build, starting from principles that have been hundreds times demonstrated wrongly-founded, still bigger and more costly machines, which will not give more results but that will burden the tax-payer still more.

But there is another way for capturing the cosmic energy, this which consists in transforming it first into heat. It's theoretically possible, it's even much simpler than the direct conversion to the electric form, and it's a way of study that, *a priori*, doesn't ask for sophisticated equipments: an independent should be able, with a minimum of intuition and a clear vision of the space, to be the first to offer to the mankind the definite and completely ecological solution to the great problem of man, the energy. We can imagine a prototype that would resemble a cylinder, a few centimetres high, that would have the property of being at a proper temperature always above the ambient. All the graduates at Polytechnique, as the spiritual heirs of Sadi-Carnot, as well as their colleagues of the Ecole Centrale, would tell you that it is impossible, because this would mean a perpetual movement. When you know the existence of the ether and its properties, the principle of the conservation of the energy, that seems to be caught out because defined in a close system, takes a new signification in an open system, and the creation of heat in the thermogenetic pile -it's the name we could give to the prototype- wouldn't pose any longer a problem of principle. We can then imagine a group of some thousands of thermogenetic piles in a “vessel”, like Descartes would have said at his time, let's rather say a watertight tank equipped with an inferior input, by which the water would enter, say at a temperature of 10°, and a superior output where it would exit at a temperature of 30°, and then a vertical stack of 5 or 6 of these tanks, of which the highest would give a water at more than 100°, in other words steam. Then, like it is made in any type of power stations, nothing easier to this equipment than make turning a steam turbine, which would drive an alternator and transform the whole into a mini-power station, each of this one having the capacity of providing with electric current one of the 36 000 villages of France. It would be the end of the oil as fuel, the end of gas, the advent of the all-electric, the end of the pollution by greenhouse effect, and a fantastic disruption of the occidental society. If that is not a project, what is it?

Good luck to the candidates!

Provisional conclusion,
in the form of an epilogue

Corps, Isère[16]. Chapelle Saint-Roch.

It's autumn. The sun is just appearing behind the orange crests, and begins to dissipate the clouds and the mists that shrouded the valleys of their wet and protecting veil. Despite the early morning hour, the weather is already warm, but some signs show that the season is on its end: on the Sautet lake, below, the south wind creases the surface of the water, like a gesture of impatience, and we have the impression that the silvered snake of the Drac[17] pains to maintain the level in the reservoir of the dam. There is in the air, adding to the strange softness, something heavy that incites to profit as quickly as possible of the passing while.

An old man, half lying on a big rock of pudding-stone, his chin in his palm and the elbow on the rock, his gaze nowhere, enjoys the minute that passes. Each of us has known this indefinable sensation of total happiness when, at a certain place and at a certain moment, not very long in general, we have the impression of being alone in the World and in complete communion with the Nature. It's probably what the old man must feel, and his half-smile, both calm and tired, wants to witness it. He often comes here to meditate, at this place that nothing seems to differentiate from others, but where there must be something special, so that he

16 French department
17 River in Isère

be so keen to come again each day, like if he was waiting for an exceptional event here, nobody knows when. Fifty years of physics, as he is a physicist, have finally taught him he knows so few, and may be he hopes, he feels, he wants to believe, that the truth could appear to him at this precise place. It's the reason for which he comes here, in this immaterial refuge, letting wander as they want thousands of ideas that turn in his old brain, full of both experience and uncertainties.

Suddenly, his eye is drawn by a furtive movement, almost imperceptible, in the middle of the path just besides, and incites him to leave his reverie. Slowly and with regret, he stands up and looks, from nearer, at what his reflexes, surprisingly intact, have moved in his tired brain, conditioned by so many years of observation, of an obsessional searching for anything, for all the things in fact. There is here a big green grasshopper, or more exactly half a grasshopper. The back part doesn't exist any longer, and the rest, that is the corselet, the head, the antennas and pieces of legs, weakly move in an anarchic way. Around and above, there are ants and flies, morning gravediggers of the probable victim of a rambler of the dawn. The ants, very organized and industrialised, practise the speciality that we know: they cut tiny pieces of the still living corpse and carry them, by the track of the ants, to the house of the ants, where they'll be stocked to feed mother-ants and baby-ants. The flies, more carefree, are here for the warm buffet and the refreshment bar. Looking at them nearer, we can see that there are at least three kinds of them: very small, about two millimetres long, middle-sized, of half a centimetre, and two golden flies four times bigger than the previous. A wasp comes at that moment and claims -please forgive us for this poor but irresistible pun- for its right to the antenna, that it seems to prefer to the blood-red meat that the flies adore. Looking still nearer, we can hardly distinguish others tiny insects, besides which the smallest flies are giants, and that take their share of the banquet, without being awkward for the others, so minute they are. At this stage, without a magnifier or a pocket-microscope, we don't see anything more, but we guess that there must be other beings, still smaller, like the corpse-eaters, so well described by Henri Barbusse and that are able to detect the dead meat at a distance of a kilometre.

The old man, abruptly pushed out of his morning reverie by this unpleasant scene, has stood up and look again at far, instinctively, toward the summits now gleaming in the ascending sun. Indeed he doesn't look at a distance, at the very peaks, but beyond, farer than the clouds, farer than the atmosphere. He had just suddenly passed, his mind still gently asleep, from a show that was like a door open to the infinitely small to another, leading to the infinitely large. And each time the circumstances put him in this situation, as it is not the first time he's being the accomplice toy of the fantasies of the Nature, he feels the same impression, both reasonable and instinctive, a bit depressing, not to have his place in that two scales Universe, in which he doesn't exactly know himself where he can be. It's well easy, once we define an infinitely small and an infinitely large, to say that we are in the middle: that has strictly no signification, as we don't know where begins and ends the one and the other. The notion of the infinity is one of the most confusing, among these that man had invented. It sets the so close frontier between what he may hope knowing one day and what he never could reach, except in his imagination or in his dreams.

Moreover, is it well reasonable to want knowing all? Is this curiosity part of the human nature, or has it appeared the day someone has decided not to accept the Nature, but to master it at his profit? Is it this way that was born the industry, at the beginning necessary handicrafts of the wood, then of the bones, and later of the metals, before becoming what it is to-day, that is a fantastic mean to access to the fortune? Can science, about which we absolutely want to think that it is a disinterested activity, of pure motivations, really handle this role of a Madonna, lighting the world and showing it the way of the progress? Obviously no. The modern occidental world, managed by the financiers and the trade corporation, has since a long time stuffed a science that is anyway of a fragile nature, diverting it for its unique profit from its essential role, the quest of the Knowledge.

The old tired man has now left his rock. Around him, the human activity has taken again its daily curse, the air begins to warm up, there are people who pass and make noise. It's time for him to find another refuge, where meditation will be possible again, as it is now the only activity that fills the most of his time, the only one that his old carcass, smashed up by the life, allows him to have. He walks up to the bushy part of the sent, but cannot avoid looking back behind him, toward a horizon masked by the mountains, toward the infinity. He closes his eyes for a while, something that has become a tic to him. Like anyone who does that, he sees the black, but he neither would say or think that he doesn't see anything, he knows what is the black. He sees the enormous mass in which his vaporous body moves without efforts, maintained in its shape by the colossal pressure of the invisible radiations, that press it from all sides. He feels himself flattened on the ground by the slow and irresistible flow of the gravity. It took him years, see tens of years, so that the nature of the physical world, of the space, becomes obvious to him. He's not happier for that, he simply feels having a huge relief, the deep satisfaction of having found the truth, at least "his" truth, the only one in which he feels easy, the only one he understands. This complicated feeling, difficult to express by the everyday words, also carries with it the sadness of someone who has the sensation of being alone in his way of thinking, and that the doubt often comes back to seize him, without completely vanishing for once. What can be the actual value of a discovery, if the others don't want of it? What is the meaning of a vision of the world that finally explains all that the mathematics gave to the physics, and still more, but that, for the moment, does only that, without bringing the technological overthrow that would be its irrefutable proof? Nonetheless, the fact of seeing the universe as it is, and not has it appears, should be of a large satisfaction for all the physicists, with the promise of an inevitable progress. But the partitioning of the so numerous facets of the physics, now split in an incredible number of specialities, maintains the physicists, searchers or professors, in tunnels that have no communication between them. And all of them would find profitable to themselves in assimilating the notion of ether, that surely could make easier their comprehension of the phenomena, all the phenomena.

To be well penetrated by this reality, it's often necessary, like for the old sitting man, to close our eyes and look at this black, that before represented nothing. Never, by the way, we'll be able to see anything else, as the normal vision, that exists only through the EM radiations named “visible” -which besides is quite stupid, as they are themselves totally invisible- makes the ether completely transparent to us. In the absence of light, it appears to us black and impenetrable, in the presence of light it disappears, while remaining present, to allow us contemplating the world such that we know it, but also such that it is not, though our sensations are not completely wrong: they simply are the inverse of the reality, what finally is something very close. Indeed, the inverse of something is always very similar to the thing itself, and can be deduced through a simple operation, let's say a simple transform if we absolutely want to model everything and use the pretentious language of the asses mathematicians.

The simplest exercise, for those who want to discover the ethered world and criticisingly test the guilty habits of our instinctive thought, consists in going one night on the side of a mountain lake, where they will be sure to be alone, and walk bare-foot on the fine sand, so as not to make the least noise. They choose a direction where there are no obstacles over a few metres, they close their eyes and they slowly go ahead, repeating to themselves, internally: “*The ether is black and I see it... The ether is black and I see it...The ether is black and I see it...*”. They must, in the same time, imagine their body as the aggregation of mass-free particles, so distant the ones from the others that it moves without difficulty in a very heavy fluid, that we may imagine being mercury to make an image of it, but several hundreds or thousands times heavier, it doesn't matter. This short experiment is not expensive, and repeated as many times as necessary, it's the ideal mean to solidly fix the ideas in the memory. That's a proceeding that wouldn't be disowned by sophrologists, and it's this way that the other image of the reality will make its nest in the deep memory, and cohabit with what is inside the brain since the birth. The fact of being alone is very important. It allows to be completely concentrated, and specially not having to dread the banters from fellows more or less motivated, or not motivated at all. If you have no mountain lake at hand, you can do the same in a quiet street, or a country track, or even in a cellar, if its dimensions allow you to do some steps without a risk of accident, but in any case it must be a lonely, silent and personal experiment. When you have done that several times, playing honestly the game, the idea of the massive ether seems less and less ridiculous, its hypothesis takes slowly but solidly its place in your brain. Your mind can then tackle in good conditions, being rich of new intellectual weapons, the new visitation of more physical questions, such as, for example, the whirling nature of the Kepler's laws, or the homogeneous repartition in the space of the hidden black mass. It's also recommended to practice this exercise regularly, so as not to lose its acquired knowledge. Luckily, it's easier and easier as the idea is progressively agreed by the subconscious, and with a minimum of willingness and perseverance we can memorize and digest the aspect, first incredible, of the true nature of the Universe, while keeping in the enormous memory that is ours the enforced model of the theoretical physics, the two visions being able to cohabit without any difficulty.

But why, will you say, make so many cerebral efforts to finally admit an hypothesis that doesn't immediately give anything more, in the matter of discoveries? What can well bring us the fact that we can't see, from our very constitution, the Universe as it really is, when science has always gone forward with that state of affairs, while nevertheless progressing with the usual means and the priority use of the mathematics? First of all, this impression of constant progress is totally contestable. If we keep aware, so few it be, of the state of advancement in the priority ways of research, like for example this of the controlled fusion, we are obliged to ascertain that, for forty years, there has not been any significant advance. Since 1970 and the Tokamaks, we are going round in circles. The Vallée episode, that could have been determining in this terrible case, has only produced internal conflicts that have been transformed into a manhunt, and the lonely man has lost, of course, carrying away his invention. The Tokamaks being then no more controllable, and the so-called parasitic emissions of energy being explained exclusively by the man who had been sacked, it was decided that the equipment had become dangerous, and was to be destroyed. And all that for what? For building again, forty years after, another Tokamak, gigantic and named ITER, ten times more expensive and that will not give anything more, apart some manipulations where were created, as it seems, impressing energetic fluxes, for very short moments. In parallel, it was carefully hidden that the global energetic efficiency ratio was still less than the unity. But, just like forty years ago, the authorities explain today to the tax-payer that it's the only way of accessing to the free and permanent energy, by imitating the sun, to which is attributed that unverifiable property to radiate its energy from the nuclear reactions of its plasma. As a result of this reasoning, we only have to create a plasma, and we try to put this plasma to the temperature of the sun, and we can't, and even if we could, in a hundred years, we couldn't directly transform it into electric energy, with an efficiency higher than one.

We have thus, here, a first practical argument to justify that we deploy some efforts to try to create a new way of thinking: the science is slowing, it's marking time, it doesn't create any longer. It's short of directing ideas, while its daughter the technology runs at a hundred mph, and swamps the market of improvements of which the utility is less and less evident, while their noxiousness on the individual behaviour is more and more obvious. There are even domains, like the biology or the astrophysics, where the denial of the ether, with the concomitant action of the fictive forces that replace it, necessary to the coherence of the hypothesis but hiding the genuine forces, leads to full and definite blind ways.

But the most important argument is not of that order. It touches to the very physics, to its role in the intellectual development of man, to its ethic, to its deep justification as indispensable tool of the progression of our knowledge. Knowledge is not a catalogue of the inventions, it's neither the technological window of the common life commodities, nor the scintillating face of the industrial production. It's the satisfaction of discovering, step by step, degree by degree, patiently and with humility, the secrets of the big machine where the random has placed us, where we are, and that was so well song by Maeterlinck, entomologist by his profession but also physicist, from heart and soul. Reading "*The great secret*" or "*The great fantasy*" is infinitely more profitable, to the individual who wants to mind and be educated, than to try to understand the Theory of Relativity, this muddy lake that we imperatively must dodge if we don't want to be horribly drowned. The attitude of the physicists in front of the ether, in general and at all the levels, from the student to the specialist in fundamental research, this manner they have to superbly and contemptuously ignore a problem that is the base of their physics, and that will come back to them, fully in the face one day or another, is intolerable. Of course, we may not reproach them to use and stand surety for a system that they were enforced to learn, and that has not, moreover, so badly worked for a so long period, but in return we may wonder where have passed, for that people, the founding values that are the curiosity, the permanent calling into question, the Cartesian doubt, the taste of learning for learning, this of the observation, the pleasure of the investigation? The physics is not dead, it will never die, but it is ill. Ill of its physicists, themselves ill of the system where they are prisoners. The specialisation to which they are submitted cut them one another, and make them resemble racehorses, equipped with their blinkers: obligation of remaining on the track and running up with the others, without the possibility to see on the right and the left.

The ether, when we accept to agree on its existence, that only narrow-minded people can now contest, gives to the vision of the world a depth and a density that cannot exist in the theoretical physics. The solar systems, supposed to move in the vacuum, unless their trajectories change, without we know, moreover, how their movements have been initialized, are an aberration and a challenge to the good sense. The same, replaced in their motor fluid, don't have mystery any longer: the planets are only the visible marks of the vortex that drives them. Those who have accepted to compare the two schemas have agreed that the whirling model is much better. The only problem is that it obliges to reset to zero all the vision of the world we had before, to change for this it implicates. It's not thinkable that a student, to whom we would give the necessary reflection time, and to whom we would explain why a solar system, as it is usually described, is unstable from its conception, could still believe in that model, specially if we propose him, besides, the whirling model. The same for the planetary model of the atom, originally inspired from the previous and also totally improbable, which couldn't stand the comparison with a multi-resonant model, the only one that justifies the quantification without needing the quantum physics, and this without any more help than this of the classical theory of the resonance.

There is as much difference, between the empty space of the theoreticians and the real space, where the ether doesn't tolerate any vacuum, except inside the particles, as between the drawing of a vehicle and the vehicle itself, or between a map and the territory that it represents. The physicists and the astronomers live in a virtual world where they have been squelching for centuries, without having doubts, but this world provides them with their usual food, and they are not keen to abandon it. The day a sufficient number of them will stand up for the promotion of the ether, the opinions will shift and the physics will start again like a rocket, and even the more refractory will be obliged to follow the movement and leave down his whims.

This being said, we must recognize that imagining a body, and all the visible matter around, as structures having no proper mass and essentially made of vacuum, asks at the beginning for a pretty considerable effort of auto-persuasion, out of the range of an average individual. He must be helped. That's why it is strongly recommended to the interested people to work in group and debate. The essential is to begin, taking the hypothesis by the side you want, under the only condition that it causes an interrogation, that it excites the curiosity, and once you have started you'll very quickly see that you obliged to play the game. This word of "game" is moreover the key-word of the venture: it's worth, to keep the motivation, taking the case as a game. Then, things link together and the game becomes more and more serious, to finally end by a deeper reflection, of a metaphysic order, that will take place in the memory and never leave it. All that can represent years of efforts and doubts, but it's the price to pay for lastly catch a fleeting glimpse of the physical truth of the World. That truth cannot be what we see, as we cannot see the ether when the light is present, and the ether "is" the World.

The light propagates in the ether, with the other EM waves, but among these last only the light cannot cross the ordinary matter, and allows us to see the things as we are accustomed to see them. If our eyes were sensitive to the X-rays, we would remain able to see the World, but with a different image, without colours and where the inside of the bodies would be more visible than their surface. If they were sensitive to the infra-red, we would see an image that would be the chart of the bodies temperatures, instead of the very bodies, but with a certain resemblance with their genuine shapes. These two examples are more expressing than others, because we know, thanks to our laboratory apparatuses, how to transpose these two sorts of images into the visible spectrum, thus we know them, and we have this way an approximate, though wrong, idea of what would be our vision in another frequency band. But we also must note that, if we had the choice of our functional conception, it's very probable that we would choose the actual way, with a vision given by the light, as it certainly is the best matched to the terrestrial life of man.

In that new vision of the World, the notion of infinity suddenly takes another aspect and another signification. It has now a groundwork, this omnipresent fluid, hidden to our eyes that only can see through it, which sprays everywhere its enormous mass, letting no place unoccupied and confirming in a peremptory manner the horror of the Nature for the vacuum. The Big Bang, already classed among the myths of the science, takes here the finishing stroke: the space being full and homogeneous, there cannot be an expansion. The only movements allowed inside a fluid are the laminar sliding and the vortex. This last is visible only when huge celestial bodies unveil it to either our glance, or to our observation instruments, under the shape of solar systems or spiral galaxies. As for the sliding, it is similar to the movement of the air in a room, that only the cigarette smoke reveals. The apparent immobility of the starred sky, from one year to the next, is indeed only the majestic slowness of an impalpable cosmic placenta, in which we are bathing and about which only a long observation, at our time scale, can reveal that it is moving. It moves slowly, in all the directions of a three dimensions Universe, that, while remaining globally stable, never rests, like someone who dreams during his sleep, moving without changing place, turning on itself. It's this way that constellations move the ones in regard to the others.

The tired old man has now left his favourite rock, after having turned over in his mind all these iconoclast ideas. He knows he's right, but also that he has not the means for convincing. There are, here and there, in this country and in others, independent searchers who, disappointed by the traditional physics, discouraged by its suspect complexity and irritated by the dictatorship if the equations, try to build a personal idea of the Universe. Some of them are, may be, as intelligent as Descartes or Huygens, they probably have their capacity of reasoning, and it is normal that, from time to time, anywhere, in a modest flat of the suburbs or in a hotel bedroom, avant-garde ideas can arise, with luminous solutions to problems that are reputed as inaccessible, beautiful and good slices of physics. But nobody will be aware of these amateur works, this word of amateur being taken in its etymological meaning, that is this "who loves". All of them take avidly notes, they all have personal libraries that often contain tens of years researches, they are all animated by the same passion, but all of them will die without seeing their works published. These last will probably be put to the dustbin, or burnt by descendants who will not understand their importance. Nobody will know if what they have found is important or no, so is life.

At the end of the track, above the valley, the old man has now disappeared. He must have reached back his house, where may be someone is waiting for him, but may be not. What is sure, is that to-morrow he will be here, on his rock, or a bit farer, whatever happens, unless it rains. If it rains he'll remain sheltered, as his old bones don't stand any longer the elements, but here or there, all the ideas that were moving in his brain will come back, like every day, to make the circle in his intact and solid memory. Untiringly, they will turn, coming each time in a different order, with different clothes, appealing other ideas, similarly surprising, that will come to take their places in the innumerable empty compartments of the great puzzle of the knowledge. To know that you are perhaps early on the others doesn't give you any advantage, if you can't share your ideas with the others. Moreover, someone who is obliged to mind lonely always ends up to mistake, sooner or later.

To understand the bases of the Universe, to finally know, even approximately, the structure of the ether and how it works, to pitilessly destroy the bookstall of the lost searchers lucubrations, all that is sure of a big personal satisfaction. But what can we do with a knowledge so much in opposition with the casual thought that nobody wants, neither to debate about it, or even study? We also can take the measure, by this way, of the more dramatic solitude of the men who have been taken in such an unjust and stupid ostracism, like Maxwell, before the well-founded of his discoveries was finally agreed, or like Tommasina or Vallée, who didn't even have this chance?

So, what can we do? Is it really worth racking one's brain to try to discover the secrets of the Cosmos? At the contrary, shouldn't we do like the majority of the citizens does, take life as it comes, with all the temptations that offers the technological civilisation, enjoying the best possible the present while and putting one's head under the sand as soon as we hear speaking about ecology and social morale? It's sure that there are other values than these that absolutely want to increase the knowledge. Nevertheless, the absolute necessity of finding other sources of energy than oil or gas, key of the survival for a world population that's inordinately swelling, makes that there always will be engineers and searchers. But besides, there are also the pleasures of the life, the simple happiness that makes the existence bearable, without pressing one's meninges, and that allows us, would it be a while, to touch the eternal and deep ambition of man: the good fortune. And if good fortune means the discovery of the radium for Pierre and Marie Curie, it's also, for a majority of people, these short moments of intense and sheer pleasure that are the meetings between friends, a nice eating around a barbecue, children laughing, or still simpler, after having rambled for a long time, in summer, through a sweet-smelling forest, your body dead-beat and full of sweat, shouting under the shower the hit of the moment, and rubbing one's intimate parts with a well harsh glove.

Which goes to show we are very little, ultimately.

Bibliography

1- General

- Abraham-Sacerdote : Recueil de constantes physiques, Gauthier-Villars, 1913.

- Arnould Henri : L'énergie, Quillet, 1920.

- Auger Léon : Gilles Personne de Roberval, Albert Blanchard, 1962.

- Bamberger Yves : Mécanique de l'ingénieur, Hermann, 1997.

- Belot Emile : Essai de cosmogonie tourbillonnaire, Gauthier-Villars, 1911.

- Belot Emile : L'origine dualiste des mondes, Payot, 1924.

- Belot Emile : La naissance de la Terre, Gauthier-Villars, 1931.

- Boll Marcel: la Science, ses progrès, ses applications, Larousse, 1933.

- Boll Marcel-Féry André : Précis de physique, Dunod, 1927.

- Bouasse Henri : Bibliothèque Scientifique de l'Ingénieur et du Physicien, 1917-1947 (45 volumes).

- Bruhat Georges : Le Soleil, Felix Alcan, 1931.

- Bureau des Longitudes: Encyclopédie scientifique de l'Univers, Gauthier-Villars, 1981.

- Castelfranchi Gaeteno : La physique moderne, Dunod, 1949.

- CNRS : Roemer et la vitesse de la lumière, Vrin,1978.

- CNRS : œuvres de Jean Perrin, CNRS, 1950

- Cochin Denys : Le Monde extérieur, Masson, 1895.

- Cornu M.A : Mémoire sur la détermination de la vitesse de la lumière.

- Couderc Paul : Univers 1937, Editions Rationalistes, 1937.

- De Broglie : Introduction à l'étude de la mécanique ondulatoire, Hermann, 1930.

- Davies Paul : Les forces de la nature, Armand Colin, 1989.

- De Heen P. : La matière, Hayez, Bruxelles, 1905.

- Descartes René : Le Monde, Chez Jacques le Gras, 1664.

- Draper John William : Les conflits de la Science et de la Religion, Germer-Baillière, 1882.

-Dreyfus F.Camille : L'évolution des mondes et des sociétés, Alcan, 1893.

- Einstein Albert : La théorie de la relativité restreinte et générale, Gauthier-Villars, 1976.

- Esclangon Ernest : La notion de temps, Gauthier-Villars, 1938.

- Fabry Charles : Physique et Astrophysique, Flammarion, 1935.

- Feuer Lewis : Einstein and the Generations of Science, Basic Books, 1974.

- Gallais-Rumeau : Chimie Générale, Delagrave, 1958.

- Gauzit J. : Les grands problèmes de l'Astronomie, Dunod, 1957.

- Gilpin Robert : La science et l'état en France, Gallimard, 1970.

- Grove W.R. : Corrélation des forces physiques, chez Leiber, 1867.

- Guyon-Hulin-Petit : Hydrodynamique physique, CNRS éditions, 2001.

- Hugolin L. : L'inertie – La force d'inertie, Blanchard, 1977.

- Lakhovsky Georges : Le grand problème, Alcan, 1935.

- Laplace (Marquis de) : Exposition du Système du Monde, Bachelier, 1824.

- Larminat (J. de) : L'éther, Plon-Nourrit, 1920.

- Llambi Campbell P.: Le grand secret de l'Univers, Hachette, 1934.

- Maeterlinck Maurice : La grande loi, Fasquelles éditeurs, 1933.

- Maeterlinck Maurice : La grande féerie, Charpentier, 1929.

- Maxwell J-C : Traité d'Electricité et de Magnétisme, Gauthier-Villars, 1885.

- Maxwell J-C : Traité élémentaire d'Electricité, Gauthier-Villars, 1884.

- Maxwell J-C : The Scientific Papers, Hermann, 1927.

- Millikan Robert-Andrews: L'électron, Alcan, 1926.

- Newton Isaac: Traité d'Optique, Gauthier-Villars, 1955.

- Nodon Albert : Eléments d'Astrophysique, Blanchard, 1926.

- Pacotte Julien : La physique théorique nouvelle, Gauthiers-Villars, 1921.

- Parenty H: Les Tourbillons de Descartes, Louis Bellet, 1903.

- Perrin Jean : Masse et Gravitation, Hermann, 1940.

- Poincaré Henri: La Mécanique Nouvelle, Jacques Gabay/ Gauthier-Villars, 1989.

- Poincaré Henri: La Théorie de Maxwell, Scientia.

- Popper Karl : La Connaissance sans certitude, PPUR, 1991.

- Romani Lucien : Théorie générale de l'univers physique, Blanchard, 1975.

- Ronchi Vasco : Histoire de la lumière, Armand Colin, 1956.

- Royer Clémence : La Constitution du Monde, Schleicher Frères, 1900.

- Sesmat Augustin: Systèmes de Référence et Mouvements, I-VII, Physique Relativiste, Hermann, 1937.

- Thellier Michel/Ripoll Camille: Bases Thermodynamiques de la Biologie Cellulaire, Masson, 1992.

-Thirring H. : L'idée de la théorie de la Relativité, Gauthier-Villars, 1923.

- Tommasina Thomas : La Physique de la Gravitation, Gauthier-Villars, 1928.

- Vincent Maxime : Les dépressions sidérales, Librairie du moniteur juridique, scientifique et littéraire, 1910.

2 – The French anti-relativists

- Bessière Gustave : Calculs et Artifices de Relativité, Dunod, 1932.

- Bouasse Henri : La Question Préalable contre la Théorie d'Einstein, Albert Blanchard, 1923.

- Bourbon B. : Pesanteur Electricité Magnétisme, Dunod, 1939.

-Brisset D. : La matière et les forces de la Nature, Dunod, 1911.

- Colliard Paul : Les deux Ethers, Chiron, 1925.

- Cornelissen Christian : Les Hallucinations des Einsteiniens, Albert Blanchard, 1923.

- Décombe L. : La Célérité des Ebranlements de l'Ether, Scientia n_o9, Gauthier-Villars, 1909.

- Destieux Jean : Incroyable Einstein, éd du Carnet critique, 1924.

- Dive Pierre : Les Interprétations physiques de la Théorie d'Einstein, Dunod, 1945

- Duport H : Critique des Théories Einsteiniennes, Imprimerie Darantière à Dijon, 1923.

- Gandillot Maurice : Véritable Interprétation des Théories Relativistes, Gauthier-Villars, 1922.

- Leredu Raymond : L'Equivoque d'Einstein, PUF, 1925.

- Leredu Raymond : La Théorie d'Einstein ou la Piperie Relativiste, Douriez-Bataille, 1928.

- Prunier F. : Essai d'une Physique de l'Ether, Blanchard, 1932.

Appendix

In physics, contesting becomes a moral right, see an obligation, since the moment we are going to ingest incomprehensible things. It's also the weapon of the unsatisfied searcher, who wants to progress and has, to do that, no other mean than disobey the scientific Establishment. Like in other domains, it's not possible to create something without destroying something else, that's a universal law. Contesting, in science, can take very various forms, but obligatorily needs first the compilation of the books written by the Ancients, of whom works and ideas has been largely forgotten or modified, and where we can dig at will. Reading them is often a subject of amazement, as it was one of the basis of this book, and it is for this reason, and also to pay homage to the pioneers of the physics, that this short anthology follows. The last pages, left white, will help the accomplice reader to add the result of his own investigations.

Paul Couderc :"*Following a certain idea, precious to the professor Langevin, it seems that the Nature takes a sadistic pleasure to present the phenomena through the wrong side: accessory phenomena most often come, in all the domains of the science, to hide us the simplicity of the fundamental fact; who would think, seeing the capricious fall of a leaf, of the elementary law of the gravitation? I see in the stellar kinematics a new example of the malice of the Nature.*"

Poincaré : "*Mathematics are sometimes a constraint, or even a danger, when, by the very precision of their language, they lead us to affirm more than we know.*"

Sturgeon: "*It seems that the dogs don't take not the slightest notice to their reflection in a mirror, because they don't smell it and trust in their nose, and not in their eyes. It's not probably the same for humans, when your brain says something and your eyes something else, we don't know which to believe.*"

Gustave le Bon : "*The new theories on the structure of the matter lead to consider it as composed of tiny whirlpools of ether, of which the rigidity and the energy are only due to the speed of their rotation movement.*

"Matter being considered as composed of rotating elements, never touching one another, we may have the presentiment of the role of the speed in the material equilibriums. In reality, matter is speed. Only the imperfection of our senses shows it immobile to us. If the constituent movement of the atoms would stop one moment only, these would vanish into an invisible dust of ether and wouldn't be anything more, absolutely nothing, not even a light vapour. The rest of the matter would be the end of things, their coming back to the nought."

"A fluid, liquid or gaseous, hindered in its translation, takes at once the whirling form...The vortices have a big stability and have a tendency to carry with them the bodies they meet, as we observe in cyclones."

Jean Perrin : "*It is said that the mass increases with the speed. This expression seems vicious to me, and the extension to follow illusory. In all cases we'll remember that: it's more and more difficult to increase the speed of a given object that goes more and more quickly.*"

Parenty : "*The understanding of Descartes couldn't conceive this invisible yarn that would keep the satellite prisoner of its planet or its sun. He has replaced the tension of this yarn by the antagonist pressure of the skies. Newton doesn't hesitate to give body to this invisible yarn, he thus may suppress the matter of the* *skies, but this rectilinear action is a fiction, a simple resultant of the forces of the unknown mechanism.*"

Descartes : "*All the planets are thus driven around the sun by the sky that contains them, the revolution is of thirty years for Saturn, two years for Mars, eight months for Venus, three months for Mercury. The opaque bodies that are the sunspots make a turn in 26 days...*"

Malebranche : "*The very scientists, and those who pride themselves on smartness, pass more than half of their life in sheerly animal actions, or such that they give to think that they take in more consideration their health, their goods and their reputation, than the perfection of their brain. They study more to acquire a chimeric size, in the imagination of the other men, than to give more strength to their mind, and more range. They make of their head a sort of furniture repository, in which they store, without discrimination and order, all that bears a character of erudition, I mean all that can appear rare and extraordinary, and excite the admiration of the other men.*"

"*It seems to me that the ration of the weight of the ether (domain of the light) to this of the atmosphere (domain of the sound) is far higher than six hundred thousands to one.*"

Leredu : "*Among all the equations that create an apparent link between his kinematics and his dynamics, only the equations (7-9) allow Einstein to incorporate in his theory certain concordances with the astronomical and experimental facts. He comes, without his knowledge, under the attraction of these accordances that come, he thinks, corroborate his ideas, and he accepts as a demonstration a proposition that, in reality, is empty of any mathematical meaning. And it's this way that the Theory of the Relativity, that was nothing, will be clothed by this deceitful splendour that, for a too long time, will comfort its existence.*

Clausius : "*All the bodies of the Nature, though they seem totally resting, are nevertheless prey to the most intense internal* *movements, and these movements are communicated to the ambient ether, so that the universal space is constantly crossed, in the most varied directions, by wave vibrations, and it is to this whole of vibrations that we give the name of temperature.*"

Grove : "*The fact that the structure or the molecular arrangement of the bodies influences, I could say in reality determines, its conducting capacity, is not at all explained in the theory that makes the electricity a fluid, whereas, if electricity is only a transmission of force or movement, the influence of the molecular state is precisely what it must be.*"

Verdet : "*...or that it exists, in the present economy of the Nature, a source of heat that instantaneously gives back to the sun all that its radiation has made it losing.*"

"*In what concerns the language, the word attraction would advantageously be replaced by aspiration, word that describes the phenomenon as well, but without suggesting the concept of attraction force. At the contrary it suggests an unbalance due to the destruction of an equilibrium.*"

"*The radiating heat is produced by the vibrations of a fluid sprayed in the whole space and to which we have given the name of ether.*"

de Broglie : "*...we know that the Brownian movement of a particle results from a continuous and random exchange of energy between that particle and a hidden medium...*"

Aristotle : "*The act from this to that and from this under the effect of that are different by the raison d'être.*"

Roberval : "*The sun, being powerfully hot, strongly heats the fluid and diaphanous matter where it is immersed, matter that is all the less rarefied as it is far from the sun. Its density will thus increase with the distance.*"

"It exists, in all the matter of the world and in each of its parts, a certain property by the force of which all this matter gathers itself into a unique and same continuous body, the parts of which move the ones toward the others, in a perpetual effort, to closely join up, at the point that they cannot be separated but with a greater force."

Roger Bacon : "*The experimental science doesn't receive the truth from the superior sciences; its it that is the mistress, and the other sciences are its servants.*

Maeterlinck : "*It's almost certain that all these movements of the asters that we believe to be circles or ellipses are only spirals, that the too short existence of the humanity has not allowed to be measured.*"

Maxwell : "*But the medium* (the ether)[18] *has other functions and other operations than to transmit the light from human to human and from world to world, and to put into evidence the absolute unity of the measurement system of the Universe. Its tiny constituents must have rotation as well as vibration movements, and their rotation axis form these magnetic lines of force that are prolonged with an uninterrupted continuity in regions that no eye has seen and that, by their action on our magnets, tell us in a not yet interpreted language, what is happening in the hidden world, from minute to minute and from century to century.*"

---

18 Author note

Made in United States
North Haven, CT
28 February 2025